IEE CONTROL ENGINEERING SERIES

SERIES EDITORS

SELF-TUNING AND ADAPTIVE CONTROL:

THEORY AND APPLICATIONS

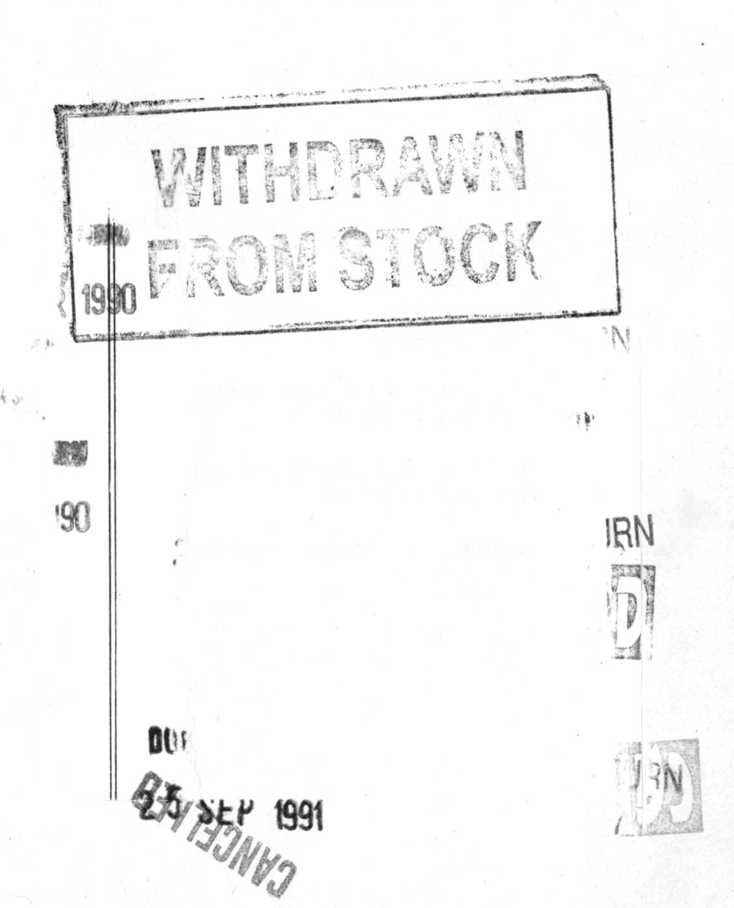

Previous volumes in this series:

Volume	1	Multivariable control theory J. M. Layton
Volume	2	Lift traffic analysis, design and control G. C. Barney and S. M. dos Santos
Volume	3	Transducers in digital systems G. A. Woolvet
Volume	4	Supervisory remote control systems R. E. Young
Volume	5	Structure of interconnected systems H. Nicholson
Volume	6	Power system control M. J. H. Sterling
Volume	7	Feedback and multivariable systems D. H. Owens
Volume	8	A history of control engineering, 1800-1930 S. Bennett
Volume	9	Modern approaches to control system design N. Munro (Editor)
Volume	10	Control of time delay systems J. E. Marshall
Volume	11	Biological systems, modelling and control D. A. Linkens
Volume	12	Modelling of dynamical systems—1 H. Nicholson (Editor)
Volume	13	Modelling of dynamical systems—2 H. Nicholson (Editor)
Volume	14	Optimal relay and saturating control system synthesis E. P. Ryan

SELF-TUNING AND ADAPTIVE CONTROL: THEORY AND APPLICATIONS

Edited by

C. J. HARRIS
Professor of Electrical and Electronic Engineering
Royal Military College of Science
Shrivenham

and

S. A. BILLINGS
Lecturer of Control Engineering
Sheffield University

PETER PEREGRINUS LTD
on behalf of the
Institution of Electrical Engineers

Published by: The Institution of Electrical Engineers, London
and New York
Peter Peregrinus Ltd., Stevenage, UK, and New York

British Library Cataloguing in Publication Data

Self-tuning and adaptive control. – (IEE control engineering series; 15)
1. Adaptive control systems
I. Harris, C.J. II. Billings, S.A.
III. Series
629.8'36 TJ217

ISBN: 0 906048 62 1

Printed in England by A. Wheaton & Co., Ltd., Exeter

Contents

List of contributors

O.L.R.JACOBS
Department of Engineering Science, University of Oxford
D.W.CLARKE
Department of Engineering Science, University of Oxford
P.E.WELLSTEAD
Control Systems Centre, UMIST, Manchester
D.L.PRAGER
Ultra Electronic Controls Ltd., London
P.C.PARKS
Department of Mathematics & Ballistics, Royal Military College of Science, Shrivenham
P.J.GAWTHROP
Department of Engineering Science, University of Oxford
H.UNBEHAUEN
Ruhr-Universität Bochum, Lehrstuhl für Elektrische Steuerung und Regelung, West Germany
B.PORTER
Department of Aeronautical & Mechanical Engineering, University of Salford
A.S.I.ZINOBER
Department of Applied Mathematics & Computing Science, University of Sheffield
D.H.OWENS
Department of Control Engineering, University of Sheffield
A.CHOTAI
Department of Control Engineering, University of Sheffield
A.J.MORRIS
Department of Chemical Engineering, University of Newcastle-upon-Tyne
Y.NAZER
Esso Chemical Company, Canada
R.K.WOOD
Department of Chemical Engineering, University of Alberta, Canada

P.M.ZANKER
Marconi Space and Defence Systems Ltd., Camberley
N.MORT
Royal Naval Engineering College, Plymouth
D.A.LINKENS
Department of Control Engineering, University of Sheffield
P.T.K.FUNG
Department of Electrical & Electronic Engineering, Sheffield City Polytechnic, Sheffield
M.J.GRIMBLE
Department of Electrical & Electronic Engineering, Sheffield City Polytechnic, Sheffield

Preface

During the past decade studies of adaptive control have rapidly increased through the advent of self tuning control and their potentially cheap and simple practical implementation using microprocessors. Simultaneously the question of global stability of the adaptive control loop for deterministic systems has been resolved for the now almost classical model reference approach. Much recent work has concentrated upon the common features and the unification of the model reference and self-tuning approaches to adaptive control. However, it is significant that by far the most effort on model reference adaptive control has concentrated upon deterministic systems via stability methods such as Liapunov's second method, whereas the self-tuning regulator approach has principally concentrated upon stochastic systems in which parameter estimation algorithms predominate. In the model reference approach to adaptive control the objective is to asymptotically drive the output of an unknown plant to that of a given reference model. In self-tuning regulation a design procedure for *known* plant parameters is initially selected and applied to the unknown plant using recursively estimated values of these parameters. In both of these approaches the question of convergence or stability of the adaptive algorithms is crucial, indeed it has been shown that adaptive control schemes which are globally stable are equivalent through the generation of identical error equations and essentially the same adaptive alws.

The recent developments in adaptive control have been accompanied by a series of international symposia on adaptive systems at Darmstadt and Bochum in Western Germany (1979), Yale (1979, 1981) and Illinois (1979) in USA and Kyoto, Japan (1981). These meetings and their proceedings have been primarily aimed at the exponents of adaptive control. The IEE Professional Group Committee on Control Theory, whilst recognising the importance of these symposia, appreciated that the increasing industrial importance of adaptive control through the recent advances in self-tuning regulation and their practical implementation through microprocessors, required to be disseminated to a wider audience at an introductory or tutorial level. Therefore the first Oxford IEE Adaptive Control Workshop was held on 26-27 March 1981 to provide an overview of the existing techniques of self-tuning and model reference control and their relevance to industrial implementation. The workshop programme was designed to give a balanced coverage of the theory of self-tuning, model reference adaptive control, variable structure control systems, and the control of badly defined systems as well as a series of typical industrial applications of adaptive control including chemical plants, turbo-charged diesel engines and surface ship control. Whilst the workshop was primarily aimed at industry with a view to encouraging further applications of adaptive control it was also designed to provide academics with a state of the art review of the subject. These two objectives are reflected in this essentially tutorial based book on the introduction to the theory and applications of adaptive control. The fourteen contributors to the workshop have each written a balanced and integrated chapter on the various aspects of modern adaptive control.

The book begins with an introduction to adaptive control systems and a brief review of the history of the subject. The general theoretical problem of nonlinear stochastic control is reviewed and suboptimal designs which correspond to the usage of the term adaptive control are considered. The concepts of separability, certainty-equivalence, caution and probing are summarised and current methods of adaptive control are reviewed. Chapter two provides a detailed derivation of the single-input single-output self-tuning algorithm and its variants. All aspects of self-tuning are discussed including parameter estimation, minimum-output-variance regulation, generalised self-tuning, implicit and explicit algorithms, pole-placement designs and some recent developments. Chapter three concentrates on multivariable self tuning systems. The chapter begins by considering a multivariable self-tuning regulator and proceeds to derive both the detuned minimum variance and pole assignment regulators for multivariable systems. Chapter four reviews the role of stability theory in adaptive control system design over the last twenty-five years. The emphasis is placed on continuous systems and the design of model reference schemes using Lyapunov and hyperstability techniques. The stability and convergence of discrete time algorithms are analysed in Chapter five. The analytic methods surveyed include the Liapunov, hyperstability, input-output and Martingale approaches. Chapter six considers the practical aspects of implementing self-tuning controllers. Robustness and integrity of the algorithms are discussed and the Fortran coding of a stable algorithm and an example of its application to the control of a chemical batch reactor is presented. A systematic design procedure for model reference adaptive systems based on Popov's hyperstability theory is presented in Chapter seven. Chapter eight considers the design of simple fixed structure controllers which provide set point tracking and disturbance rejection for multivariable plant where the detailed dynamical properties of the processes involved are unknown. Chapter nine reviews the theory of variable structure systems where the structure of the feedback controller is altered as the state crosses discontinuity surfaces in the state-space to yield insensitivity to plant parameter variations and disturbances. The design of fixed structure two-term controllers for unknown or badly-defined multivariable systems is considered in Chapter ten. The design yields a robust controller which ensures closed loop stability and insensitivity to the unknown components of the system dynamics. Chapter eleven considers applications of single and multivariable self-tuning controllers. A multivariable generalised self-tuning algorithm is derived which includes both set-point following and feedforward control action. The performance of the controllers are evaluated by application to the regulation of distillation column terminal compositions. The remaining three chapters discuss the application of self-tuning to engine control, surface ship course keeping and manoeuvring, and ship positioning systems. Each application is studied in great detail and all aspects of the controller design and implementations are discussed.

Finally the editors would like to express their appreciation to all the contributing authors, to Miss Julie Waite of the IEE for her efficient organisation of the Workshop, to Professor J H Westcott and Professor R W H Sargent for acting as Chairman at the Workshop and to Mrs J Swann for the excellent typing of the text.

May 1981

C J Harris
S A Billings

Chapter One

Introduction to adaptive control

O L R JACOBS

1.1 *Introduction*

Adaptive control has been an identifiable topic in control engineering for at least a generation. Its history, measured in numbers of papers abstracted by the IEE INSPEC service, is illustrated in Fig 1.1. It has mainly been concerned with feedback systems having the classical structure shown in Fig 1.2. The description 'adaptive' signifies that the controller in Fig 1.2 performs two simultaneous functions; it learns about the controlled process whilst, at the same time, controlling its behaviour.

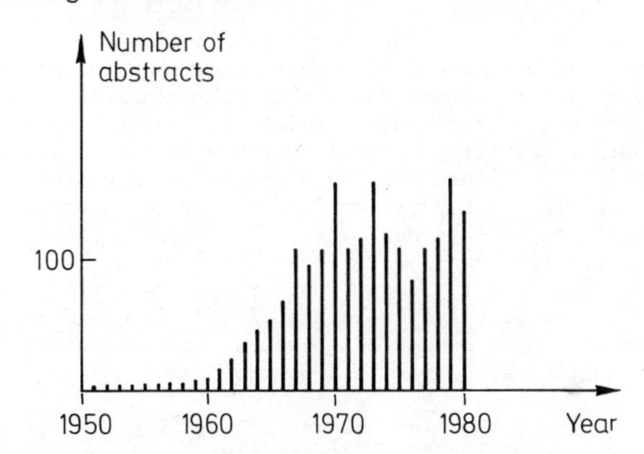

Fig 1.1 Number of INSPEC abstracts on adaptive control.

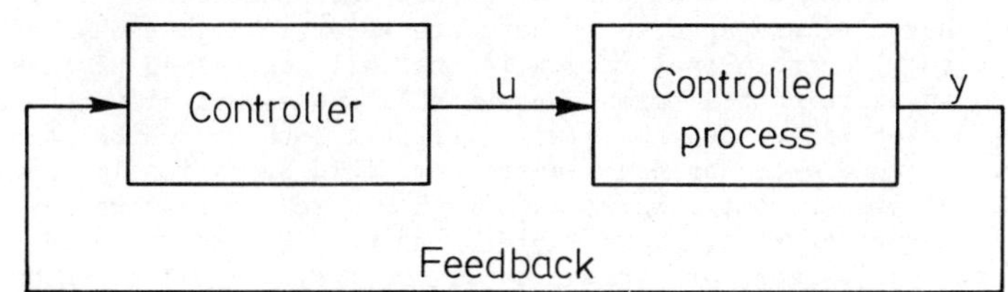

Fig 1.2 Feedback control system.

Dr Jacobs is with the Department of Engineering Science, University of Oxford.

Interest in adaptive control originated primarily with aerospace problems (Gregory, 1959). It was found that classical linear controllers in the configuration of Fig 1.2 did not always give satisfactory control of the attitude of high-performance aircraft and rockets. The difficulty arose because response characteristics of these controlled processes varied significantly in flight, and the classical controller could be matched only to a single flight condition: under other conditions within the flight envelope the fixed classical controller would be unsatisfactory. There seemed to be a need for a more sophisticated controller which could automatically adapt itself to the changing characteristics of the controlled process. It was recognised that such a controller would have many other valuable applications, for example in industrial process control or in controlling economic systems, and that it might be analogous to some biological control systems.

There are two controversial questions about adaptive control:

(i) Is there some precisely definable distinction between adaptive control and other feedback controls?

(ii) When is adaptive control useful?

The first question is controversial only if too wide a definition of adaptive control is attempted, as was done in the early days of the subject. It was argued that any broad definition, such as 'self-adjustment in accordance with changing conditions' (Aseltine et al, 1958), would not exclude classical linear feedback which has the well-known property of making closed-loop behaviour insensitive to operating conditions. The controversy can be avoided by taking Truxal's (1963) definition, that an adaptive controller is one 'designed from an adaptive viewpoint', and then stating a specific 'viewpoint'. Section 1.2 describes the viewpoint corresponding to current usage. It is that 'adaptive control' represents one particular approach to designing sub-optimal controllers for non-linear stochastic systems. This approach consists of assuming that the controlled process satisfies equations which would be linear if only the values of certain coefficients were known, and designing a controller which estimates coefficients and uses the resulting estimates in a control law.

The second question, about usefulness of adaptive control, remains very much alive (Jacobs, 1980 (a): Åström, 1980). It may seem surprising that, after the thousands of papers indicated by Fig 1.1, the question should remain open, but there are several reasons why this should be so. The principal reason is that all controlled processes for which adaptive control might be suitable are essentially both non-linear and stochastic, and therefore difficult both to control and to analyse. If they were not non-linear they could be optimally controlled by classical linear controllers; and if they were not in some way uncertain, or stochastic, there would be no need for learning in the form of self-adjustment or estimation of coefficients. Non-linear, stochastic problems are difficult because, almost by definition, there can be no general analytic solutions to them. In particular there are no general design procedures for designing controllers to match non-linear stochastic controlled processes; nor are there any general results about performance under optimal control to provide a point of comparison for performance under sub-optimal control. Each application may need to be the subject of a special case-study and there is no guarantee that the results of one study will carry over to any other situation.

In the absence of a usable general theory of adaptive control, research and development can only be assessed against the background of practical applications. Ideally no report would be regarded as complete without some indication of an appropriate application where the adaptive control could give better performance than classical control, which is usually cheaper to implement. That this counsel of perfection is not met by most of the literature of Fig 1.1 is the second reason why the question about usefulness remains open. A third reason, discussed further in Section 1.4.4, is that most published simulation studies have avoided the question.

The literature on adaptive control includes several useful survey papers. Early surveys by Aseltine et al (1958) and by Stromer (1959) characterise much of the literature: they attempt to classify adaptive control systems according to whatever 'viewpoint' was used to design the controller, but make little reference to specific applications and no reference to relative performances. Later surveys by Jacobs (1961) and by Wittenmark (1975) are of a similar nature. Asher et al (1976) give a bibliography classifying a large number of publications (about 700) in the same way and also according to nominal field of application. None of the above papers attempts critical assessment of the work surveyed.

A recent survey of applications by Parks et al (1980) is less uncritical. It includes the following summary: 'the number of significant applications is really quite small ... only very few reports deal with ... the continuous use of such control systems for more than a short pilot-scale experimental period ... reports on applications are thin and performance data from such systems even thinner'. Critical assessment has also been expressed about a NASA-sponsored program which continued the original aerospace interest by developing adaptive auto-pilots for one particular jet fighter aircraft. The assessment appears (IEEE, 1977) to have been that adaptive control is normally of little value because variations in response characteristics of the controlled process can be adequately compensated by gain-scheduling (or feed-forward control) based on available measurements.

A minority of the literature is not subject to the criticisms implied above. At least six papers have reported successful applications of adaptive control to real controlled processes. Gilbart and Winston (1974) used model-reference adaptive control to improve the tracking accuracy of a 24" optical tracking telescope by a factor of as much as six. Borrison and Syding (1974) used self-tuning adaptive control to increase the throughput of a 200 KW ore-crusher by about 10%. Cegrall and Hedquist (1975) used a similar adaptive scheme to improve the performance of the moisture control loop in a 130,000 tonne per year paper-making machine; they report that improvements, both to steady-state and to transient performance, were so marked that the adaptive scheme was permanently installed on the machine. Dumont and Belanger (1978) achieved similar results on a full-scale titanium dioxide kiln; they reported a 10% improvement in steady-state performance, and also improved transient performance which gave 'by far the most successful grade change performed in the plant'. Adaptive autopilots for ships have been more successful than for aircraft; Källström et al (1979) used self-tuning control of steering to achieve speed increases of order 1% in a 355,000 ton tanker, and Amerongen (1980) used model-reference methods to achieve smaller increases of average speed in an ocean-going survey vessel.

The above examples indicate that adaptive control can be useful and thus go some way to justify the continuing volume of research and development shown in Fig 1.1. Further stimulus is provided by the spread of inexpensive mini and micro computers which could easily be used to implement adaptive control laws of considerable complexity. A great variety of potential fields of application is indicated in the surveys by Asher et al (1976) and by Parks et al (1980). The history and current status of the subject is usefully summarised in the Yale 1979 workshop on Applications of Adaptive Control (Narendra and Monopoli, 1980).

Although assessment of work on adaptive control must ultimately be in terms of practical applications, control theory has much to offer to the design of adaptive controllers. The remainder of this Chapter is devoted to a review of adaptive control from the point of view of stochastic control theory. Its objective is to provide a conceptual foundation which can lead to successful practical applications and to elimination of controversial questions about adaptive control. Section 1.2 describes the general theoretical problem of non-linear stochastic control and specifies the particular approach to designing sub-optimal controllers which corresponds to current usage of the term 'adaptive control'. Section 1.3 summarises some relevant control-theoretic concepts: neutrality, separability, certainty-equivalence, caution and probing. Section 1.4 reviews the current state of adaptive control in the light of the foregoing. A concluding Section 1.5 summarises Chapter One.

1.2 *Approaches to controller design*

Adaptive control is just one sub-optimal approach to designing algorithms to control non-linear stochastic processes. Specification of the sub-optimal is introduced here by a summary of the optimal and of why it does not generally lead to usable algorithms. In the discussion three levels of optimality are distinguished: 'optimal' control, which is the best that could possibly be achieved; 'sub-optimal' control, which gives performance sufficiently close to optimal to be a practical proposition; 'non-optimal' control which gives performance significantly worse than optimal.

The whole discussion refers to a general mathematical model of controlled processes which is outlined in Section 1.2.1. Optimal stochastic control theory is summarised in Section 1.2.2 and classical linear control, which can be regarded as the simplest sub-optimal approach, is described in Section 1.2.3. Section 1.2.4 summarises estimation procedures, known as 'system identification', which can be combined with classical control laws to give the 'adaptive control' described in Section 1.2.5. An alternative approach which used to be described as adaptive but is now known as 'extremal' control is mentioned in Section 1.2.6.

1.2.1 Mathematical model of controlled process

For purposes of control theory it is assumed that the controlled processes of Fig 1.2 belong to a general class which can be represented by discrete-time equations relating states x and outputs y to inputs u and noises ξ as follows

$$x(i+1) = G_1 (x(i), u(i), \xi_1 (i), i) \quad (1)$$

$$y(i) = G_2 (x(i), \xi_2(i), i) \quad (2)$$

Equation (1) specifies how the dynamic states x (n in number) evolve in time i (an integer) under the influence of controls u (m in number) and of random variables ξ_1 (up to n in number). Equation (2) specifies how observable outputs y (p in number) depend on the state x and on random variables ξ_2 (up to p in number). The arguments of the vector functions G include time i to show that the functions could be time-varying. The noise variables ξ are 'white' in the sense that their values at any one point in time are independent of their values at any other time. The specification of any particular controlled process includes specifications of the functions G, of probability distributions for the noises ξ and of a probability distribution for the state $x(i_1)$ at some starting time i_1.

Two principal assumptions in the representation of equations (1) and (2) are:

(i) The process is to be controlled in discrete-time, that is by a digital controller rather than by an analog controller.

(ii) The number of state variables is finite. This implies that the controlled process is regarded as a lumped parameter process rather than as a distributed process.

These assumptions avoid some technical mathematical difficulties and do not cause serious practical problems.

1.2.2 Optimal control

Optimal stochastic control theory (Bellman, 1961; Fel'dbaum 1965; Aoki, 1967; Bertsekhas, 1976) is concerned with minimising summed cost functions of the form

$$I(i_1) = \sum_{i=i_1}^{i_2} H(x(i+1), u(i), i) \quad (3)$$

where H(i) is a scalar function specifying how the cost at each point i in a time interval starting at i_1 depends on the current controls u(i) and on the states x(i+1) resulting from those controls. The minimisation is subject to the dynamical constraint of equation (1) and also to equation (2) which constrains the information available about current states x. The random variables ξ ensure that current and future values of the states x are always uncertain, so that future values of the controls u are uncertain, and current and future values of the scalar H are also uncertain. The cost function I, when summed over future times, is thus a random variable and consequently unsuitable as a criterion of optimality. An appropriate non-random criterion is provided by the expected value J of I (Gessing, 1980)

$$J = \underset{?}{E}\,[I] \tag{4}$$

where the query indicates that it is necessary to specify the set of random variables with respect to which expectation is to be taken.

The relevant set of random variables varies with time. At the starting time i_1 the set of random variables affecting $I(i_1)$ consists of the starting state $x(i_1)$ together with the sequence of all noises ξ. At some later time i, during the interval (i_1, i_2), the set would consist of the uncertain current state x(i) together with the sequence of current and future noises ξ, which can conveniently be written ξ^i where

$$\xi^i = \{\xi_1(i), \ldots, \xi_1(i_2-1), \xi_2(i+1), \ldots, \xi_2(i_2-1)\}$$

This set $(x(i), \xi^i)$ can be used in (4) to give a non-random performance criterion

$$J(i) = \underset{x(i),\xi^i}{E}\,[I(i)] \tag{5}$$

against which the control u(i) could be optimised at time i provided that probability distributions were available for the random variables $x(i),\xi^i$. Probability distributions for the random variables in (5) are implicit in what was specified about the controlled process. Distributions for the sequence ξ^i are assumed to be known. The relevant distribution for x(i) is a conditional distribution which is, in principle, determined by combining prior information about the controlled process with the known sequence of past controls u and outputs y, which can conveniently be written y^i where

$$y^i \equiv \{u(i_1), \ldots, u(i-1), y(i_1), \ldots, y(i)\}.$$

The conditional distribution is written $p(x(i)|y^i)$ and this is sometimes abbreviated to $p(x|y)$. Algorithms for combining prior information with observed sequences of inputs and outputs so as to update the conditional distribution are the subject of estimation theory (Jazwinski, 1970).

Optimal control theory specifies the optimal value of control u(i) as part of the solution to equations characterising the optimal (minimal) value of the cost function. The main results of optimal stochastic control theory are:

(i) The optimal value, written f(i), of the expected value J(i) of the cost function I(i) satisfies a functional recurrence equation

$$f(i) = \min_{u(i)} \{ \underset{x(i),\xi^i}{E}[H(i) + f(i+1)]\} \tag{6}$$

This equation is notoriously insoluble because it is non-linear, and because its solution must satisfy mixed boundary conditions, some specified at the current time i and some at the terminal time i_2.

(ii) The optimal control u(i), which minimises the right hand side of equation (6), is a function of the conditional probability distribution $p(x(i)|y^i)$ for the current state x(i). This contributes further to the intractability of equation (6) because the estimation theory which specified how $p(x(i)|y^i)$ should be updated is itself a major topic of about the same difficulty as optimal control theory. A further difficulty is that probability distributions are generally functions having infinite dimensionality so that the solutions f and u to equation (6) would be functionals $(f(p(x|y)), u(p(x|y))$ of the probability distribution rather than functions $(f(x), u(x))$ of the finite number of states. In a few special cases the conditional distribution is characterised by a finite number of sufficient statistics (Striebel, 1965) so that this further difficulty is eliminated; the small repertoire of problems for which equation (6) has been solved are all of this nature.

The above results do not lead to specific optimal solutions to specific practical problems, but they do have at least one important practical conclusion. It is that a controller for a stochastic system should perform two separate functions at each point i in time when control is exercised:

(i) update $p(x(i)|y^i)$

(ii) use the resulting information about x(i) to determine u(i).

Fig 1.3 shows the structure of such a controller in the feedback control system of Fig 1.2.

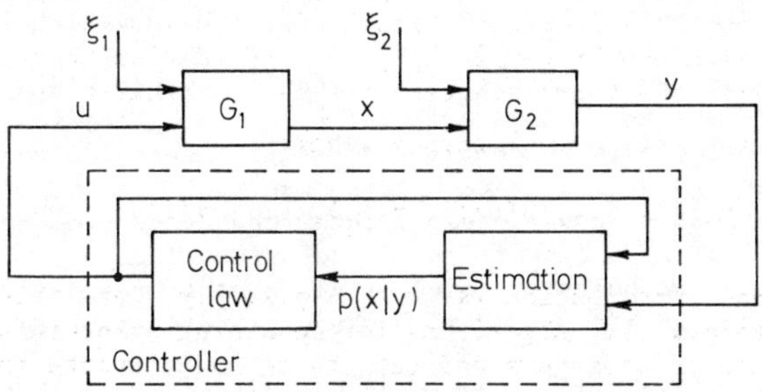

Fig 1.3 Structure of optimal controller.

The optimal way to perform these two functions can be found by analysis only where equation (6) can be solved to specify an optimal control function and where estimation theory also leads to explicit algorithms for optimal estimation. The only significant class of problems where these conditions are met is where the functions G are linear, the cost function H is quadratic and the random variables ξ and $x_1(i_1)$ are all Gaussian. This class, known as LQG (linear, quadratic, Gaussian), includes all the linear problems of classical feedback control theory; it is the subject of a large literature (for example IEEE 1971, Kwakernaak and Sivan, 1972). For almost all other problems either or both of the functions of estimation and control must be done sub-optimally.

Practical problems are generally non-LQG. Real controlled processes almost always include some more or less significant non-linearity; real performance objectives are seldom identical to that of a scalar quadratic cost function; real random variables, disturbances and noises are often non-Gaussian. Practical controllers are therefore usually sub-optimal and can be classified according to the sub-optimal approach used in their design, as follows.

1.2.3 Classical control

The simplest sub-optimal approach is to assume that the problem is LQG and to design a corresponding linear controller. Because every linear controller is optimal for some LQG problem (Kalman, 1964) classical control theory, which is concerned with the design of linear controllers, can be regarded as embodying this approach.

The starting point for controller design is then an explicit assumption that the controlled process can be represented by a linear mathematical model in the form of a transfer function or corresponding difference equation (or differential equation for continuous-time processes) having known constant coefficients a, b, c, d and time-delay k ($k \geq 1$),

$$\sum_{j=0}^{n} a_j y(i-j) = \sum_{j=0}^{n} b_j u(i-k-j) + \sum_{j=0}^{n} c_j \xi(i-j) + d. \qquad (7)$$

In (7), and in most of the rest of this chapter, u and y are the scalar input and output of a single-input single-output (SISO) controlled process. The discussion here is restricted to SISO systems, as is most work on adaptive control, in the interests of simplicity. The classical approach often achieves further simplification by neglecting the bias d and the white Gaussian random variables ξ in (7). Classical design seeks to compromise between conflicting requirements for good transient response and for good steady-state performance of linear feedback control systems. Transient response, associated with the complementary function term in the solution to linear dynamic equations (Jacobs, 1974), can be characterised by the poles or modes of closed-loop response. Steady-state performance is associated with the particular integral.

The above simple classical approach has been found to lead to satisfactory performance in a great variety (Dorf, 1980) of real non-LQG control problems. The main reason for this is that classical control is robust. The performance of a well-designed classical feedback system is insensitive to values of parameters of the controlled process (Horowitz, 1963) and so the linear mathematical model of equation (7) can sometimes be quite inaccurate without much serious effect on performance.

1.2.4 System identification

Information about the model (7) is nevertheless essential as a starting point for controller design. Much of this information is implicit in the behaviour of the controlled process and some of it can actually be extracted from observed time-histories of input and output. The development of algorithms to perform this task is a recognised topic in control engineering, known as 'system identification' (Eykhoff, 1974; Isermann, 1981). It draws on classical methods of estimation theory and has a history similar to that of adaptive control. Most identification algorithms generate estimates of coefficients of linear equations, such as the a, b, c, d of equation (7). The algorithms are usually designed on the assumption that structural information about the process to be identified is available a priori. Such structural information includes:

(i) the order n of the dynamics

(ii) the magnitude k of the time-delay

(iii) the location with respect to the unit circle of factors of the polynomials† A(z), B(z), C(z) which specify poles and zeros, and hence stability, of the corresponding transfer functions.

(iv) whether or not there is a non-zero bias term such as d in (7).

Unfortunately it is the structural information which is usually most important for purposes of controller design. The information needed to design an identification algorithm might also be used to design a fixed-parameter classical feedback controller having performance which could be insensitive to precise values of the coefficients a, b, c, d. Any estimates subsequently generated by identification would contribute little more to control performance. Thus, from the point of view of control, the information provided by system identification is liable to suffer from what might be caricatured as the 'bank-loan' effect: it is normally only available under conditions which make it inessential.

1.2.5 Adaptive control

The sub-optimal approach which has come to be known as 'adaptive' originated as a scheme of simultaneous identification and control. It combined the identification algorithms of Section 1.2.4 with the classical or LQG design procedures of Section 1.2.3. The 'bank-loan' effect, that information obtainable by identification might have little value for purposes of control, was not considered. This oversight could be a factor contributing to the controversies mentioned in Section 1.1.

The approach assumes that the controlled process can be described by a mathematical model such as equation (7) in which the values of coefficients a, b, c, d may be uncertain. Real-time algorithms are used to estimate the uncertain coefficients and the resulting estimates affect a control law designed to match the assumed model of equation (7) which would be linear if all the coefficients had known values. The approach could therefore be regarded as being based on a linearisation of, or Taylor series approximation to, the underlying non-linear controlled process. However equation (7) itself is not linear when the values of some of its coefficients are uncertain, so the approach does not actually amount to linearisation. The non-linearity becomes

†The notation here is $X(z) = \sum_j x_j z^{-j}$ where z is the z-transform variable or forward shift operator. When the summation is from j=0 to j=∞ then X is the z-transform of x (Jacobs, 1974).

apparent in a state-space representation of equation (7) in which the uncertain coefficients are represented by additional state variables. The point is illustrated here by giving the state-space representation of a simple first-order (n=1) version of equation (7)

$$y(i) + ay(i-1) = bu(i-1) + \xi(i) + c\xi(i-1) + d \qquad (8)$$

Four state variables are needed to represent the four uncertain coefficients a, b, c, d

$$x_1 \equiv -c,\ x_2 \equiv b,\ x_3 \equiv c - a,\ x_4 \equiv d,$$

and a fifth state

$$x_5 \equiv y - \xi$$

represents dynamics of the first-order equation (8). The non-linearities show up in the updating equation for x_5 which, by substitution in (8), is

$$x_5(i+1) = x_1(i)\, x_5(i) + x_2(i)\, u(i) + x_3(i)\, y(i) + x_4(i) \qquad (9)$$

The principal non-linearity in (9) is the term x_1x_5 which represents uncertainty about c. The term x_2u which represents uncertainty about b is bilinear (Mohler, 1973; Jacobs and Potter, 1976): it makes the design of control laws a non-linear problem, because control u is multiplied by uncertain state x_2, but does not affect the linearity of state estimation because uncertain past states x_2 are multiplied by known past controls u. The term x_3y which represents uncertainty about a is similarly benign, and the term x_4 which represents uncertainty about d is linear.

The above categorisation of non-linearities does not depend on the order of equation (8) and so is also applicable to equation (7). It shows that in the assumed mathematical model, which is the basis of the adaptive approach, the non-linearities range from severe to non-existent as follows:

(i) Uncertainty about the coefficients c, which model noise effects, introduced genuine non-linearity. This will make it difficult to design both of the two functions, estimation and control, to match a controlled process satisfying equation (7).

(ii) Bilinearity is introduced by uncertainty about the coefficients b which model the gain and the zeros of controlled process dynamics, and by uncertainty about the coefficients a which model the poles of dynamics. This will make it difficult to design a control law, but not an estimator, to match equation (7).

(iii) Uncertainty about the bias d introduces no non-linearity but can cause observability problems as will be mentioned in Section 1.4.3.

Adaptive control is always concerned with problems where there is uncertainty about some or all of the coefficients a, b, c and thus with problems where the assumed equation (7) is non-linear. There is therefore as little likelihood of designing an optimal controller to match the assumed model (7) as there was of designing an optimal controller for the underlying non-linear controlled process. What the adaptive approach does is to replace the problem of designing sub-optimal algorithms to match non-linear processes in general, by the problem of designing sub-optimal algorithms to match the standardised non-linearities of equation (7). Fig 1.4 illustrates the structure of sub-optimal controllers designed using this approach. Such controllers risk being non-optimal in two different ways:

(i) The controller intended to match equation (7), which is non-linear because of the uncertain coefficients, may be non-optimal even for equation (7).

(ii) Modelling errors in using equation (7) to represent the actual controlled process may be so severe that even the optimal controller to match equation (7), if it could be found, would be non-optimal as a controller for the actual process.

Nothing general is known about the extent of these non-optimalities, which is one of the reasons why the usefulness of adaptive control is still questioned.

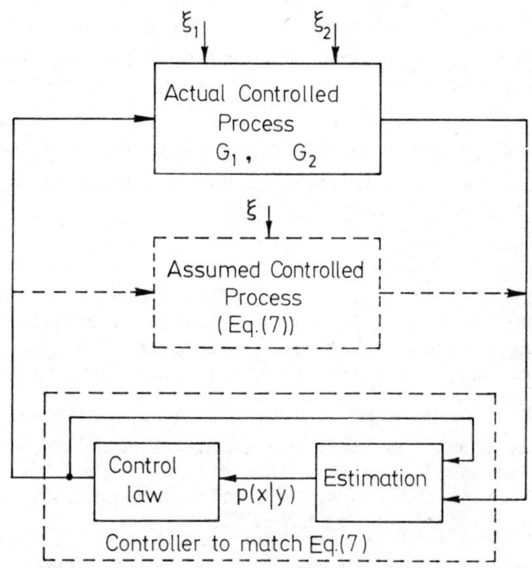

Fig 1.4 Structure of adaptive controllers.

1.2.6 Extremal control

The above approaches to controller design; optimal, classical and adaptive, can mainly be described as implicit with respect to the cost functions H, I and J. They are implicit in the sense that no explicit measurement of costs is required for controller operation; the cost function affects control action implicitly, by way of the control law which is derived for the specified cost function.

An alternative, explicit, approach would be actually to measure the cost H and to use this measurement to drive a feedback controller specially designed to adjust the controls u so as to minimise H. Such a controller would be designed to search for an extreme† value in the relationship between inputs u and outputs H. The approach is therefore known as extremal control (Jacobs and Langdon, 1970). It has the structure shown in Fig 1.5(a) and differs from the general optimal structure of Fig 1.3 as follows:

(i) No attempt is made to estimate any of the dynamic states x of equation (1) nor any of the coefficients of an input-output model such as equation (7).

(ii) The approach concentrates explicitly on the relationship between input u and cost H rather than on that between u and the states x or their measured values y. This relationship, being extremal, is inherently non-linear, as illustrated by the simple static characteristic shown in Fig 1.5(b). Controller design is thus inherently more difficult than in other approaches which are based on linearisable relationships.

These differences could be expected to have an adverse effect on performance.

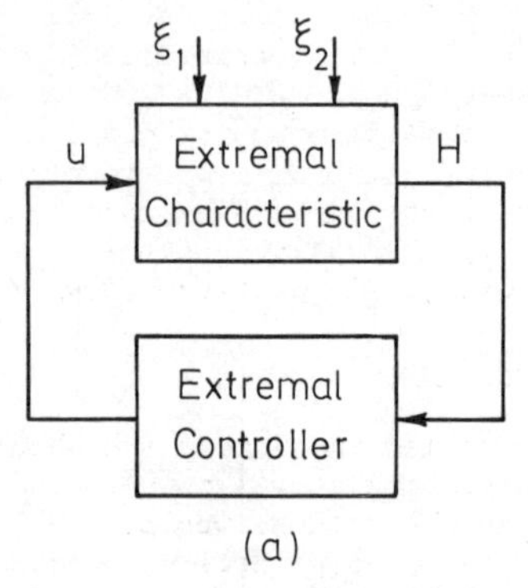

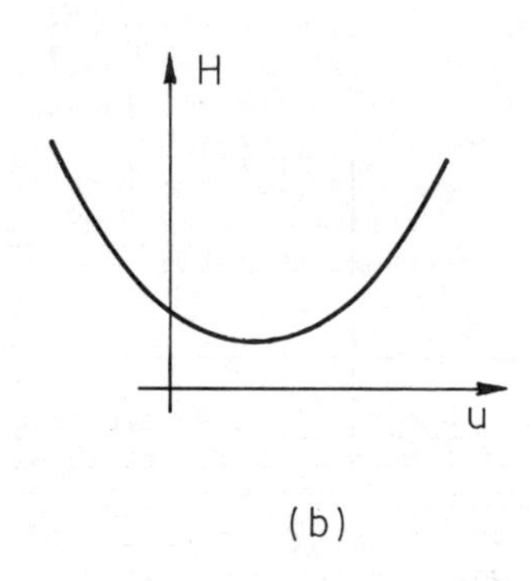

Fig 1.5 Extremal control
(a) Structure of control system (b) Static extremal characteristic

Extremal control is mentioned here mainly because it is part of the history of adaptive control. Many of the papers during the first decade shown in Fig 1.1 were about extremal control, but its impact on control engineering has been negligible and it seems no longer to be an active research topic. Its closest descendant is model-reference adaptive control, discussed in Section 1.4.2, which is explicit in the same sense as is extremal control and which uses adaptation mechanisms bearing some relationship to extremal control algorithms.

†'Extreme' means either maximum or minimum. Here the output H is to be minimised, but extremal control can also be designed for maximisation.

1.3 *Structural properties of stochastic control*

The known results of optimal control theory, summarised in Section 1.2.2, may not lead directly to usable general algorithms but they have given rise to theoretical concepts which are relevant to the design of all stochastic control algorithms, including adaptive algorithms. These concepts characterise interactions between the two functions of estimation and control shown in Fig 1.3. Two types of interaction are possible:

(i) Uncertainty in the current estimate of the current state could result in cautious control which exercises control less strongly than if there were no uncertainty.

(ii) Control might introduce some special probing action in order to reduce future uncertainty more rapidly than would otherwise happen.

Both interactions happen to be absent in the class of LQG problems, which is one reason why problems of optimal LQG control are soluble. It is also a reason why LQG results cast little light on interactions in other problems.

The relevant theoretical concepts are described here as 'structural properties' of stochastic control. Three such properties of a stochastic control problem (Patchell and Jacobs, 1971), neutrality, separability and certainty-equivalence are defined in Section 1.3.1, and some sufficient conditions under which they are known to hold are summarised in Section 1.3.2. Two corresponding properties of stochastic control laws (Jacobs and Patchell, 1972), caution and probing, are defined in Section 1.3.3. Their applicability to adaptive control is discussed and illustrated by a theoretical example in Section 1.3.4.

1.3.1 Structural properties of control problems

1.3.1.1 Neutrality. A stochastic control problem is neutral (Feldbaum, 1965) if the rate of reduction of uncertainty about states x is independent of the control u. In a neutral problem there would therefore be no reason to introduce any special probing action, because there would be no possibility of affecting the rate of reduction of uncertainty.

Neutrality is a property of the estimation function in a stochastic controller. Its distinguishing feature is that the shape of the conditional probability distribution $p(x(i)|y^i)$ is independent of past controls u. This shape could be characterised by central moments (Bar-Shalom and Tse, 1974) but is more economically summarised by the information-theoretic measure of negative entropy

$$s(i) = -\int_{x(i)} p(x(i)|y^i) \log(p(x(i)|y^i))\, dx(i) \qquad (10)$$

The resulting definition is that

a stochastic control problem is neutral when equations (1) and (2) are such that s(i) is independent of all past controls u(j) (j<i), for all i.

For a Gaussian conditional distribution $p(x|y)$ the negative entropy of equation (10) would be proportional to the covariance matrix Σ according to

$$s = \tfrac{1}{2} \log |\Sigma| + \text{constant} \tag{11}$$

A control problem having Gaussian conditional distribution would therefore be neutral if the conditional covariance matrix Σ were independent of the controls, as it is in LQG problems.

1.3.1.2 Separability. All stochastic control problems can be loosely described as separable in the sense that the controller performs the two separate functions of Fig 1.3. The property defined here refers more specifically to the amount of information about the current state $x(i)$ which needs to be transmitted from the estimation function in order to determine control action. A stochastic control problem is separable (Joseph and Tou, 1961) if the only information which needs to be transmitted is a point estimate $\hat{x}(i)$ of the current state, no information about accuracy being required. Separability is thus a property of the optimal control laws and indicates that the optimal control is independent of the accuracy of information about current states. It implies that neither probing, to improve accuracy, nor caution, to allow for inaccuracy would be needed.

The definition is that

a stochastic control problem is separable when equations (1), (2) and (3) are such that the optimal control law has the form

$$u(i) = \text{function of } (\hat{x}(i)). \tag{12}$$

This definition may be useless for practical purposes because it is not generally possible to determine optimal control laws. Its value here is in establishing the concept of separability. Fig 1.6 shows the structure of the optimal controller for a separable problem: it differs from the general controller of Fig 1.3 only in the information transmitted from the estimator to the control law.

In LQG problems the optimal control is proportional to the conditional mean m of $p(x(i)|y(i))$ and is independent of the covariance Σ. The mean m is one form of estimate $\hat{x}$, and so LQG problems are separable.

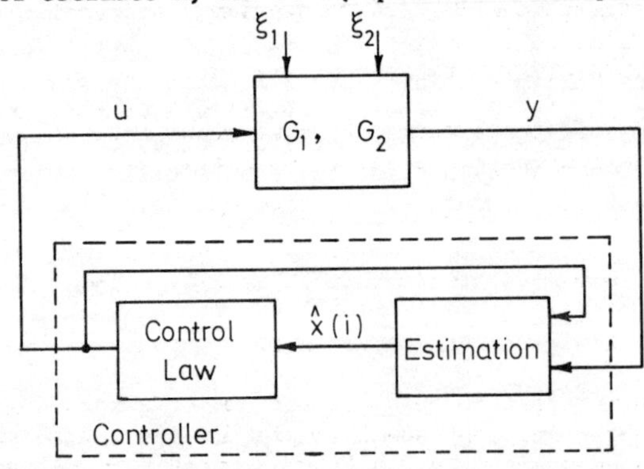

Fig 1.6 Structure of controller for separable problem.

1.3.1.3 Certainty-equivalence. Certainty-equivalence is a stronger property than either neutrality or separability. It refers only to separable problems and relates to the form of the control law in equation (12). A separable problem is certainty-equivalent (Simon, 1956; Theil, 1957) when the form of this function is identical with that of the function specifying the optimal feedback control law for an equivalent deterministic problem, defined below, having no uncertainties about the state x. Certainty-equivalence implies that the control law can be designed without any consideration for stochastic effects.

The equivalent deterministic problem for the general stochastic problem of Sections 1.2.1 and 1.2.2 has dynamic equation similar to equation (1) except that the random variables ξ, are replaced by their mean values of zero and that there is assumed to be perfect information (y=x) about the state x which satisfies

$$x(i+1) = G_1(x(i), u(i), 0, i). \tag{13}$$

The deterministic control problem is then to minimise the cost function I of equation (3) subject to the constraint of equation (13). The solution to this problem can be expressed as a feedback control law of the form

$$u^c(i) = \text{deterministic feedback of } (x(i)) \tag{14}$$

as shown in Fig 1.7.

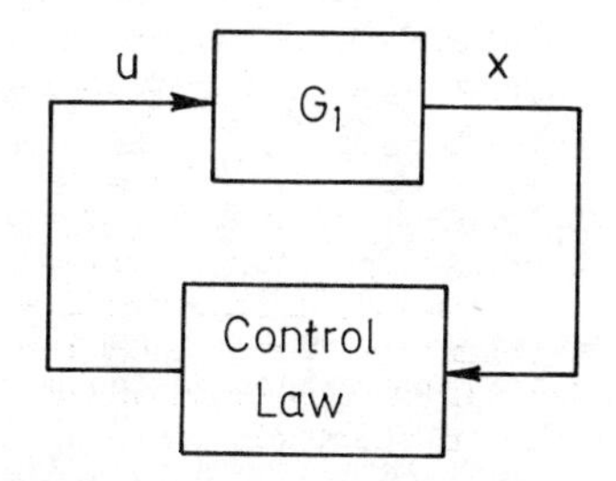

Fig 1.7 Deterministic equivalent of system in Fig 1.3.

The control law of equation (14) and Fig 1.7 could be used within the separable structure of Fig 1.6 to provide a stochastic control law

$$u^c = \text{deterministic function of } (\hat{x}(i)) \tag{15}$$

This control is said to be certainty-equivalent and is not necessarily optimal. The definition of certainty-equivalence is that

A stochastic control problem is certainty-equivalent when equations (1), (2) and (3) are such that its optimal control is the certainty equivalent control law of equation (15).

LQG problems are certainty-equivalent.

The three definitions of this Section cannot generally be used to determine whether or not any specific problem has the defined structural properties. To determine separability or certainty-equivalence it would be necessary to determine the optimal control law: to determine neutrality would require expressions in closed form for the conditional probability density or its estimator. These results are not generally obtainable for non-linear stochastic problems such as those to which adaptive control may be a sub-optimal solution. The significance of the defined properties is that they provide a conceptual basis for discussing interactions between the dual functions of a stochastic controller. The concepts can be useful for the design of sub-optimal approaches such as adaptive control.

1.3.2 Some sufficient conditions

Necessary and sufficient conditions for problems in the general class of problems (1), (2) and (3) to be neutral, or separable, or certainty-equivalent are not known. What is known is that LQG problems possess all three properties and that some sufficient conditions for each of the three are as follows

A problem is:

(i) *neutral* if the functions G of equations (1) and (2) are linear (Bar-Shalom and Tse, 1974)

(ii) *separable* if, in addition to (i), all the random variables are Gaussian (Striebel, 1965; Wonham 1968)

(iii) *certainty-equivalent* if, in addition to (i), the function H of equation (3) is quadratic (Akashi and Nose, 1975; Alspach and Sorenson, 1971; Bar-Shalom and Tse, 1974; Root, 1969).

The properties of neutrality and separability are preserved (Wonham, 1968) when equation (1) has a more general linear form

$$x(i+1) = Ax(i) + B(u(i)) + \xi_1(i) \tag{16}$$

in which the control u enters as a, possibly non-linear, additive term such as arises when there are constraints on the magnitude of u.

The above conditions are summarised on a Venn diagram in Fig 1.8.

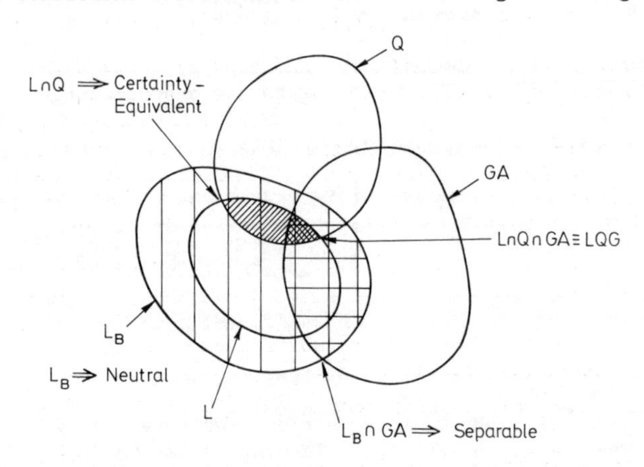

Fig 1.8 Venn diagram summarising sufficient conditions.

The diagram shows the set L of all problems having linear functions G, the set Q of all problems having quadratic function H, and the set GA of all problems having Gaussian random variables. It also shows the set L_B of problems where function G_1 has the form of equation (16) and function G_2 is linear. L is then a subset of L_B and LQG is a subset of $L \cap Q$, the known set of certainty-equivalent problems.

Linearity is an essential part of all the above conditions. It follows that adaptive control, which is an approach to non-linear problems, is concerned with problems which do not satisfy the known sufficient conditions for neutrality, separability or certainty-equivalence and which are very unlikely to enjoy any of the three properties. Interactions between the dual functions of estimation and control are therefore to be expected in adaptive control and adaptive algorithms should be designed to allow for these interactions. The properties of caution and probing, defined in Section 1.3.3 below, provide a conceptual basis for such designs.

1.3.3 Properties of stochastic control laws

For a given stochastic control problem the optimal control law would make optimal allowance for any interactions between the dual functions of estimation and control. The certainty-equivalent control, defined in Section 1.3.1.3, makes no allowance for any interactions. The difference between optimal control and certainty-equivalent control can therefore be regarded as some indication of the optimal allowance to be made. It can be written $(u^* - u^c)$ where u^* represents optimal control and u^c represents certainty-equivalent control.

In allowing for interactions there are two possible effects; caution and probing as described at the beginning of Section 1.3. These effects are identified here by arbitrarily assuming them to be additive, so that the difference between optimal and certainty-equivalent control is written

$$u^* - u^c \equiv \text{caution} + \text{probing} \tag{17}$$

The single equation (17) cannot identify two separate properties of caution and probing. The definitions are completed by introducing a further hypothetical control law, the neutral control u^n (Jacobs and Patchell, 1972). This is the control which would be optimal if all interaction between current control and future uncertainty were ignored. The neutral control thus excludes probing, and would be optimal in a neutral problem. No general definition to specify neutral control laws is known, but single-stage control problems ($i_2 = i_1$ in equation (3)) are always neutral because they have no future time when future uncertainty could be of consequence. In some specialised simple examples, including one to be summarised in Section 1.3.4, this single-stage control can be identified as the neutral control.

The significance of neutral control is that it provides a conceptual definition of probing as the difference between optimal control and neutral control

$$\text{probing} \equiv u^* - u^n. \tag{18}$$

A corresponding definition of caution is also provided by combining equations (17) and (18) to give

$$\text{caution} \equiv u^{n} - u^{c}. \tag{19}$$

Caution and probing, as defined here, are theoretical concepts which cannot generally be evaluated in non-LQG problems, where they may be non-zero. Their significance is that they suggest structure for the control law in the general feedback controller of Fig 1.3 when the controlled process is non-linear. Equation (17) can be rewritten

$$u^{*} = u^{c} + \text{caution} + \text{probing}$$

in a form which shows that stochastic control can be regarded as the sum of three components; certainty-equivalent control, caution and probing. The suggestion is that in non-linear problems, where it may be impossible to implement either optimal control or optimal estimation, a useful suboptimal control u^{so} might be constructed by summing three corresponding suboptimal components to give

$$u^{so} = u^{c,so} + \text{caution}^{so} + \text{probing}^{so} \tag{20}$$

This suggestion is as relevant to non-linear problems which are to be the subject of adaptive control as to any other problems.

1.3.4 Caution and probing in adaptive control

One well-known form of probing is the use of test signals in system identification (Section 1.2.4). Successful estimation of process parameters requires that the input signal u be 'persistently exciting' (Åström and Bohlin, 1965); when this condition is unlikely to be satisfied special test signals which may be sine waves or pseudo-random sequences are added to the input u in order to probe the process (Godfrey and Brown, 1979).

Many of the adaptive controls described in the literature have used sub-optimal certainty-equivalent control laws matched to equation (7) with no attempt to introduce either probing or caution. This could be one reason why there have been so few successful applications of adaptive control. That caution and probing may be needed for adaptive control is indicated by the results of an optimal control study of a simple example (Jacobs and Patchell, 1972) which is summarised below.

The study was of a controlled process for which equations (1) and (2) had the form

$$x_1(i+1) = x_1(i) + x_2(i)\,u(i) \tag{21a}$$

$$x_2(i+1) = x_2(i) \tag{21b}$$

$$y(i) = x_1(i) + \xi(i) \tag{21c}$$

This represents a process satisfying the first-order input-output equation (8) with uncertainty about the coefficient b and with other coefficients having the values

$a = -1, \ c = -1, \ d = 0$

The only non-linearity here is the bilinear term $x_2 u$ in equation (21a). The example was chosen more for its solvability than for its practical significance, but it approximately represents control of an integrator having unknown gain and noisy output. The probability distributions for $\xi(i)$ and for the initially uncertain states $x(1)$ are assumed Gaussian so that the conditional distribution $p(x(i)|y^i)$ is also Gaussian with mean m and covariance matrix Σ providing sufficient statistics which can be updated by a Kalman filter. The elements σ_{11}, σ_{12}, σ_{22} of the covariance matrix here depend on past controls, so the problem is not neutral. The resulting equations (6) of optimal control are of sufficiently low dimension that they could be solved numerically (Patchell, 1971) and optimal controls to minimise N-stage quadratic cost functions of the form

$$J_N(i) \;=\; \underset{x(i),\xi^i}{E}\left[\sum_{j=i}^{i+N-1} x_1^2(j+1)\right]$$

have been computed for values of N up to six. Fig 1.9 shows how the optimal control u_N^* was found to depend on the conditional mean m_1 of the state x_1 and on the number N of stages-to-go for some fixed combination of values of the other sufficient statistics (m_2, σ_{11}, σ_{12}, σ_{22}). The figure also shows:

(i) the certainty-equivalent control

$$u^c = -m_1/m_2 \tag{22}$$

(ii) the neutral control, which is identified with the single-stage (N-1) control,

$$u^n = u_1^* = -\frac{m_1 m_2 + \sigma_{12}}{m_2^2 + \sigma_{22}} \tag{23}$$

(iii) the caution $u^n - u^c$

(iv) the probing $u^* - u^n$

It can be seen that both caution and probing are present. These are attributable to the effect of uncertainty about state x_2; if there were no such uncertainty (σ_{12}=0, σ_{22}=0) the neutral control of equation (23) would be identical with the certainty equivalent-control of equation (22), which would be optimal for all N. Simulations showed that, at least for small values of N, the performance of sub-optimal controls which neglected caution or probing could be significantly worse than performance under optimal control. The certainty-equivalent control, which had neither caution nor probing, could be particularly unsatisfactory.

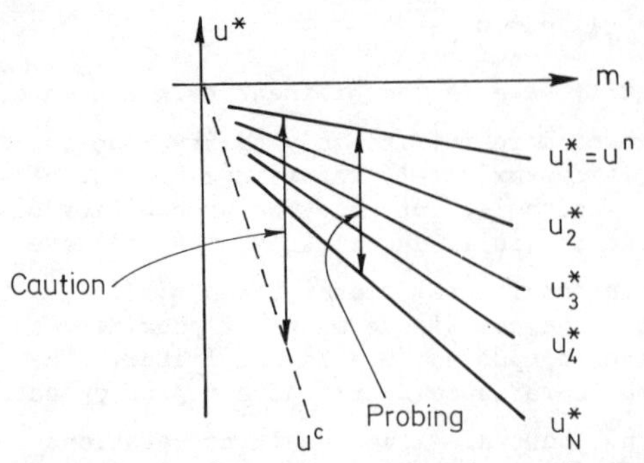

Fig 1.9   Example of caution and probing in optimal control.

It is not known whether the above comparisons would hold for very large N in this example. Some other simple examples have been investigated (Åström and Wittenmark, 1971; Jacobs and Hughes, 1974) where probing was, under certain circumstances, always necessary in order to avoid an observability problem described as 'turn-off'. Neither is it known to what extent caution and probing are necessary in sub-optimal control laws to match the general class of equation (7) with uncertain coefficients.

1.4 *Review of adaptive control*

The literature on adaptive control consists of a large number of papers discussing algorithms, a smaller number reporting simulation studies and, as mentioned in Section 1.1, about half a dozen papers reporting successful applications. References to many of the individual contributions can be found in the survey papers by Wittenmark (1975), Asher et al (1978) and Parks et al (1980). The current state of the literature on algorithms is reviewed here by discussing three identifiable classes of adaptive control algorithm; self-tuning, model-reference, and sub-optimal, which are the subjects of Sections 1.4.1, 1.4.2, 1.4.3 respectively. The role of simulation is discussed in Section 1.4.4 and reference is made to Section 1.1 where the successful applications have already been summarised.

Self-tuning algorithms and model-reference controls share the structure shown in Fig 1.10. Their control u is the certainty-equivalent sub-optimal control u^c for an assumed controlled process, such as equation (7), which is linear with uncertain coefficients. It is generated by a feedback control law having adjustable coefficients which are tuned or adapted by the algorithms to be discussed in Sections 1.4.1 and 1.4.2. These two classes were originally derived starting from very different points of view, but have recently come to be recognised as closely related (Gawthrop, 1977; Ljung and Landau, 1978). The structure of Fig 1.10 differs appreciably from the general adaptive structure of Fig 1.4; it would not be optimal even if the controlled process satisfied equation (7) without any modelling errors. Whether or not the differences significantly affect performance is not, in general, known but could be investigated as suggested in Section 1.4.4.

The sub-optimal adaptive controls to be discussed in Section 1.4.3 have the structure shown in Fig 1.4, which would be optimal if there

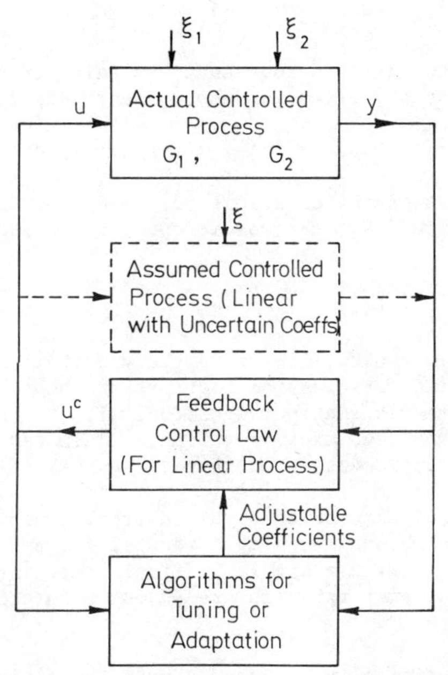

Fig 1.10 Structure of self-tuning and model-reference controllers

were no modelling errors. This structure can be used as a basis for designing sub-optimal control laws, which could include caution and probing according to equation (20), matched to the assumed equation (7).

1.4.1 Self-tuning control

Self-tuning control is, in a sense, the simplest possible adaptive control algorithm derivable from the point of view of the discrete-time stochastic control theory which is the basis of this Chapter. The simultaneous algorithms which it uses for identification and for control, as described in Section 1.2.5, are both made as simple and easily implementable as possible, rather than optimal for the assumed controlled process of equation (7).

The identification algorithm is simplified by making the additional assumption that all the coefficients c in equation (7) are zero, except for c_o which has the value unity. In the notation of the footnote in Section 1.2.4 this assumption is that

$$C(z) = 1 \tag{24}$$

The effect of the assumption is to eliminate uncertainty about the coefficients c so that, as discussed in Section 1.2.5, the problem of estimating the remaining coefficients is linearised. This linearisation is demonstrated by rewriting equation (7) using a vector θ to represent the remaining uncertain coefficients†

†It is further assumed, without loss of generality to the linearisation, that $a_o = 1$ and $d = 0$.

$$\theta \equiv \{a_1, \ldots, a_n, b_o, \ldots, b_n\}^T \tag{25}$$

together with a vector X(i) to represent the relevant past sequence y^i of inputs u and outputs y affecting the current output y(i)

$$X(i) \equiv \{-y(i-1), \ldots, -y(i-n), u(i-k), \ldots, u(i-k-n)\}^T \tag{26}$$

Then, using the notations of equations (25) and (26) together with the assumption of equation (24), the assumed process of equation (7) can be written

$$y(i) = X^T(i)\ \theta + \xi(i). \tag{27}$$

This equation (27) is clearly linear in the uncertain coefficients θ. It is a starting point for classical estimation theory (Rao, 1965) which leads to least-squares estimation algorithms with the nice properties that they generate unbiased minimal-variance estimates $\hat{\theta}$, and that they can conveniently be implemented recursively in real time.

Self-tuning control assumes that the controlled process satisfies equation (27). It combines the simple recursive least-squares estimation of coefficients θ with a similarly simple certainty-equivalent control law having parameters which are adjusted in accordance with the resulting estimates $\hat{\theta}$.

Self-tuning was originally proposed, under the title 'Self-optimising', by Kalman (1958) who described a mid-1950s hybrid computer implementation using multiturn potentiometers and thermionic valves. The proposal had little immediate consequence because of two difficulties:

(i) Computing technology was not sufficiently advanced.

(ii) Equation (27) seemed to be too specialised to represent any very general class of practical controlled processes.

In recent years both the above difficulties have been much reduced and self-tuning has become a very popular development area in adaptive control. Difficulty (i) has been virtually eliminated by the widespread availability of mini- and micro-computers. Reduction of difficulty (ii) has been due to Åström and his co-workers at Lund in Sweden.

There were two stages to this reduction. The first was to establish that equation (7) is a canonical input/output representation for a large class of SISO linear systems. It is thus about as general as a linear model can be, which equation (27) is not. The second stage was to prove (Åström and Wittenmark, 1973) that if a controlled process really did satisfy equation (7) then self-tuning control, designed on the assumption that the controlled process satisfied equation (27), could actually converge to the correct certainty-equivalent control to match equation (7). This property of convergence, sometimes known as the self-tuning property, refers to the control system shown in Fig 1.11. Taken together these two stages meet the above difficulty (ii) that self-tuning seemed only to match a very specialised class of controlled processes; Åström's results establish that self-tuning control can converge to the certainty-equivalent control for the general class of linear controlled processes satisfying equation (7). Valuable though these results have been in stimulating developments in self-tuning control, they do not claim to guarantee that self-tuning would be successful in practical applications. For this purpose the results are incomplete in several ways, as follows:

(i) The established convergence is asymptotic convergence as time goes to infinity. Nothing is said about the rate of convergence even though, in practice, this would need to be fast compared to the rate of any changes in the uncertain coefficients.

(ii) Results are proved mainly for simple cases.

(iii) Even under the most favourable conditions, convergence is not guaranteed. What is proved is only that if the algorithm converges at all it will converge to the correct control law.

(iv) The results assume that the controlled process actually does satisfy equation (7), as shown in Fig 1.11. Nothing is said about performance when the controlled process is really non-linear as shown in Fig 1.10.

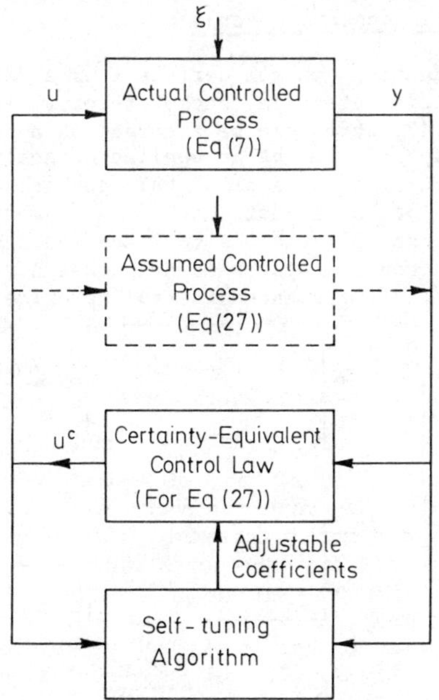

Fig 1.11 Control system having the self-tuning property.

Development work on self-tuning control has not been inhibited by the incompleteness of established theoretical results. The proliferation of techniques to implement and tune estimation algorithms, of derivations of certainty-equivalent control laws, of studies of stability and convergence, of robustness studies aimed to meet point (iv) above, of simulation studies, of real applications, and of generalisations to multi-input multi-output (MIMO) systems and to continuous-time systems are described in other chapters of this book. The popularity of self-tuning control can be attributed to three factors:

(i) It uses algorithms which are as simple as possible.

(ii) It is derived from discrete-time stochastic control theory in a form which can readily be implemented on available mini- and micro-computers.

(iii) It has been found possible to establish some theoretical results which, although incomplete, indicate that satisfactory stable performance can sometimes be expected.

This popularity is thought to be the main reason why as many as four out of the six successful applications mentioned in Section 1.1 were achieved by self-tuning.

There are no guarantees that self-tuning control, with its deliberately simplified algorithms for estimation and control, is necessarily the best sub-optimal adaptive control that could be implemented. A simulation study which indicated situations where other adaptive controls might perform significantly better is summarised in Section 1.4.4.

1.4.2 Model-reference adaptive control

Model-reference adaptive control derives from a different viewpoint than that of optimal discrete-time control theory. This viewpoint, illustrated in Fig 1.12, which can be regarded as a modified version of Fig 1.10, is that adaptation might be applied directly to the certainty-equivalent linear controller of a classical feedback control system. Adaptation takes the form of adjustment of one or more of the coefficients of such a controller so as to force the response of the resulting closed-loop control system towards that of some desired model. This model is explicitly implemented in real hardware and is subjected

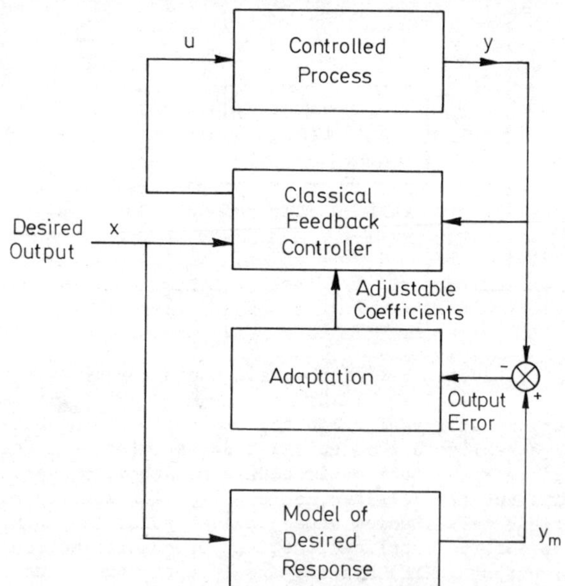

Fig 1.12 Structure of model-reference adaptive control.

to the same desired value signal x as is the feedback controller. Its output y_m is the reference signal of the name 'model-reference'. The adaptation mechanism must somehow be designed so as to reduce to zero the measured output error $(y_m - y)$.

Model-reference adaptive control was one of the methods originally suggested for aerospace problems in the 1950s (Whitaker et al, 1958). It was designed for implementation using analog hardware and so its subsequent development has not had to wait on advances in digital technology, although these advances now stimulate further work. The main differences between model-reference adaptive control and the other classes reviewed here are:

(i) It is explicit, as is extremal control (Section 1.2.6), in that it uses a signal, the output error, which explicitly indicates quality of performance.

(ii) The actuating signal u is not explicitly made available to the adaptation mechanism.

(iii) It was intended primarily for servo problems having time-varying target values x for the output y. In regulator problems, where the target value is constant, probing may need to be added at the input x in order to generate the necessary reference excitation.

(iv) It was originally developed to be implemented using analog hardware.

The structure illustrated in Fig 1.12 is by no means the only way to use model-reference methods within a control system. Various structures have been proposed where the model-reference adaptation may operate on a control law, as in Fig 1.12, or on an estimator, and may be driven by the output error $(y_m - y)$, as in Fig 1.12, or by some other difference between corresponding signals in the model and the system. Such structures have been reviewed by Landau (1979). In addition to structure, the main features of a model-reference design are the feedback control law and the adaptation mechanism. Most work has been concerned with designing the adaptation mechanism. This mechanism must be able, on the basis of the observed output error, to determine which way to adjust the controller coefficients and must also remain stable under all operating conditions. It appears to be a disadvantage of the method that there is no obvious general theoretical basis for such designs. A variety of mechanisms have nevertheless been developed which are, to some extent, described in other chapters of this book.

Most analysis of model-reference systems has been of deterministic continuous-time systems and has concentrated on stability (see also Chapters 4 and 7). The results indicate that under some conditions stability can be achieved. It is the more recent of these stability results which have been recognised as closely related to corresponding results about self-tuning control. From the practical point of view the theoretical results about model-reference methods have the same status as those about self-tuning; they give no guarantee of stability in practice nor any indication of what is the best adaptive control to implement. As with self-tuning control, the popularity of the method is little inhibited by incompleteness of theoretical results. This popularity can partly be attributed to some correspondence between the structure of Fig 1.12 and the steps which a control engineer might himself take to adjust the coefficients of a classical feedback controller.

1.4.3 Sub-optimal adaptive controls

Adaptive controllers can be, and perhaps should be, based on the optimal structure of Fig 1.4 rather than on the simplified structure of Fig 1.10. They can then be designed to take sub-optimal account of some or all of the non-linearities and interactions, described in Section 1.2 and 1.3, which are known to be inherent to adaptive control. Three possible features of a sub-optimal controller are:

(i) Non-linear estimation, to give better estimates of the states x than can be achieved by assuming linearity.

(ii) Caution in the control law, to give better control than can be achieved by assuming separability.

(iii) Probing in the control law, to give better control than can be achieved by assuming neutrality.

One popular sub-optimal non-linear estimator is the extended Kalman filter (Jazwinski, 1970) which can be used to estimate jointly the dynamic state and the uncertain coefficients of equation (7). The first-order example of equations (8) and (9) in Section 1.2.5 has indicated how an appropriate set of state variables x can be defined (Cox, 1964). Time-varying coefficients are, for this purpose, assumed to satisfy first-order state equations of the form

$$x_j(i+1) = g_j\ x(i) + d_j + \xi_j(i) \tag{28}$$

in which g_j and d_j are known and ξ_j is a white noise of known variance σ_j. Here g_j represents the dynamics of coefficient change and, provided $|g_j|<1$, the state x_j has mean value

$$\bar{x}_j = d_j/(1-g_j) \tag{29a}$$

and variance

$$\text{var}(x_j) = \sigma_j/(1-g_j^{\,2}) \tag{29b}$$

The extended Kalman filter assumes that all random variables are Gaussian and generates estimates $\hat{m}$, $\hat{\Sigma}$ of the mean and covariance matrix of the conditional distribution $p(x|y)$. These estimates can be transmitted on to the control law of Fig 1.4. If the control law takes account of the covariance $\hat{\Sigma}$, which indicates the accuracy of $\hat{m}$ as an estimate of x, it is being cautious.

One sub-optimal control law which has been proposed for use in conjunction with an extended Kalman filter is the single-stage (N=1) neutral control u^n introduced at equation (23) in Section 1.3.4 (Jacobs and Hughes, 1975). The variance terms in (23) show that this control is cautious. For the first-order example of equations (8) and (9) the neutral control u^n has been the subject of a comparative study (Jacobs and Saratchandran, 1980) summarised in Section 1.4.4.

Many other sub-optimal estimators and sub-optimal control laws could be combined within the structure of Fig 1.4. These could use simpler estimators, for example the parallel filter (Nelson and Stear, 1976) in which states and coefficients are separately estimated by parallel Kalman filters each of which takes the estimates generated by other as values for its own parameters (Saratchandran, 1980). They could use simpler control laws, for example the certainty-equivalent control. Or they could use more complicated algorithms which might be more nearly optimal. Saridis (1977) summarises several such combinations, although he does not explicitly refer to Fig 1.4. All these sub-optimal schemes are more expensive to implement, in terms of the computing power needed, than is the simple self-tuning scheme of Section 1.4.1. Whether the extra expense would be justified by superior performance is not, in general, known: the comparative study summarised in Section 1.4.4 suggests that significant differences in performance can sometimes exist.

Sub-optimal probing has been found particularly expensive to introduce on any systematic basis (Tse and Bar-Shalom, 1973). It may be that probing need only be introduced on an ad hoc basis as a perturbation signal to provide the persistent excitation needed for identifiability or observability of all state variables. Examples of such probing have already been mentioned in Section 1.3.4 and, in Section 1.4.2, for model-reference regulation. Observability problems can arise in the sub-optimal schemes of Fig 1.4 when there is a non-zero bias d to be estimated in equation (7) and when operating conditions are such that the control u is approximately constant. Without a perturbation signal added to u it is impossible to distinguish between what otherwise would be two synchronised terms, for example bu and d on the right hand side of equation (8).

1.4.4 Simulation and comparative studies

In the design of any adaptive control system there are two outstanding questions about the non-optimalities mentioned in Section 1.2.5:

(i) Would the proposed adaptive algorithm give a good sub-optimal control of the assumed controlled process satisfying, for example, equation (7)?

(ii) Will an algorithm for which the answer to question (i) is 'yes' also give good control of a real controlled process in the control system of Fig 1.4?

The non-linear and stochastic nature of adaptive control makes it virtually impossible to answer either of these questions by analysis. Many investigators have therefore turned to simulation as the most practicable available tool for studying adaptive control systems.

As indicated in Section 1.1, most published simulation studies do not give wholly satisfactory answers to either of the above questions. What is frequently missing in any attempt to make critical comparisons between performance under a proposed adaptive control and performance under any other controls (Sawaragi and Yoshikawa, 1969).

Two obvious candidate controls for such comparisons are classical control which makes no attempt at adaptation, and the optimal control which would achieve the best possible performance for a given controlled process. It is only when there is significant difference between performance under classical control and performance under optimal control that there would be any need for adaptive control. It is only when performance under adaptive control is significantly closer to optimal performance than to performance under classical control that the adaptive control can be said to be 'good'.

Such comparisons could, in principle, answer the question 'when is adaptive control useful?'. In practice the stated comparisons cannot be made because the optimal control and its resulting performance are, almost by definition, not known; adaptive control was defined as one approach to sub-optimal control of processes for which the optimal control cannot be found. An alternative control, which can be used to give some, possibly optimistic, indication of the best performance achievable, is the linear control which would be optimal if the true values of the uncertain coefficients of equation (7) were known. This control, the certainty-about-coefficients (CAC) control, usually is known in simulations designed to answer question (i) and so can provide a basis for systematic investigations of the quality of sub-optimal control of processes satisfying equation (7). This type of investigation could answer some of the controversial outstanding questions about adaptive control. It is surprising that few such investigations have been reported; one is summarised below.

This study (Jacobs and Saratchandran, 1980) consisted of Monte Carlo simulations in which each of several control algorithms controlled a process satisfying a version of equation (8) in which the bias coefficient d was zero. The controllers simulated were:

(i) The CAC control defined above, which provided a best bound on performance.

(ii) A classical certainty-equivalent control based on initial estimates of the uncertain coefficients and making no attempt to reduce uncertainty about them. This control, the no-learning (NOL) control, provides a worst bound on performance.

(iii) A self-tuning control, as described in Section 1.4.1, referred to as ST control.

(iv) A combination of extended Kalman filter and single-stage control, as described in Section 1.4.3, referred to as "non-linear" (NL) control (Jacobs 1980 (b)).

Monte Carlo simulations were used because initial uncertainty about values of the coefficients a, b, c is an important part of the problem. It is simulated by using repeated trials in which initial values of the coefficients are drawn at random from a probability distribution characterising initial uncertainty. The objective of control was to regulate the output y to zero and the quality of control at each trial was measured by a summed quadratic cost function.

$$I_N = \sum_{i=1}^{N} (y^2(i+1) + q\, u^2(i)) \tag{30}$$

in which q is a specified weighting factor on cost of control.

The results were presented graphically on logarithmic scales, as idealised in Fig 1.13, showing how the average of costs I_N increased

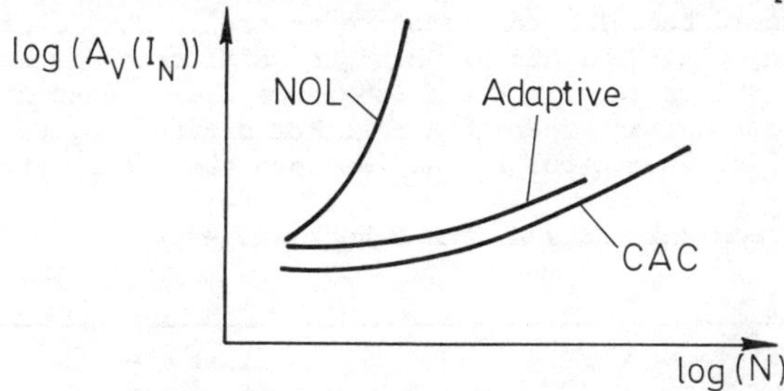

Fig 1.13 Idealised simulation results showing adaptive behaviour

with N, the length of time simulated. The figure shows behaviour of an adaptive control whose costs migrate with time from large values associated with classical (NOL) control to much lower values associated with certainty about coefficients (CAC). This reduction of cost is the reward for reducing uncertainty about the coefficients a, b, c. Few results had the idealised behaviour of the figure; sometimes NOL costs were close to CAC costs, indicating little need for adaptation, and sometimes NOL costs were so high as to be off-scale for all but the smallest values of N. It was found that whether or not adaptive control was necessary and whether or not the simplest (ST) algorithm would be adaquate were related to the values of the uncertain coefficients, to the initial uncertainty about them, and to the rate at which they varied in time. The coefficient values and the initial uncertainty were found to fall into eight significant classes, as shown in Table 1.1, depending on two of the structural features mentioned in Section 1.2.4, whether the controlled process was stable ($|a| \lessgtr 1$) and whether the noise model was minimum-phase ($|c| \lessgtr 1$), and on whether the sign of the gain b was initially known with certainty. Three rates of coefficient variation were simulated:

(i) constant unknown coefficients ($g_j = 1$ in equation (28))

(ii) time-varying coefficients ($g_j = 0.8$)

(iii) more rapidly varying coefficients ($g_j = 0.6$).

Table 1.1 Classification of controlled processes

Class	1	2	3	4	5	6	7	8
$\lvert a\rvert<1$?	✓	✓	x	x	✓	✓	x	x
$\lvert c\rvert<1$?	✓	x	✓	x	✓	x	✓	x
sign of b certain?	✓	✓	✓	✓	x	x	x	x

The simulation results are summarised in Table 1.2 which quotes incremental average costs (slopes from Fig 1.13) at the thousandth stage for those controls which could give stable behaviour. These results go some way towards indicating what characterises a control problem which would benefit from adaptive control. They confirm that situations exist, Classes 1 and 5, where the difference between the two bounds, CAC and NOL is so slight that there may be no need for adaptation. They also show that there can be other situations, Classes 7 and 8, where even adaptive control fails. Between these two extremes the adaptive algorithm may be more or less successful and may even approach CAC performance, as in Class 2 and, for constant coefficients, in Classes 3 and 4. Performance generally deteriorated as the rate of variation of coefficients increased, and the non-linear algorithm then gave better performance than the self-tuner.

Table 1.2 Incremental costs at thousandth stage

Control	Coefficient	Class							
	rate	1	2	3	4	5	6	7	8
CAC	(i) - (iii)	1	17	1.1	20	1	17	-	-
NL	(i)	1	17	1.1	20	1	18	-	-
	(ii)	1	17	2	560	1.1	19	-	-
	(iii)	1	17	4.6	-	1.1	21	-	-
ST	(i)	1	17	1.1	23	-	-	-	-
	(ii)	1	24	3	-	-	-	-	-
	(ii)	1	93	25	-	-	-	-	-
NOL	(i) - (iii)	1	-	-	-	1.1	-	-	-

There is considerable scope for further investigations along the lines summarised above. Some possible topics are:

(i) Do the results carry over to other controlled processes in the general class of equation (7) with less simple dynamics than the first-order equation (8)?

(ii) How are the results affected by non-zero bias d?

(iii) Are there quantitive relationships between average cost and rate of change of coefficients?

(iv) Are there problems which are stochastically uncontrollable in the sense that not even the optimal control could be stable? This question (Athans et al, 1977; Sternby, 1979) may not be answerable by simulation.

(v) How do the many simulation results which have been published piecemeal fit into the systematic classification?

Research along these lines should lead to a valuable body of knowledge about sub-optimal control of processes satisfying linear equations having uncertain coefficients. This would answer question (i) at the beginning of the Section.

Question (ii) about the performance of real processes under adaptive control is not likely to yield to any similar systematic investigation. Two major difficulties are:

(i) The certainty-about-coefficients control would not, indeed could not, be known. In the presence of modelling errors there would be no controlled process actually satisfying a linear equation (7) and so no way of assigning true values to its coefficients nor of designing a controller to match them.

(ii) Modelling errors are specific to each application. With non-linear controlled processes and non-linear controllers there is little hope that general results will be obtainable. To answer question (ii) with reference to a real application will therefore require an investigation specific to the application. The real controlled process can be initially simulated, but final assessment of adaptive control must depend on full-scale trials, as stated in Section 1.1.

1.5 *Conclusions*

Adaptive control has a history spanning a generation and a literature of thousands of papers. Although there appear to be many practical problems which need adaptive control and although recent developments in computer technology make it possible to implement many proposed algorithms, only about half a dozen successful applications have been reported.

The subject has been bedevilled by two controversial questions: 'what is adaptive control?' and 'when is it useful?'. The first question is answered here by specifying current usage, which is that adaptive control is the name of one particular approach to designing sub-optimal controllers for non-linear stochastic systems. It is to assume that the controlled process satisfies equations which would be linear if only the values of all its coefficients were known; a sub-optimal controller is then designed to match this assumed non-linear stochastic process. The approach has the structure shown in Fig 1.4. There are two ways in which the resulting control could be non-optimal (bad):

(i) It could be non-optimal even for the assumed controlled process.

(ii) Modelling errors may be such that what would be a sub-optimal control for the assumed process becomes non-optimal for the real process.

Stochastic control theory cannot solve non-linear problems but it offers theoretical concepts, such as certainty-equivalence, caution and probing as described in Section 1.3, which can help with the design of sub-optimal controllers. Much of the reported work on adaptive control has concentrated on certainty-equivalent controllers which are, from the theoretical point of view, the simplest possible.

There has been a shortage of critical studies to investigate (i) above, the need for and performance of adaptive control as a sub-optimal control for the assumed control process. Simulation methods could be used, as indicated in Section 1.4.4, to build up a systematic body of knowledge on this point.

The question 'when is adaptive control useful? really relates to (ii). It can properly be answered only in the context of full-scale practical applications.

References

Akashi H and Nose K, (1975), On certainty-equivalence of stochastic optimal control problem, Int J Cont, 21, 7, pp 857-863.

Alspach D L and Sorenson H, (1971), Stochastic optimal control for linear but non-Gaussian systems, Int J Cont, 13, 6, pp 1169-1181.

Aoki M, (1967), Optimization of stochastic systems, Academic Press.

Amerongen J van (1980), Model reference adaptive control applied to steering of ships, in Methods and applications in adaptive control, Lect notes in Cont and Inf Sciences, 24, Springer.

Aseltine J A, Mancini A R and Sarture C W, (1958), A survey of adaptive control systems, IRE Trans AC, 6, pp 102-108.

Asher R B, Andrisani D and Dorato, P, (1976), Bibliography on adaptive control systems, Proc IEEE, 64, 8, pp 1226-1240.

Åström K-J, (1970), Introduction to stochastic control theory, Academic Press.

Åström K-J, (1980), Why use adaptive techniques for steering large tankers?, Int J Cont, 32, 4, pp 689-708.

Åström K-J and Bohlin T (1965), Numerical identification of linear dynamic systems from normal operating records, 2nd IFAC Symposium on Theory of self-adaptive control systems, Plenum Press.

Åström K-J and Wittenmark B (1971), Problems of identification and control, J Math An and Apps, 34, pp 90-113.

Åström K-J and Wittenmark B, (1973), On self-tuning regulators Automatica, 9, pp 185-199.

Athans M, Ku R and Gershwin S B, (1977), The uncertainty threshold principle: some fundamental limitations of optimal decision making under dynamic uncertainty, IEEE Trans AC, AC 22, 3, pp 491-495.

Bar-Shalon Y and Tse E, (1974), Dual effect certainty-equivalence and separation in stochastic control, IEEE Trans AC, AC 19, 5, pp 494-500.

Bellman R, (1961), Adaptive control processes, Princeton.

Bertsekhas, D P (1976), Dynamic programming and stochastic control, Academic Press.

Borisson U and Syding R, (1974), Self-tuning control of an ore crusher, IFAC Symposium on Stochastic Control, Budapest, preprints, pp 491-497.

Cegrell T and Hedquist T, (1975), Successful adaptive control of paper machines, Automatica, 11, 1, pp 53-59.

Cox H, (1964), On the estimation of state variables and parameters for noisy dynamic systems, IEEE Trans AC, AC 9, pp 5-12.

Dorf R C, (1980), Modern control systems, 3rd ed, Addison-Wesley.

Dumont G A and Belanger P R, (1978), Self-tuning control of a titanium dioxide kiln, IEEE Trans AC, AC 23, 4, pp 532-538.

Eykhoff P, (1974), System Identification, Wiley.

Feldbaum A A, (1965), Optimal control systems, Academic Press.

Gawthrop P J (1977), Some interpretations of the self-tuning controller, Proc IEE, 124, 10, pp 889-894.

Gessing R, (1980), Stochastic optimal control and its connection with estimation, Ox Univ Eng Lab report 1328.

Gilbart J W and Winston G C, (1974), Adaptive compensation for an optical tracking telescope, Automatica, 10, 2, pp 125-131.

Godfrey K R and Brown R F, (1979), Practical aspects of the identification of process dynamics, Trans Inst MC, 2, pp 85-95.

Gregory P C - ed Proc Self Adaptive Flight control Symposium, (1959), WADC Report No 59-49.

Horowitz I, (1963), Synthesis of feedback systems, Academic Press.

IEEE, (1971), The linear quadratic Gaussian problem, IEEE Trans AC, AC 16, 6, special issue.

IEEE, (1977), Mini-issue on NASA's advanced control law program, IEEE Trans AC, AC 22, 5, pp 752-806.

Isermann R, ed, (1981) System identification tutorials, Pergamon.

Jacobs O L R, (1961), A review of self-adjusting systems in automatic control, J Elect and Cont, 10, 4, pp 311-322.

Jacobs O L R, (1974), Introduction to control theory, O.U.P.

Jacobs O L R (1980a), When is adaptive control useful? Proc IMA Conf on Control Theory at Sheffield, Academic Press.

Jacobs O L R (1980b), On non-linear adaptive control, Ox Univ Eng Lab report, 1323.

Jacobs O L R and Hughes D J (1974), Turn-off escape and probing in non-linear stochastic control, IFAC Symposium on Stochastic Control, Budapest, preprints, pp 343-352.

Jacobs O L R and Hughes D J (1975), Simultaneous identification and control using a neutral control law, 6th IFAC Congress, Boston, paper 30-2.

Jacobs, O L R and Langdon S M, (1969), An optimal extremal control system, Automatica, 6, pp 297-301.

Jacobs O L R and Patchell J W, (1972), Caution and probing in stochastic control, Int J Cont, 16, pp 189-199.

Jacobs O L R and Potter R V, (1978), Optimal control of a stochastic bilinear system, in Recent theoretical developments in control, ed Gregson M J, Academic Press.

Jacobs O L R and Saratchandran P, (1980), Comparison of adaptive controllers, Automatica, 16, pp 89-97.

Jazwinski, A H, (1970), Stochastic processes and filtering theory, Academic Press.

Joseph P D and Tou J T, (1961), On linear control theory, Trans AIEE (Apps and Ind), 80, pp 193-196.

Källström C G, Åström K J, Thorell N E, Erikson J and Sten L, (1979), Adaptive autopilots for tankers, Automatica, 15, pp 241-254.

Kalman R E (1958), Design of a self-optimising control system, Trans ASME, 86D, pp 51-60.

Kalman R E (1964), When is a linear control system optimal?, Trans ASME, 86D, pp 51-60.

Kwakernaak H and Sivan R, (1972), Linear optimal control systems, Wiley.

Landau Y D (1979), Adaptive control, the model reference approach, Marcel Dekker.

Ljung L and Landau Y D, (1978), Model reference adaptive systems and self-tuning regulators - some connections, 7th IFAC Congress, Helsinki.

Mohler R R, (1973), Bilinear control processes, Academic Press.

Narendra, K S and Monopoli, R V, (1980), Applications of Adaptive Control, Academic Press.

Nelson N W and Stear E, (1976), The simultaneous on-line estimation of parameters and states in linear systems, IEEE Trans AC, AC 21, pp 94-98.

Parks P C, Schaufelberger W, Schmid C and Unbehauen H, (1980), Applications of adaptive control systems, in Methods and Applications of adaptive control, Lect notes in Cont and Inf Sciences, 24, Springer.

Patchell J W, (1971), On the structure of stochastic control laws, Ox Univ D.Phil Thesis.

Patchell J W and Jacobs O L R, (1971), Separability neutrality and certainty equivalence, Int J Cont, 13, pp 337-342.

Rao C R (1965), Linear statistical inference and its applications, Wiley.

Root J G, (1969), Optimum control of non-Gaussian linear systems with inaccessible state variables, SIAM J Conf, 7, pp 317-323.

Saratchandran P (1980), On the performance of an adaptive controller under different estimators, JACC, San Francisco.

Saridis G N, Self-organizing control of stochastic systems, Marcel Dekker.

Sawaragi Y and Yoshikawa T, (1969), Performance bounds for discrete time stochastic optimal control problems, Int J Cont, 13, 337-342.

Simon H, (1956), Dynamic programming under uncertainty with a quadratic function, Econometrica, 24, pp 74-81.

Sternby J, (1979), Performance limits in adaptive control, IEEE Trans AC, AC 24, 4, pp 645-647.

Streibel C, (1965), Sufficient statistics in the control of stochastic systems, J Math An and Apps, 12,3, pp 576-592.

Stromer P R, (1959), Adaptive or self-optimizing control systems - a bibliography, IRE Trans AC, May, pp 65-68.

Theil H, (1956), A note on certainty equivalence in dynamic planning, Econometrica, 25, pp 346-349.

Truxal J G, (1963), Adaptive control, Proc 2nd IFAC Congress, Basle, pp 386-392, Butterworths.

Tse E and Bar-Shalom Y V, (1973), An actively adaptive control for linear systems with random parameters via the dual control approach, IEEE Trans AC, AC 18, 2, pp 109-117.

Whitaker H P, Yamron J and Kezer A, (1958), Design of model reference adaptive control systems for aircraft, MIT Instrument Lab Report, R-164.

Wittenmark B, (1975), Stochastic adaptive control methods - a survey, Int J Cont, 21, 5, pp 705-730.

Chapter Two

Introduction to self-tuning controllers

D W CLARKE

2.1 *Introduction*

The automatic tuning of controllers for industrial processes has excited both theoretical and practical interest for many years. Although it is generally recognised that the conventional PID regulator is remarkably effective in practice, its initial tuning and the maintenance of good tuning on a plant with many control loops can be a time-consuming acitivity, particularly if the process dynamics are slow. Furthermore, with the increasing demands for tighter quality control and for energy conservation it may well be that some critical loops will require controllers with many parameters whose values have to be chosen in a more systematic way. Such loops may contain, for example, significant dead-time or they may be subjected to known disturbances which are best regulated using properly-tuned feedforward action.

Self-tuning control is one approach to the automatic tuning problem. It can be viewed either as a tuning aid for control laws that are more complex than PID but which have fixed parameters, or as a means by which a time-varying process can be controlled in a consistent way. Although it is not optimal in the sense of dual control, it can - if properly used - give good tuning and control performance for a wide class of practical systems. Moreover, unlike dual control, the self-tuning algorithms require modest and predictable computing power. When one considers that the PID algorithm takes only a small fraction of the computational resources of a typical industrial microprocessor-based regulator, it is seen that there is scope with current technology to implement a self-tuner in a microprocessor at a cost not greatly exceeding that of the PID-based system. This combination of performance and cost means that the self-tuner could become accepted by industry on grounds of convenience even for loops where there is no need for adaptive capability.

A self-tuning controller has three main elements. There is a standard feedback law in the form of a difference equation which acts upon a set of values such as the measured output and feedforward signals, the current set-point etc, and which produces the new control action. A recursive parameter estimator monitors the plant's input and output and computes an estimate of the plant dynamics in terms of s set of parameters in a prescribed structural model. The parameter estimates

Dr Clarke is with the Department of Engineering Science, University of Oxford.

are fed into a control-design algorithm which then provides a new set of coefficients for the feedback law. The general structure of a self-tuner is shown in Fig 2.1. The control-design algorithm simply accepts current estimates and ignores their uncertainties (unlike dual control algorithms); such a procedure is called *certainty-equivalent*. (See section 1.3.1.3 for example). The rationale for this approach is that, although there may be poor control during the tuning phase, this can be minimised by taking suitable precautions, but the overall algorithm is considerably simplified. A further simplification is possible by omitting the control-design stage in an *implicit* self-tuner where the process model equations are reformulated such that the estimator directly produces the coefficients of the required control law; a self-tuner which includes both an estimator of a 'standard' process model and a control-design stage is called *explicit*.

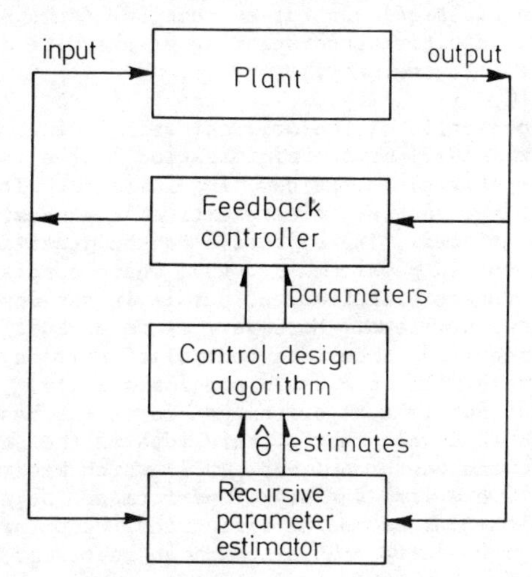

Fig 2.1 Structure of an explicit self-tuner.

The idea of self-tuning is, of course, not new. Kalman (1958) derived the first recognisable self-tuner and even attempted to realize the algorithm in a special purpose computer. However, the theory and the technology were insufficiently developed and the technique was not actively pursued compared with the then emerging state-space methods. The theory was revived and extended to cover stochastic aspects by Peterka (1970), but it was not until the key paper of Åström and Wittenmark (1973) that the current great interest in the subject was initiated.

This paper led directly to a series of practical applications, which themselves engendered new theoretical developments as the problems involved with implementing adaptive control in practice become clarified. The field is now moving very rapidly and many algorithms have been proposed from different points of view. It is relatively easy to derive a new algorithm; it is more difficult to derive convergence conditions or to show that any given method is 'better' than others. It is to be hoped that a more unified framework and criteria for choosing the best algorithms will be developed in the future.

As Åström (1980) puts it, self-tuners are *performance-oriented*. A PID regulator has three coefficients - K T_i and T_d - which the engineer can adjust to give acceptable control performance (as he sees it) for the current plant conditions. This performance may, of course, deteriorate later as the plant dynamics change, or the tuning may take so long that the 'acceptable' performance is inferior to that ultimately possible on the plant. In a self-tuner, on the other hand, the user specifies the *closed-loop performance* he would like, and the algorithm sets out to attain this performance despite unknown plant parameters or drift. This implies, of course, that the desired performance is in fact achievable on the plant with the given saturation characteristics of the control actuator, and it is in this area that the skill of the plant engineer would still be important. Similarly, recent self-tuning theory has attempted to broaden the range of possible control objectives, and to interpret them in classical control engineering terms so that a user would have a greater intuitive understanding of the ultimate closed-loop performance under self-tuning.

The performance objective of the original self-tuning regulator of Åström and Wittenmark (1973) is the minimisation of the variance of the measured process output y(t) at the sampling instants. It is in some ways the stochastic analogue of a discrete-time 'dead-beat' controller, and shares the same defects. The objective can be justified in cases such as quality control in paper-making, say, where consistent thickness is desired despite process disturbances, but it is not appropriate for a broad range of control problems. Moreover, as no account is taken of the control effort required, excessive control signals may be generated and in some cases there will be closed-loop instability. It is, however, the simplest implicit algorithm and forms the basis of more general self-tuners. A development of this idea is the self-tuning controller of Clarke and Gawthrop (1975, 1979) which was interpreted by Gawthrop (1977) to give a wide variety of performance objectives. This implicit algorithm includes set-point variations, hence providing tracking as well as regulatory control, and can be viewed, for example, as minimising a combination of control and output variances. This means that control effort can be traded against output variations to suit the available actuator characteristics; in most cases the trade-off is extremely favourable in that a large reduction in control variance may imply only a modest increase in output variance.

Simple implicit self-tuners are based on predictive control theory which depends on knowledge of the system time-delay Δ. It has been argued that explicit methods, though involving more computation, do not require this knowledge as the delay can be estimated as part of the process dynamics. Such an approach has been advocated by Wellstead et al (1979), based on earlier work by Edmunds (1976), in which the associated design procedure is the placement of the closed-loop poles at prescribed locations while leaving the zeros in their open-loop positions. This has been extended to cover tracking and regulation, and Åström and Wittenmark (1980) have produced a similar design procedure which also places 'safe' process zeros in the deterministic servo case. A further extension is in the use of state-space methods, in which the estimated model is transformed into its equivalent state-space representation, and the design procedure becomes that used for linear-quadratic-Gaussian (LQG) control. Early work in this area is due to Peterka and Åström (1973). Although the LQG approach requires the most computation, various 'tricks' can be used to minimise the computing load, such as performing only one iteration of the associated Riccati equation per cycle; details are discussed in Lam (1980).

An algorithm is called *self-tuning* if, as the number of input and output samples tends to infinity, the control signal generated becomes that which would be produced by the corresponding feedback law designed on the basis of *known* process dynamics. This property, and its proof, depends upon various assumptions about the process, such as time-invariant dynamics, and most of the methods described in this chapter have been shown theoretically and experimentally to possess self-tuning behaviour. However, in practice self-tuners are used for more general problems, such as for time-varying processes, and the algorithms are adjusted so that effective control is still produced. Appropriate modifications to the basic algorithms, and their relative robustness to the assumptions made about the process, are still matters of current research.

This chapter is intended to be mainly tutorial in nature, and organised as follows. Section 2.2 describes the underlying continuous-time process model assumed by the self-tuners, and how the corresponding discrete-time model is obtained. Of particular interest here are models written in 'predictive' form and the locations of the discrete-time poles and zeros. Section 2.3 derives the standard recursive-least-squares parameter estimation algorithm, and its generalization to cover certain types of correlated noise. Section 2.4 derives the prototype self-tuning regulator which minimises the process output variance, and comments on its properties. Section 2.5 describes a generalised self-tuning controller and gives various interpretations of its performance. Section 2.6 outlines the derivation of explicit self-tuners based on the placement of closed-loop poles in prescribed locations. Section 2.7 briefly described some recent developments such as explicit LQG-based self-tuners and 'hybrid' self-tuning. In view of its tutorial nature, the development of the theory will concentrate on single-input/single-output systems though most of the algorithms can be extended to the general case.

2.2 *System Models*

Although all practical processes are non-linear, good regulation can generally be achieved using models which are transfer-functions given by local linearization around the current operating point. A process model appropriate for our discussion is therefore:

$$y(t) = \frac{B(s)}{A(s)}\, u(t-\Delta) + d(t) \tag{1}$$

where A(s) and B(s) are polynomials in the differential operator $s \triangleq \frac{d}{dt}$. The transfer-function relating the control input u(.) and the plant output y(.) is:

$$G(s) = e^{-s\Delta}\,\frac{B(s)}{A(s)} \tag{2}$$

where s can be interpreted as the Laplace-transform variable, and Δ is the time delay which is found in many processes. The signal d(t) is a 'disturbance' which may have several components:

a) a constant d_1 reflecting the fact that zero control does not in general lead to zero plant output; the value of this constant would change if the operating point changed, as would the parameters of G(s);

b) a slowly-varying value d_2 depending on load-disturbances acting upon the process;

c) a feedforward element $d_3(t)$ which can be represented by:

$$d_3(t) = G_1(s)\ v(t) \tag{3}$$

where v(t) is some auxiliary measurable signal- such a term may be reduced by an appropriately chosen feedforward compensator;

d) a stochastic signal $d_4(t)$ given by:

$$d_4(t) = G_2(s)\ \xi(t) \tag{4}$$

where $\xi(t)$ is a white-noise process, and $G_2(s)$ gives the appropriate spectral-density (assumed to be rational) to $d_4(t)$ as $G_2(jw)G_2(-jw)$.

The treatment of the first three cases above will be considered later as simple extensions to the basic self-tuning methods, and we concentrate instead on the stochastic component $d_4(t)$. By augmenting A(s) to include the poles of $G_2(s)$, the model becomes:

$$y(t) = \frac{B(s)}{A(s)}\ u(t-\Delta) + \frac{C(s)}{A(s)}\ \xi(t) \tag{5}$$

where by the representation theorem (Åström, 1970) no root of C(s) lies in the right-half plane; it is further assumed that no root lies on the imaginary axis.

All self-tuners are implemented digitally, so the process model as seen by the controller is the z-transform (Franklin and Powell, 1980) of (5) after the usual zero-order-hold (ZOH) has been inserted in the forward path and a sampler placed on the output, as shown in Fig 2.2.

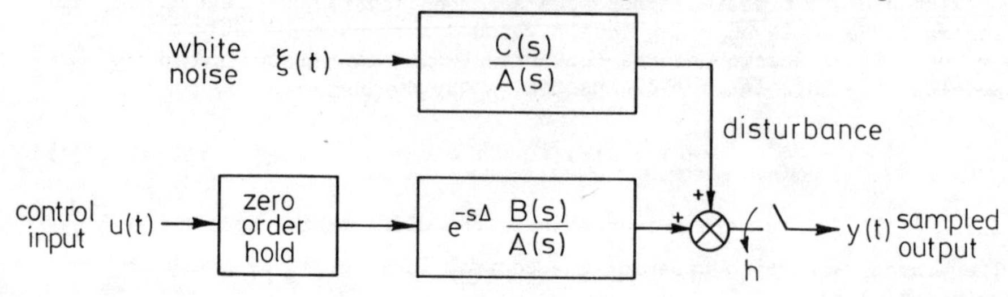

Fig 2.2 The open-loop process.

Suppose the sample interval is h. Then the basic procedure for obtaining the discrete-time equivalent of G(s) together with its ZOH is, assuming distinct poles:

(i) find k, δ where k is integral and $0<\delta\leq h$, such that:

$$\Delta = (k-1)h + \delta \tag{6}$$

and write $e^{-s\Delta}$ as $e^{-skh} \cdot e^{s(h-\delta)}$.

(δ is called the 'fractional delay')

(ii) expand $e^{s(h-\delta)} \dfrac{B(s)}{sA(s)}$ to give $\dfrac{G(0)}{s} + \Sigma_1^n \dfrac{\gamma_i}{s+p_i}$ (7)

where the γ_i are the residues at the poles p_i:

$$\gamma_i = \frac{e^{-p_i(h-\delta)}B(-p_i)}{\left[\dfrac{d}{ds}(sA(s))\right]_{s=-p_i}}$$

(iii) take z-transforms term-by-term of this expansion, giving:

$$\frac{G(0)}{1-z^{-1}} + \Sigma_1^n \frac{\gamma_i}{1-q_i z^{-1}}, \text{ with } q_i = e^{-p_i h} \tag{8}$$

(iv) recombine and premultiply by $(1-z^{-1})z^{-k}$ to give:

$$G^*(z^{-1}) = z^{-k} \frac{B^*(z^{-1})}{A^*(z^{-1})} \tag{9}$$

where A^* and B^* are polynomials in the backward-shift operator z^{-1}. Note that the terms in $(1-z^{-1})$ cancel and that the leading coefficient of A^* (i.e. $A^*(0)$) is 1. Moreover, the steady-state system gain $G(0)$ is also given by $G^*(1)$. If the system order (degree of $A(s)$) is n, it is clearly seen that the order of G^* (degree of A^*) is also n, as is the degree of B^*. Equation (8) shows that the leading coefficient of B^* is given by $B^*(0) = G(0) + \Sigma_1^n \gamma_i$, and the expansion in (7) shows that, for $\delta=h$, this equals the coefficient of s^n in $B(s)$. For real systems the number of continuous-time zeros is less than the number of poles, so this value is zero.

Hence the z-transfer-function G^* may be written as:

$$G^*(z^{-1}) = z^{-k} \frac{(b_o+b_1 z^{-1} + \ldots +b_n z^{-n})}{1 + a_1 z^{-1} + \ldots +a_n z^{-n}} \tag{10}$$

where the degree of the numerator polynomial is one less than that of the denominator if there is no fractional delay. The input-output relationship at times $t = ih$ can be written in the difference-equation form:

$$y(t) + \Sigma_1^n a_i y(t-i) = \Sigma_o^n b_i u(t-i-k) \tag{11}$$

Of crucial importance in discrete-time control is the position of the poles and zeros of $G^*(z^{-1})$ in the z-plane. For each s-plane pole there is a unique z-plane pole given by the mapping $z = e^{sh}$, so that stable poles in the left-half s-plane map into discrete-time poles in the unit disc. There is no such simple relationship between the zeros, as the following lemmas indicate:

Lemma 2.2.1

If the fractional delay δ is zero, and if the degree of $B(s)$ equals the degree of $B^*(z^{-1})$, the zeros of B^* are at:

$$z_i = 1 + h\beta_i + O(h^2) \tag{12}$$

where β_i is the ith zero of $B(s)$.

This lemma shows that for minimum-phase systems where $\beta_i < 0$ the discrete-time system has zeros within the stability region if degree (B^*) = degree (B). However, the discretization process makes degree (B^*) equal to n-1, so that this result only holds for systems with one more pole than zero in the s-plane. (The result can be extended to cover equality of poles and zeros). Few practical systems possess this property, as it implies an asymptotic phase lag of only 90°. It is the excess s-poles that lead to extra z-plane zeros, and it is the location of these extra zeros that causes difficulty in some self-tuning algorithms. They may in fact lie outside the stability region (hence loosely termed 'non-minimum phase'), as implied by the second lemma:

Lemma 2.2.2

If $G(s)$ has l more poles than zeros, then $G^*(z^{-1})$ becomes, as $h \to 0$, proportional to that obtained if $G(s)$ were simply l integrators:

$$G^*(z^{-1}) \to b_m(1-z^{-1}) \frac{D^l}{l!} \left\{\frac{1}{1-z^{-1}}\right\} \text{ as } h \to 0, \tag{13}$$

where D is the operator $-hz \frac{d}{dz}$.

For $l = 1$ to 3 this formula shows that the numerator polynomials are proportional to:

$$B_1^* = 1;\ B_2^* = 1 + z^{-1};\ B_3^* = 1 + 4\,z^{-1} + z^{-2},$$

giving unstable zeros at -1 for l=2 and -3.73 for l=3. For larger values of pole-excess the number and magnitude of the unstable B^* zeros increase continually.

Hence many stable, minimum-phase continuous-time systems will, if sampled rapidly enough, have discrete-time models with zeros outside the stability region. This has the following practical consequence:

when a self-tuning controller is implemented on a process by plant engineers with PID experience, there is a great temptation to minimise the sample-interval h on the grounds that the controller will respond more rapidly to disturbances. This will lead to the migration of the discrete-time zeros, so that it is important to ensure that the control design is relatively insensitive to their location.

The above lemmas are in Gawthrop (1980) and Gawthrop and Clarke (1980); further discussion is in Åström, Hagander and Sternby (1980).

The position of the fractional delay δ also influences the zeros, as is shown in the following example:

Example 2.2.1

Find $G^*(z^{-1})$ corresponding to $\frac{e^{-s\delta}}{s+a}$ where $0 < \delta \leq h$

Following the above procedure, expand $\frac{e^{s(h-\delta)}}{s(s+a)}$, giving

$$\frac{1}{as} - \frac{e^{a(\delta-h)}}{a(s+a)} \rightarrow \frac{1}{a(1-z^{-1})} - \frac{e^{a(\delta-h)}}{a(1-qz^{-1})}$$

where $q = e^{-ah}$. On recombining and multiplying by $(1-z^{-1})\,z^{-1}$:

$$G^*(z^{-1}) = z^{-1} \cdot \frac{b_o + b_1 z^{-1}}{(a(1-qz^{-1})}, \text{ where:}$$

$$b_o = 1 - e^{a(\delta-h)} \quad \text{and}$$

$$b_1 = -q + e^{a(\delta-h)} = e^{-ah}(e^{a\delta}-1) = 0 \text{ if } \delta = 0.$$

The zero of G^* is at $z_i = \frac{-b_1}{b_o} \simeq \frac{\delta}{\delta-h}$ for small h.

For $\delta < h/2$ this lies within the unit circle, but moves outside for $\delta > h/2$. Note that part of the fractional delay can, in some cases, be attributed to the computation time of the algorithm (i.e. the time taken between sampling $y(t)$ and generating $u(t)$). As this time is generally fixed, it becomes an increasing fraction of h as h is reduced.

In the development of the self-tuning algorithms, we shall assume that the plant model is in the standard discrete-time form and drop the * henceforth, giving the input-output model at sample instants $t = ih$:

$$y(t) = \frac{B(z^{-1})}{A(z^{-1})} u(t-k) + \frac{C(z^{-1})}{A(z^{-1})} \xi(t) \tag{14}$$

where the order of the polynomials is n, the leading coefficients of A and C are 1, $b_o \neq 0$ so that k, the delay, is > 1. The disturbance is due to an uncorrelated sequence $\xi(t)$, whose variance is σ^2, passing through dynamics C/A where all roots of C are assumed to lie within the unit circle. Here, as in the sequel, z is interpreted as the forward-shift operator and polynomials such as $A(z^{-1})$ are written simply as A where convenient. The model (14) can be written in difference-equation form as:

$$y(t) + \Sigma_1^n a_i y(t-i) = \Sigma_o^n b_i y(t-i-k) + \xi(t) + \Sigma_1^n c_i \xi(t-i)$$

2.2.1 Predictive Models

Many self-tuning strategies, particularly implicit methods, are based on predictive control designs, where the prediction horizon is the system delay k. The argument for predictive control is as follows: As the delay is k, the first output that can be influenced by the current control u(t) is y(t+k). Meanwhile disturbances are acting upon the process, so that if an optimal prediction (Yaglom, 1973) of their effect at t+k is available at time t, the control u(t) can be chosen to neutralise them. The effectiveness of this control depends on the accuracy of prediction, which itself clearly depends on the characteristics of the disturbance and the prediction interval k.

Rewrite (14) to give:

$$y(t+k) = \frac{B}{A} u(t) + \frac{C}{A} \xi(t+k) \qquad (15)$$

By expanding $\frac{1}{A}$ as an infinite power series in z^{-1}, it is seen that the disturbance term has components $\xi(t+i)$ in the 'future' and $\xi(t-j)$ in the 'past'. These past values of $\xi(t)$ can be reconstructed using (14) and measured input-output data {u(t), y(t)}, but future values are unpredictable as ξ is an uncorrelated sequence. This development is in Clarke and Hastings-James (1971). Now C/A can be resolved in an identity:

$$\frac{C}{A} = E(z^{-1}) + z^{-k} \frac{F(z^{-1})}{A} \qquad (16)$$

where the E and F polynomials can be obtained uniquely by comparing coefficients of powers of z^{-1} if the degree of E is k-1. It is seen that E(0) = 1 and the degree of F is n-1.

Equation (15) and (16) may be rewritten as

$$Ay(t+k) = Bu(t) + C\xi(t+k) \qquad (17)$$

$$C = EA + z^{-k}F \qquad (18)$$

((18) is a Diophantine equation - see Kučera (1979)).

Now (17) is premultiplied by $E(z^{-1})$, giving:

$$EA\ y(t+k) = EB\ u(t) + EC\ \xi(t+k)$$

or $$Cy(t+k) - Fy(t) = Gu(t) + EC\xi(t+k)$$

where $G(z^{-1}) = EB$, giving:

$$y(t+k) = \frac{Fy(t) + Gu(t)}{C} + E\xi(t+k) \tag{19}$$

Defining the optimum prediction to be $y^*(t+k/t)$ and the error to be $\tilde{y}(t+k/t)$, (19) can be written as $y = y^* + \tilde{y}$, where

$$\left.\begin{aligned} Cy^*(t+k/t) &= Fy(t) + Gu(t) \\ \tilde{y}\,(t+k/t) &= E\xi(t+k) \end{aligned}\right\} \tag{20}$$

Note that y^* and $\tilde{y}$ are orthogonal, as would be expected with an optimum predictor (in the Wiener sense), and the prediction accuracy can be obtained from:

$$\text{Var }\{\tilde{y}\} = \sigma^2\,(1+e_1^2 + e_2^2 + \ldots + e_{k-1}^2) \tag{21}$$

This variance increases with k.

Example 2.2.2

Rewrite the system of (22) in prediction form (19).

$$(1-0.9\,z^{-1})\,y(t) = 0.5u(t-2) + (1+0.7\,z^{-1})\xi(t) \tag{22}$$

Equation (18) is for this system where $n = 1$ and $k = 2$:

$$(1+0.7\,z^{-1}) = (1+e_1z^{-1})(1-0.9z^{-1}) + z^{-2}\,f_o$$

Comparing coefficients of z^{-1} gives $e_1 = 1.6$, and then comparing coefficients of z^{-2} gives $f_o = 0.9 \times 1.6 = 1.44$, so that the prediction model is:

$$y(t+2) = \frac{1.44\,y(t) + (0.5 + 0.8\,z^{-1})\,u(t)}{1 + 0.7\,z^{-1}} + (1 + 1.6\,z^{-1})\xi(t+2)$$

The prediction error variance is $\sigma^2(1 + 1.6^2) = 3.56\sigma^2$. Note that if k were 1, the 1-step variance would be σ^2; this shows how the prediction deteriorates as k increases. For the uncontrolled case, $u(t) = 0$, the variance of $y(t)$ can be shown to be

$$\text{Var }(y(t)) = \sigma^2\,\frac{1 + 2 \times 0.7 \times 0.9 + 0.7^2}{1 - 0.9^2} = 14.47\,\sigma^2$$

so that the prediction 'explains' roughly 3/4 of the output variance.

Some versions of self-tuning such as the controller of Clarke and Gawthrop (1975, 1979) involve the prediction of an *auxiliary* output $\psi(t)$ defined in terms of a user-specified transfer-function $P(z^{-1})$:

$$\psi(t) \triangleq P(z^{-1})\ y(t) = \frac{P_N(z^{-1})}{P_D(z^{-1})}\ y(t) \tag{23}$$

In this case, the identity (16) becomes:

$$\frac{CP_N}{AP_D} = E + z^{-k}\ \frac{F}{AP_D} \tag{24}$$

The degree of $E(z^{-1})$ is still k-1, although the coefficients are different from those of E in (18), and the degree of F is n-1 + degree of P_D (at least in the useful case where P corresponds to the discretization of a continuous-time transfer-function). Repeating the analysis of equations (17) to (20), it is found that the optimal prediction of $\psi(t+k)$ is given by:

$$\left.\begin{aligned} C\psi^*(t+k/t) &= Fy^f(t) + Gu(t) \\ \psi(t+k) &= \psi^*(t+k/t) + E\xi(t+k) \end{aligned}\right\} \tag{25}$$

where $y^f(t)$ is the known signal obtained by passing y(t) through the filter $1/P_D$, so that $P_D(z^{-1})y^f(t) = y(t)$.

2.3 *Recursive Parameter Estimation*

In regression analysis (Plackett, 1960) an observation ('output') is hypothesised to be a linear combination of explanatory variables ('inputs'), and a set of observations is used to estimate the weighting on each variable such that some fitting criterion is optimised. The common criterion chooses the model parameters such that the sum-of-squares of the errors between the model outputs and the observations is minimised ('least-squares').

Consider the model that is linear-in-the-parameters:

$$\left.\begin{aligned} \phi(t) &= \theta_1 x_1(t) + \theta_2 x_2(t) + \ldots + \theta_n x_n(t) + \varepsilon(t) \\ \text{or}\quad \phi(t) &= x'(t)\ \theta + \varepsilon(t) \end{aligned}\right\} \tag{25}$$

where: θ is a vector of n unknown parameters (assumed constant)

x(t) is a vector of *known* data

$\varepsilon(t)$ is an error statistically independent of the elements $x_i(t)$

$\phi(t)$ is an 'output' observation.

Suppose N observations are made, where $N > n$. The collection of equations like (25) can be stacked to give:

$$\phi_N = X_N\ \theta + \varepsilon_N \tag{26}$$

where ϕ_N, ε_N are N-vectors, and X_N is a Nxn matrix. The estimates $\hat{\theta}$ are obtained by minimising the loss-function $L = \Sigma_1^N e^2(t)$, where e(t) is the model error $\phi(t) - x'(t)\hat{\theta}$, with respect to $\hat{\theta}$. This is readily shown to give the 'normal equations of least-squares':

$$\hat{\theta} = (X_N' X_N)^{-1} X_N' \phi_N \tag{27}$$

Hence $\hat{\theta} = (X_N' X_N)^{-1} X_N' (X_N\theta + \varepsilon_N)$, using (26)

$$= \hat{\theta} + (X_N' X_N)^{-1} X_N' \varepsilon_N$$

Now as $x_i(t)$ and $\varepsilon(t)$ are independent, by assumption, the expected value of $\hat{\theta}$ depends on the expected value of ε_N. If it is further assumed that the noise has zero mean (if not, $x_{n+1}(t)$ can be defined as 1 and the mean of ε estimated) then $E\{\hat{\theta}\} = \theta$, the true parameter vector, so that the estimates are *unbiased*. Moreover, for the case where $\varepsilon(t)$ is an uncorrelated (white-noise) sequence, the least-squares estimates can also be shown to have minimal variance compared with all other linear unbiased estimates, and whose accuracy depends on the covariance matrix:

$$E\{(\hat{\theta}-\theta)(\hat{\theta}-\theta)'\} = \sigma^2 (X_N' X_N)^{-1} \tag{28}$$

where σ^2 is the variance of $\varepsilon(t)$.

The above equations were derived in a form suitable for batch processing, in which the amount of data storage and computation increases with N; this is obviously undesirable for a method such as self-tuning where the computational requirements should not increase with time. This property can be achieved by reformulating the equations in *recursive* form in which a fixed computation is performed for each new observation. Noting that the dimension of X'X do not increase with N, define

$$S(t) \triangleq (X_t' X_t) \tag{29}$$

where X_t is the matrix of known data acquired up to time t. With a slight abuse of notation, write the equation (26) in partitioned form as:

$$\begin{pmatrix} \phi_{t-1} \\ \phi(t) \end{pmatrix} = \begin{pmatrix} X_{t-1} \\ x'(t) \end{pmatrix} \theta + \begin{pmatrix} \varepsilon_{t-1} \\ \varepsilon(t) \end{pmatrix}$$

Letting $\hat{\theta}(t)$ be the n-vector of estimates using all data up to time t, (27) can be written as:

$$\theta(t) = S(t)^{-1} [X_{t-1}' \phi_{t-1} + x(t)\phi(t)]$$

$$= S(t)^{-1} [S(t-1)\hat{\theta}(t-1) + x(t)\phi(t)]$$

but $\quad S(t) = S(t-1) + x(t)\, x'(t) \tag{30}$

$$\hat{\theta}(t) = S(t)^{-1}\,[S(t)\hat{\theta}(t-1) - x(t)x'(t)\hat{\theta}(t-1) + x(t)\phi(t)]$$

$$= \hat{\theta}(t-1) + S(t)^{-1}\,x(t)\,[\phi(t) - x'(t)\hat{\theta}(t-1)]$$

giving, together with (3.6), the recursive updating of $\hat{\theta}$:

$$\hat{\theta}(t) = \hat{\theta}(t-1) + K(t)\,[\phi(t) - x'(t)\hat{\theta}(t-1)] \tag{31}$$

where K(t) is an n-vector (the 'Kalman gain'). Note that the estimation update for each parameter is of the feedback form:

$$\begin{bmatrix}\text{new}\\ \text{estimate}\end{bmatrix} = \begin{bmatrix}\text{old}\\ \text{estimate}\end{bmatrix} + \text{gain x (prediction error of old model)}$$

Although suitable for convergence analysis these equations are still not entirely useful as they involve the inverse of S(t) at each stage. Further simplification is possible by invoking a 'matrix inversion lemma' which involves P(t), the inverse of S(t):

$$P(t) = (S(t-1) + xx')^{-1} = P(t-1) - \frac{P(t-1)\,xx'\,P(t-1)}{1 + x'P(t-1)x}$$

Using this lemma, recursive-least-squares (RLS) becomes (31) together with:

$$K(t) = \frac{P(t-1)x(t)}{1 + x'(t)P(t-1)x(t)} \tag{31a}$$

$$P(t) = (I - K(t)\,x'(t))P(t-1) \tag{31b}$$

where it is seen that no matrix inversion is required. Now S^{-1} exists only after n observations have been taken, but it is convenient to initialise the algorithm by a suitable choice of P(0). Typically it is taken to be a diagonal matrix αI where a large value of α (10^4, say) implies little confidence in $\hat{\theta}(0)$ and gives rapid initial changes in $\hat{\theta}(t)$, whilst a small value of α (0.1, say) implies that $\hat{\theta}(0)$ is a reasonable estimate of θ and $\hat{\theta}(t)$ changes only slowly. Note that the covariance matrix of the estimates is given by $\sigma^2 P(t)$ (for large t at least); σ^2 can itself be estimated using the model prediction error e(t).

Now $S(t) = \Sigma x(i)x'(i)$, so that its norm $\|S\|$ is 'small' initially but tends to infinity, provided that x(t) is 'sufficiently exciting'. Hence $\|K\|$ tends to zero, and $\hat{\theta}(t)$ tends to a constant vector θ. This behaviour is acceptable if the 'true' parameter were in fact constant, but in practice we would want the algorithm to *track* slowly-varying parameters (so that the self-tuner stays in tune). There are many ways in which (31) can be modified to achieve this, the most popular being the use of a 'forgetting factor' β, where $0 < \beta \le 1$.

If, instead of giving equal weights to the errors in the least-squares criterion, more weighting is given to the more recent data giving:

$$I = \Sigma_1^t \beta^{t-j} e^2(j)$$

to be minimised, the RLS equations (31a) and (31b) become:

$$K(t) = \frac{P(t-1)\ x(t)}{\beta + x'(t)P(t-1)x(t)} \tag{32a}$$

$$P(t) = (I-K(t)x'(t))P(t-1)/\beta \tag{32b}$$

In this case $||P||$ and $||K||$ do not tend to zero, so that $\hat{\theta}(t)$ can vary for large t. Although $\hat{\theta}$ is affected by *all* past data, the 'asymptotic sample length' : ASL $\underline{\Delta}$ $1/(1-\beta)$ indicates the number of 'important' previous samples contributing to $\hat{\theta}(t)$. Typical values of β are in the range 0.95 (fast variations) to 0.999 (slow); the corresponding ASLs are 20 to 1000.

There are many ways in which recursive estimation algorithms can be derived. For example θ could be considered to be the 'state' of a system whose 'output measurement' is given by (25), and the estimation equations (31), (32) could be obtained directly from Kalman filtering theory. In this approach σ^2 would need to be known (or estimated), but the parameter variations could be described by a 'state-equation' such as the random-walk model:

$$\theta(t) = \theta(t-1) + \psi(t);\ \psi(t)\text{ an uncorrelated sequence.}$$

In this case the updating of P(t) would read:

$$P(t) = (I - K(t)x'(t))P(t-1) + R$$

where R is the covariance matrix of $\psi(t)$; this gives one alternative approach to the estimation of drifting parameters.

2.3.1 Application to Dynamic Models

The model (25) is general - by allocating different variables to $x_i(t)$ and different interpretations to the elements of θ a variety of model structures can be formulated. For example, consider the difference equation:

$$y(t) + a_1y(t-1) \ldots + a_ny(t-n) = b_ou(t-k) + \ldots + b_nu(t-k-n) + \varepsilon(t) \tag{33}$$

which can be rewritten as:

$$y(t) = -a_1y(t-1) \ldots + b_nu(t-k-n) + \varepsilon(t).$$

Taking $\phi(t) = y(t)$, and:

$$\theta' = (a_1, a_2 \ldots a_n, b_o, \ldots, b_n)$$

$$x'(t) = (-y(t-1), \ldots -y(t-n), u(t-k), \ldots u(t-k-n))$$

as elements of (25), the RLS algorithm can be used to estimate the parameters of a dynamic model. It $\varepsilon(t)$ is an uncorrelated random sequence where $\varepsilon(t)$ is also uncorrelated with $u(t-i)$, $y(t-i)$, the above theory also shows that RLS gives optimal, unbiased estimates.

Consider, however, the general system difference equation (15), where the 'error' $\varepsilon(t) = \xi(t) + \Sigma_1^n c_i \xi(t-i)$ is no longer uncorrelated unless $c_i = 0$, all i. Moreover, as $\varepsilon(t)$ contains past values of ξ, it is correlated with $y(t-1) .. y(t-n)$. Hence a principal property of the least squares method - that the estimates are unbiased - no longer holds. This is not important for high signal/noise ratios, but the bias of the estimates is substantial when the signal/noise ratio is low (<10/1, say). It is surprising, therefore, that many self-tuning algorithms can in fact use RLS such that either there is no bias, or the bias has no effect on the self-tuning property. Each case, however, must be investigated separately to show this; examples of this will be given later. Other self-tuners are affected by bias, and estimation algorithms such as those described below must then be used.

The system model (15) can be written in the form of (25) as:

$$y(t) = -a_1 y(t-1) - \ldots + b_o u(t-k) + \ldots + c_1 \xi(t-1) + \ldots + c_n \xi(t-n) + \xi(t) \quad (34)$$

where RLS could be used if the variables $\xi(t-1) \ldots \xi(t-n)$ were in fact 'known'. The idea of the extended-least-squares method (Panuska, 1969) is to *proxy* $\xi(t-i)$ by estimates $e(t-i)$ obtained using the current model parameter estimates $\hat{\theta}(t-i)$, giving:

$$\theta' = (a_1 \;..\; a_n,\ b_1 \ldots b_n,\ c_1 \quad c_n)$$

$$x'(t) = (-y(t-1) \ldots -y(t-n),\ u(t-k) \ldots u(t-k-n),\ e(t-1) \ldots e(t-n))$$

An ordinary RLS iteration is performed using these extended vectors, and a new estimate of the residuals $e(t)$ is obtained from:

$$e(t) = y(t) - x'(t)\hat{\theta}(t) \quad (35)$$

This estimate is then placed into the ring-buffer of the e variables for the next iteration; initial values $e(0)$, $e(1)$ etc are usually taken to be zero.

Extended-least-squares (ELS) is a straight forward extension of RLS, but is not as robust as RLS in practice (a comment which could be made about most algorithms). Its one theoretical shortcoming is that convergence cannot be proved for all C polynomials (Ljung, 1978), though this has not been found to be a serious problem in many applications. A method which has been shown to converge for all useful C is the recursive maximum-likelihood (RML) algorithm. This method notes that the estimation of all parameters in (34) is the non-linear hill-climbing problem of finding the parameter set $\hat{\theta}$ which minimises the loss-function $L = \frac{1}{2}\Sigma\xi^2(t)$ (see Eykhoff, 1974). The function $K(t)$ in (31) can in this context be interpreted as an approximate gradient of the loss-function with respect to $\hat{\theta}(t-1)$, and the overall algorithm as a recursive approximation to the standard Newton-Raphson iterative method for minimising a sum-of-squares. The RML method, therefore, is superficially the same as ELS except that the x(t) vector of (31a) and (31b) becomes the vector $x^f(t)$ with *filtered* data components, where the filter used in $1/\hat{C}(z^{-1})$. For example:

$$\hat{C}(z^{-1})u^f(t-k-i) = u(t-k-i), \text{ giving:}$$

$$u^f(t-k-i) = u(t-k-i) - \hat{c}_1 u^f(t-k-i-1) \dots - \hat{c}_n u^f(t-k-i-n)$$

and: $$x^f(t) = \text{col } (-y^f(t-1) \dots u^f(t-k) \dots u^f(t-k-n) \dots$$

The RML algorithm needs more computations than ELS due to the extra filtering, and the ring-buffers containing filtered as well as raw data add to the data storage requirements.

RLS, ELS and RML belong to a class of algorithms called 'prediction error identification methods', whose structure is of the form (31) and whose convergence properties have been studied by Ljung (1979). Now suppose that a system:

$$y(t) = -ay(t-1) + bu(t-1) + \varepsilon(t)$$

is in closed-loop with a proportional controller $u(t) = \alpha y(t)$. It is seen that the input/output relationship can be described by:

$$y(t) = - (a+\lambda)y(t-1) + (b+\lambda\alpha)u(t-1) + \varepsilon(t)$$

Hence closed-loop estimation based on minimising $L = \Sigma\varepsilon(t)^2$ does not give unique parameter estimates, as $\hat{\theta}' = [b+\lambda\alpha,\ a+\lambda]$ give the same value for the loss-function L. This problem is surveyed in Gustavsson et al (1977), in which it is shown that *unique* closed-loop estimates are obtained if:

(i) there is a separate, persistently exciting signal such as a set-point, acting in the loop, or

(ii) the feedback law is time-varying.

This second case is important in adaptive control, whilst the first is important when monitoring an existing controlled process in order to derive new control parameters. It may not, in fact, be necessary to have unique estimates, provided that the resultant controller design produces the same control signal for all possible estimates (as in self-tuning). In practice, however, uniqueness is useful as it avoids the possibility of numerical overflow or underflow.

2.4 *Minimum-Variance Control and the Self-Tuning Regulator*

Recall that the system model (15) can be written in the predictive form (20):

$$Cy^*(t+k/t) = Fy(t) + Gu(t)$$

$$y(t+k) = y^*(t+k/t) + \tilde{y}(t+k/t) = y^*(t+k/t) + E\xi(t+k)$$

Consider a control law which chooses u(t) to minimise the variance $I = E\{y^2(t+k)\}$, where the expectation is conditioned on all data over the range $(-\infty, t)$ and the ensemble is all realizations of the random processes $\xi(t+i)$ affecting the output after time t. The argument for the choice of I as a control criterion is as follows: in quality control it is desirable that only a specified small percentage of the product has a quality (e.g. paper thickness) outside some given range (e.g. to control the tearing strength). If there are large fluctuations in the quality, the set-point giving the average quality may need to be set at a relatively high value, which may affect the average product cost. If the fluctuations are minimised, this set-point can be reduced giving the same reject rate at a lower average cost, as shown in Fig 2.3.

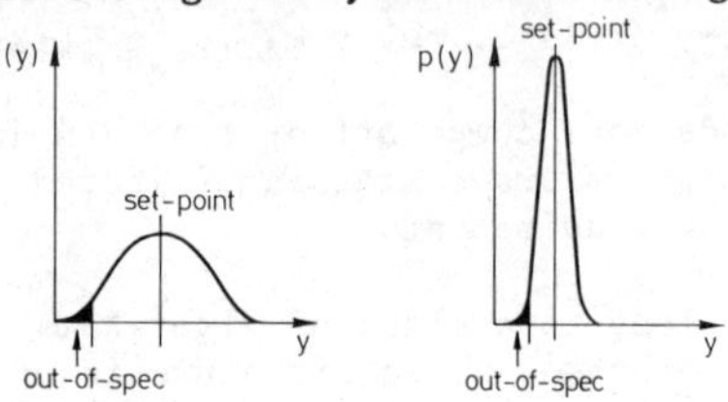

Fig 2.3.1 Non-minimum-variance control

Fig 2.3.2 Minimum-variance control

Now $I = E\{y^2(t+k)\} = E\{(y^*(t+k/t) + \tilde{y}(t+k/t))^2\}$

$$= (y^*(t+k/t))^2 + E\{(\tilde{y}(t+k/t))^2\}$$

using the fact that y^* and $\tilde{y}$ are orthogonal and y^* is known at time t. As $\tilde{y}$ is unaffected by u(t), the minimum I is clearly when y^* is set to zero by the choice of control:

$$Fy(t) + Gu(t) = 0 \tag{36}$$

or $$u(t) = -\frac{Fy(t)}{BE}, \text{ a feedback law} \tag{37}$$

Using this control the minimal variance I_{min} is simply that of the k-step-ahead prediction error $\sigma^2(1+e_1^2 + .. + e_{k-1}^2)$. From example (2.2.2) the above results give a minimum-variance controller:

$$u(t) = \frac{-1.44\ y(t)}{0.5 + 0.8\ z^{-1}}, \text{ or in difference-equation form:}$$

$$u(t) = -1.6\ u(t-1) - 2.88\ y(t)$$

which reduces the uncontrolled output variance of $14.47\ \sigma^2$ to the controlled variance $3.56\ \sigma^2$.

Note that the closed-loop characteristic equation is:

$$1 + \frac{F}{BE}\frac{z^{-k}B}{A} = 0, \text{ or } B(EA + z^{-k}F) = 0,$$

which, using the identity (18) becomes:

$$B(z^{-1})\ C(z^{-1}) = 0 \tag{38}$$

As C is a stable polynomial it is seen that the closed-loop stability depends on the sampled-process zeros. Section 2.2, however, shows that there are many cases (e.g. with fast sampling or large fractional delay) where B has roots outside the stability region. Hence minimum-variance control should be used with caution; ways of overcoming this difficulty are described later.

The original self-tuning regulator of Åström and Wittenmark (1978) used the minimum-variance objective function. Consider, first, equation (20) for the case where $C(z^{-1}) \equiv 1$, and at time t rather than t+k:

$$y(t) = F(z^{-1})y(t-k) + G(z^{-1})u(t-k)+E(z^{-1})\xi(t) \tag{39}$$

Compare (39) with (25) and make the following correspondences:

$$\theta' = (f_o,\ f_1,\ ..f_{n-1},\ g_o,\ g_1\ ...g_{n+k-1})$$

$$x'(t) = (y(t-k),\ ...,\ u(t-k),\ ...\ u(t-k-n+1))$$

$$\varepsilon\ (t) = \xi(t)+e_1\xi(t-1)\ +\ ...\ +\ e_{k-1}\xi(t-k+1)$$

$$\phi\ (t) = y(t)$$

Here $\varepsilon(t)$ is an autocorrelated process, but is independent of all elements of the 'known data' vector x(t). Hence the RLS algorithm of (31) *et seq* can be used to obtain unbiased (though not optimal) estimates of $\hat{\theta}$. These estimates are then used in the certainty-equivalent control law:

$$\hat{F}(z^{-1})\ y(t) + \hat{G}(z^{-1})\ u(t) = 0 \tag{40}$$

This is an *implicit* self-tuner, as the required feedback parameters are estimated directly rather than via a control-design calculation.

The parameter estimates in (40) are not unique, for this equation may be multiplied by an arbitrary constant without affecting the calculation of u(t). This means that the estimates will lie on a linear manifold - 'wandering in unison' - and may lead to excessively large or small values with possible numerical problems developing. A unique estimation can be achieved by 'fixing' one parameter, such as $g_o = b_o$ the first sample on the system's pulse response. This can be done in two ways. One is to omit g_o and u(t-k) in θ and x and to make $\phi(t) = y(t) - \hat{b}_o u(t-k)$, where $\hat{b}_o$ is the fixed value of g_o. The other method is to set the corresponding diagonal element in P(0) - which would otherwise be α I - to zero, indicating that the parameter is considered to be known exactly. This second approach, though involving slightly more computation, has the advantage that the same basic self-tuner software can be used with any chosen parameter fixed. Though $\hat{b}_o$ need not be close to b_o, the convergence rate depends on the ratio $b_o/\hat{b}_o$, and a 'small' value of $\hat{b}_o$ may lead to initially excessive control signals, as seen intuitively by solving (40) for u(t):

$$u(t) = -\frac{1}{\hat{b}_o}(\hat{f}_o y(t) + \hat{f}_1 y(t-1) + .. \hat{g}_1 u(t-1) + ..)$$

An alternative is simply to ignore lack of uniqueness in the estimation and not to fix any parameter; although this complicates convergence analysis the self-tuning performance is still effective in most cases.

Consider now what happens when the self-tuner is used with a system where $C(z^{-1}) \neq 1$ but is a general polynomial. The discussion of recursive estimation would indicate that RLS is not sufficient and ELS or RML should be used, but this is not in fact the case. This can be seen by expanding $1/C(z^{-1})$ as an infinite polynomial:

$$C^{-1} = 1 + \gamma_1 z^{-1} + \gamma_2 z^{-2} + \ldots$$

and rewriting the predictor model as:

$$y(t) = Fy(t-k) + Gu(t-k) + \gamma_1(Fy(t-k-1) + Gu(t-k-1)) + \ldots + E\xi(t)$$

Suppose the algorithm has converged such that $\hat{\theta} = \theta$; then the control (40) sets all the terms on the RHS to 0 so that this equation reduces to (39) as if $\gamma_i = 0$ and hence as if $C \equiv 1$. This implies that $\hat{\theta} = \theta$ is a fixed-point of the algorithm, but does not of course *prove* that it is a *stable* fixed-point - this must be done using alternative methods. In the initial tuning stage, however, $\hat{\theta}$ may be remote from θ and the γ_i's may then have a significant effect on the *rate* of convergence. Intuitively, the dynamics of 1/C and the convergence rate are related.

There is one subtle point concerning the implication of the above argument in practice, as it depends on (40) being satisfied by the control u(t). In many systems there are hard limits on the control that can be exercised and there may be occasions in which the desired u(t) cannot be used so that (40) no longer holds. Although in principle RLS then fails to give unbiased estimates of F and G, it is found that provided u(t) is not always clipped the algorithm is still effective.

2.5 *Generalized Minimum-variance and the Self-tuning Controller*

A control design for minimising the output variance is often inappropriate as it ignores both the 'practicable controllability' of the process and the variance of the control effort required; moreover it is highly sensitive to the positions of the process zeros in discrete-time which, as seen in section 2.2, bear little relationship with their continuous-time counterparts for many systems. To overcome these disadvantages a general class of control laws was developed (Clarke and Gawthrop, 1975, 1979), based on prediction theory, which broadened the class of performance objectives yet retained the implicit self-tuning property.

Recall that the auxiliary output $\psi(t) \triangleq P(z^{-1})y(t)$, where P is a prescribed transfer-function, admits the predictive model given by (25):

$$\left.\begin{aligned} C\psi^*(t+k/t) &= Fy^f(t) + Gu(t) \\ \psi\ (t+k) &= \psi^*(t+k/t) + E\xi(t+k) \end{aligned}\right\} \qquad (41)$$

The development of section 2.4 can be directly applied to these equations to derive the (self-tuned) minimum-variance control of $\psi(t)$. This can be viewed as generating an augmented system $G^*(z^{-1})P(z^{-1})$ whose control is in some sense 'easier' than the control of the original system G* alone. For example, P could be a phase-advance element which compensates for the excessive phase-lag of G* itself.

The full predictive control law, however, has other features. Let w(t) be the system set-point at time t and $Q(z^{-1})$ a further prescribed transfer-function. Then the predictive feedback law is defined to be:

$$u(t) = \frac{w(t) - \psi^*(t+k/t)}{Q(z^{-1})} \qquad (42)$$

This has the structure of a classical feedback control except that the prediction ψ^* is fed back rather than the system output, and it has various interpretations (Gawthrop, 1977) according to the user's choice of P and Q. Note that w(t) need not be the signal demanded by the plant operator w'(t), but could be 'tailored' by a unity-gain transfer-function $R(z^{-1})$, giving w(t) = Rw'(t). This could, for example, be used to smooth out sudden changes in demand. Combining (41), (42) and the system model (14) gives the conceptual feedback system shown in Fig 2.4.

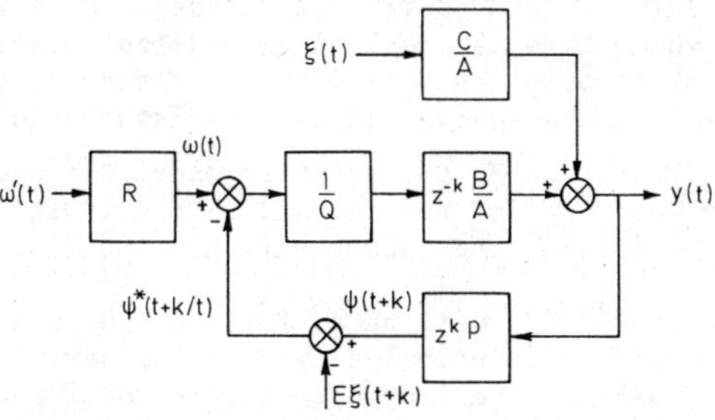

Fig 2.4 Conceptual feedback system

Note that not all elements in Fig 2.4 are causal though the relationships between the system variables of interest are causal, as the identity (24) gives a causal approximation to the element z^kP. To find the closed-loop behaviour u(t) has to be eliminated from the equations, giving:

$$y(t) = \frac{B}{PB + QA}\, w(t-k) + \frac{EB + QC}{PB + QA}\, \xi(t) \qquad (43)$$

which, on using identity (24), can be shown to require a control:

$$u(t) = \frac{A}{PB + QA} w(t) - \frac{F/P_D}{PB + QA} \xi(t) \tag{44}$$

This shows that the characteristic equation of the closed-loop is:

$$PB + QA = 0 \tag{45}$$

where it is seen that the system time-delay z^{-k} is absent - this is a useful feature of predictive control. Moreover the roots of the characteristic equation no longer depend solely on B; for example if $Q(z^{-1}) = \lambda$, say, the closed-loop poles are the roots of:

$$P_N B + \lambda P_D A = 0,$$

which, for large enough λ, is dominated by A. Hence for stable open-loop systems there is also closed-loop stability even if a root of $B(z^{-1})$ is outside the stability region, thus desensitizing the control to the position of the system's discrete-time zeros. Note that λ can be viewed as a root-locus parameter.

2.5.1 Some Interpretations

Consider the predictive control in closed-loop (43) for Q = 0:

$$y(t) = \frac{1}{P} w(t-k) + \frac{E}{P} \xi(t), \text{ with control} \tag{46}$$

$$u(t) = \frac{A}{PB} w(t) - \frac{F}{P_N B} \xi(t) \tag{47}$$

Defining $M(z^{-1}) \triangleq 1/P(z^{-1})$, the control law is seen to be a 'model-reference' type in which M is the desired closed-loop model specifying how y(t) responds to set-point changes w(t-k). Moreover, the model specifies to some extent the equivalent output disturbance: $M(z^{-1})E(z^{-1})\xi(t)$. Clearly as $M \rightarrow$ the identity model the system becomes 'faster', requiring more control action, and in the limit the control law reduced to one minimising the output variance.

A typical design procedure would be as follows. Suppose the open-loop plant is considered to be principally second-order with typical time-constants T_p. Then a convenient closed-loop model would also be of second-order with unity gain (clearly) and 'faster' than the open-loop:

$$\text{i.e. } M(s) = \frac{1}{(1+sT)^2}, \quad T < T_p.$$

(A more suitable model could have its poles in a Butterworth configuration at a radius 1/T). This model would be discretized assuming impulse samplers and a zero-order-hold to give $M(z^{-1}) = Z(M(s))$, and hence $P(z^{-1})$. This approach is more satisfactory than simply specifying $M(z^{-1})$ directly, as it positions the closed-loop zeros at convenient locations. This formulation is easy to use, as the desired model T can be chosen on *engineering* grounds to suit the time-scale of the process; in a self-tuning context the closed-loop behaviour would be *consistent* even if the open-loop process changes.

However, (47) shows that the model-reference control is obtained by cancellation of process dynamics, and (45) implies the same sensitivity to process zeros as the simple minimum-variance law. To overcome this problem a *detuned* model-reference law is used with $Q \neq 0$, and where the use of $Q = \lambda$ trades 'closeness of model following' against control effort. In practice, Q is a subsidiary parameter which can be set to zero during the initial tuning and increased later so as to reduce control effort.

Another way to view $Q(z^{-1})$ is by comparing (42) with a conventional controller. Taking $P = 1$ and defining $L(z^{-1}) = 1/Q$, where L has a PID-like structure, (42) is like a Smith-predictor (1959) controller in that the predicted rather than the currently measured output is fed back. As with Smith's method, the closed-loop characteristic equation is $LB + A = 0$ instead of $z^{-k}LB + A = 0$ (which it would be if predictions were not used, as in conventional control). This means that the gain of L can be increased without causing instability, thus giving tighter control of systems with time-delay. The advantage of the above method is that, unlike Smith's predictor, it takes account of disturbances and can work on open-loop unstable processes.

Consider the variance:

$$\begin{aligned} I_1 &\triangleq E\{(\psi(t+k) + Qu(t) - w(t))^2\} \\ &= E\{(\psi^*(t+k/t) + Qu(t) - w(t) + E\xi(t+k))^2\} \qquad (48) \\ &= E\{(\psi^*(t+k/t) + Qu(t) - w(t))^2\} + E\{E(z^{-1})\xi(t+k)^2\} \end{aligned}$$

as $\xi(t+i)$ is orthogonal to other terms on the RHS. It is seen that the control (42) minimises I_1 as it sets the first bracket to zero. The terms in I_1 are not, however, in the usual quadratic form. A more conventional measure of performance is:

$$\begin{aligned} I_2 &\triangleq E\{(\psi(t+k) - w(t))^2 + \mu(Qu(t))^2\} \qquad (49) \\ &= E\{(\psi^* + \tilde{\psi} - w)^2 + \mu(Qu)^2\} \\ &= E\{(\psi^* - w)^2 + \mu(Qu)^2\} + E\{\tilde{\psi}^2\} \end{aligned}$$

If the expectation is now interpreted to be conditioned on data acquired up to time t the first term is deterministic, so I_2 is minimised by minimising $(\psi^* - w)^2 + \mu(Qu)^2$ with respect to $u(t)$:

$$\frac{\partial I_2}{\partial u(t)} = 2(\psi^*(t+k/t) - w(t)) \frac{\partial \psi^*}{\partial u(t)} + 2\,\mu q_o Qu(t)$$

where q_o is the first term of $Q(z^{-1}) = q_o + q_1 z^{-1} + \ldots$ Using (41) $\partial\psi^*/\partial u(t) = g_o$, so I_2 is minimised when:

$$\psi^*(t+k/t) - w(t) + \frac{\mu q_o}{g_o} Qu(t) = 0$$

Hence the control (42) minimises I_2 of (49) for $\mu = g_o/q_o$, where the expectation is interpreted as *conditioned* on currently available data. For example if $Q = \lambda$ and $P = 1$, I_2 becomes:

$$I_{21} = E\{(y(t+k) - w(t))^2 + g_o \lambda u^2(t) | t\}$$

so that λ can be used to trade error against control variance; in practice the trade-off is very favourable (Clarke and Hastings-James, 1971). Again, defining $Q = \lambda(1-z^{-1})$, I_2 becomes:

$$I_{22} = E\{(y(t+k) - w(t))^2 + g_o \lambda(u(t) - u(t-1))^2 | t\}$$

in which differences in control are weighted. Note that this is equivalent to inserting a forward-path integrator into the loop.

A final interpretation involves a useful way of implementing the control law. Recall that $\psi^*(t+k/t)$ is a function of $u(t)$ as all other terms are known at time t, so define $u^1(t)$ as the exact model-following control which would make ψ^* equal $w(t)$. Equation (41) gives:

$$w(t) + (C-1)\psi^*(t+k/t) = Fy^f(t) + (G-g_o)u(t) + g_o u^1(t) \quad (50)$$

where $(C-1)$ and $(G-g_o)$ are polynomials in z^{-1} with no leading constant - (50) can be used to calculate $u^1(t)$. But the actual $\psi^*(t+k/t)$ is given by (41) and the current control $u(t)$ as:

$$\psi^*(t+k/t) + (C-1)\psi^*(t+k/t) = Fy^*(t) + (G-g_o)u(t) + g_o u(t)$$

Subtracting from (50) gives:

$$w(t) - \psi^*(t+k/t) = g_o(u^1(t)-u(t)) \quad (51)$$

which, using (42) to define the predictive-control, becomes:

$$Qu(t) = g_o(u^1(t)-u(t)), \text{ or}$$

$$u(t) = \frac{u^1(t)}{1 + Q/g_o} \quad (52)$$

This again shows the effect of non-zero Q on reducing the control amplitude. In this formulation u^1 is calculated from (50), then u from (52). If there are hard limits on $u(t)$ they can be tested and $u(t)$ clipped if required. In any event (51) can be used to update ψ^* ready for the next sample, as the equations are valid no matter the *actual* $u(t)$ exercised; indeed manual control could be in operation.

Further discussions are in Gawthrop (1977).

2.5.2 Self-tuning Aspects

There are many ways in which a self-tuner can be designed based on the general predictive control law. The approach of Clarke and Gawthrop (1975) was to define a function $\phi(t)$:

$$\phi(t) \triangleq \psi(t) + Qu(t-k) - w(t-k)$$

Clearly $\phi^*(t+k/t) = \psi^*(t+k/t) + Qu(t) - w(t)$, so the control (42) sets ϕ^* to zero. Moreover, (41) gives:

$$C\phi^*(t+k/t) = Fy^f(t) + Gu(t) + CQu(t) - Cw(t)$$

$$= Fy^f(t) + (G+CQ)u(t) - Cw(t) \tag{53}$$

Defining F, (G+CQ), - C to be $F'(z^{-1})$, $G'(z^{-1})$, $H'(z^{-1})$, comparing (53) with (20), and recalling the development of the self-tuning regulator, a regression model of the form:

$$\phi(t) = F'y^f(t-k) + G'u(t-k) + H'(t-k) + \varepsilon(t) \tag{54}$$

can be proposed.

RLS can be used to obtain $\hat{F}'$, $\hat{G}'$ and $\hat{H}'$ using data vectors x(t) with components $y^f(t-i)$, $u(t-i)$ and $w(t-i)$, and the control becomes:

$$\hat{F}'y^f(t) + \hat{G}'u(t) + \hat{H}'w(t) = 0 \tag{55}$$

Note that in this case $h_o = -c_o = -1$, so this is clearly a convenient parameter to fix.

Though a straight-forward method, the above suffers from certain disadvantages: For example if w(t) is not persistently exciting, the estimation becomes ill-conditioned and as it stands (54) can only be used if Q is a polynomial, not a transfer-function. In particular, as $G' = G + CQ$, the G' parameters vary if Q is adjusted on-line, as advocated in the interpretation of the predictive law. Hence other developments are preferred, and Clarke (1980) discusses various possibilities. The important considerations are that $C(z^{-1})$ needs to be explicitly estimated (to get good servo performance) and that RLS can be justified only if u(t) of (55) can actually be transmitted to the process (e.g. no actuator saturation). Hence it is more appropriate to use ELS (or RML) on the predictor model for ψ given by (41) to give $\hat{F}$, $\hat{G}$ and $\hat{C}$, and then employ (42) to give a control:

$$u(t) = (w(t) - \hat{\psi}^*(t+k/t))/Q \tag{56}$$

where ψ^* is given by:

$$\hat{C}(z^{-1})\hat{\psi}^*(t+k/t) = \hat{f}_o y^f(t) + \ldots + \hat{g}_o u(t) + \ldots \tag{57}$$

In practice it is most convenient to use equations (50), (52) and then (51) to produce u^1, u(t) and then $\hat{\psi}^*(t+k/t)$.

A typical self-tuning algorithm using the generalized-minimum variance objective is therefore:

1) the user initially chooses $P(z^{-1}) = 1/M(z^{-1})$, where M is the discrete-time version of the desired closed-loop model M(s), chooses $Q(z^{-1}) = \lambda Q'(z^{-1})$ where $1/Q' = L(z^{-1})$ is some suitable PI algorithm with arbitrary gain, and puts $\lambda = 0$.

2) at each sample-instant: ELS or RML estimates parameters of the predictor model (41), (50) generates $u^{1}(t)$ (which can be monitored by the user), and (52) generates the actual control u(t), which can be modified by the user in his choice of λ. In this way, closeness of model-following can be traded by the user against control effort.

2.5.3 Feedforward

Many processes have some disturbances which are measurable, such as feed composition into a distillation column. With interacting processes the disturbances in one loop may be the controlled variables of other loops. Suppose the effect of a measured disturbance v(t) on the output is given by:

$$A(z^{-1})y(t) = B(z^{-1})u(t-k) + C(z^{-1})\xi(t) + D(z^{-1})v(t) \tag{58}$$

Then, repeating the analysis of section 2.2, the prediction model is:

$$Cy(t+k) = Fy(t) + Cu(t) + EC\xi(t+k) + EDv(t+k)$$

Writing $ED = E_1C + z^{-k}F_1$, (20) becomes:

$$\left.\begin{aligned} Cy^*(t+k/t) &= Fy(t) + Gu(t) + F_1v(t) \\ \tilde{y}(t+k/t) &= E\xi(t+k) + E_1v(t+k) \end{aligned}\right\} \tag{59}$$

Here the prediction y*, and the corresponding control laws, includes a term $F_1(z^{-1})v(t)$. For example, the self-tuned minimum-variance law (40) becomes $\hat{F}y(t) + \hat{G}u(t) + \hat{F}_1v(t) = 0$.

The interpretation of this result and the potential benefits of feedforward depends on the assumed nature of v(t). For example, it may be a 'deterministic' signal whose effect on y(t) is delayed by the same amount as u(t). In this case D would be of the form $z^{-k}D_1(z^{-1})$ and, provided (58) is an adequate description of its effect, E_1 would be 0 and the minimum-variance law would exactly cancel the effect of v(t). In another case v(t) could be a 'white-noise' disturbance affecting y(t) much as $\xi(t)$ does; here feedforward simply gives a better predictive model and hence reduces the variance of $\tilde{y}$. In either event, the controlled behaviour is improved. As many feedforward terms as required can be included; by using variables from other loops the interaction between loops can be reduced.

2.6 *Explicit Pole-placement Algorithms*

To achieve its performance, the minimum-variance controller cancels the system dynamics and is highly sensitive to the positions of system zeros. The use of Q reduces this sensitivity at the cost of having its model-following performance detuned. In either event the system delay k must be known so that a k-step-ahead predictor model can be used. In an *explicit* algorithm, on the other hand, the standard system model (14) is estimated; moreover a *range* of time-delays can be accommodated by overparameterizing the $B(z^{-1})$ polynomial at the cost of extra computations. Suppose k is known to lie in the range $[k_1,k_2]$; then if $\beta(z^{-1})$ is a polynomial with $n+k_2-k_1$ terms, the identified model is:

$$A(z^{-1})y(t) = \beta(z^{-1})u(t-k_1) + C(z^{-1})\xi(t) \tag{60}$$

If k were in fact k_1, the first n coefficients of $\hat{\beta}$ would be non-zero, whereas if k were k_2 the last n coefficients would be non-zero. Hence $\hat{\beta}$ can be used in a self-tuning algorithm provided that the control design is insensitive to the resultant gross variations in the zeros of $\hat{\beta}$; this insensitivity can be at the cost of ultimate performance.

2.6.1 Pole-assignment Regulators

The first work in this area was in a thesis by Edmunds (1976); since then it has been extensively developed by Wellstead and his co-workers (1979a, 1979b). The idea is to concentrate entirely on the closed-loop pole positions, leaving the zeros alone. A model (60) is estimated, but for simplicity here (14) will be used.

A standard feedback controller is:

$$F(z^{-1})\ y(t) + G(z^{-1})u(t) = 0 \tag{61}$$

where $g_0 = 1$. When used with the system (14) the closed loop is:

$$(AG + z^{-k}\ BF)y(t) = CG\xi(t) \tag{62}$$

Choose the coefficients of F and G using the identity:

$$AG + z^{-k}BF = CT \tag{63}$$

where $T(z^{-1})$ is a polynomial in the form say, $\Pi_i(1-\alpha_i z^{-1})$ and the α_i correspond to stable poles. Then the closed-loop is characterized by:

$$y(t) = \frac{G}{T}\xi(t) \text{ and } u(t) = -\frac{F}{T}\xi(t) \tag{64}$$

Hence the user's specification of T influences, via (63) and (64), the variance of y(t). Note that B is not a factor of the characteristic equation.

Example 2.6.1

Consider the system of example 2.2.2, equation (22), and let $T = 1-0.5z^{-1}$; this giving a relatively fast pole. Then (63) is

$$(1-0.9z^{-1})(1+g_1 z^{-1} + \ldots) + z^{-2}\,0.5\,(f_o + \ldots) = (1+0.7z^{-1})(1-0.5z^{-1})$$

Comparing coefficients of increasing powers of z^{-1} gives:

$$g_o = 1;\; g_1 = 1.1;\; f_o = 1.28 \text{ and } g_2 \ldots,\; f_1 \ldots \text{ are zero.}$$

Hence the control is $1.28\, y(t) + (1+1.1z^{-1})u(t) = 0$, and

$$y(t) = \frac{1 + 1.1\,z^{-1}}{1 - 0.5\,z^{-1}}\,\xi(t);\; u(t) = -\frac{1.28}{1-0.5\,z^{-1}}\,\xi(t) \text{ from (64)}$$

The variance of y(t) is 5.59 σ^2, approximately double that attained by the minimum-variance laws.

To derive a self-tuner based on pole-assignment, let $A^1(z^{-1})$ and $B^1(z^{-1})$ be the solutions of the identity:

$$A^1(z^{-1})\,G + z^{-k}\,B^1(z^{-1})\,F = T \qquad (65)$$

where F and G are given by (63). Note that (65) differs from (63) in that C is omitted. Consider now the sequence:

$$A^1 y(t) - z^{-k}\,B^1 u(t) = \frac{A^1 G + z^{-k}\,B^1 F}{T}\,\xi(t), \text{ using (64)}$$

$$= \xi(t), \text{ using (65) again.} \qquad (66)$$

This is a model of the structure of (33) in which RLS can be used to get unbiased estimates of $A^1(z^{-1})$, $B^1(z^{-1})$. The self-tuning algorithm is therefore as follows:

(i) the model structure (33) is assumed, and RLS is used to obtain parameter estimates $\hat{A}^1$, $\hat{B}^1$

(ii) the identity (65) is resolved at *each sample instant* to obtain $\hat{G}$ and $\hat{F}$ given the estimated $\hat{A}^1$, $\hat{B}^1$ and the pre-specified T

(iii) the control $\hat{F}\,y(t) + \hat{G}\,u(t) = 0$ is exercised.

The above simply demonstrates that F and G are fixed points of the algorithm; Wellstead et al (1979) give further details. As with the minimum-variance self-tuner the 'fixed-point' argument does not constitute a convergence proof; unlike the minimum-variance case a proof has not yet been established, although good simulation experience is reported.

One possible difficulty with the algorithm is the numerical sensitivity of the resolution of (65) when there is overparameterization of the model. For example, if A^1 and B^1 share a common root the matrix which is used to solve for $\hat{F}$ and $\hat{G}$ is singular. It can be argued that this is 'unlikely' and that simulation experience has been good. However, initial simulation experience with the minimum-variance regulator was also 'good' and it was not until convergence analysis was undertaken that non-convergent classes of system were discovered.

The method has been extended to the servo case (Wellstead et al, 1979) by augmenting (66) to include two extra signals, one corresponding to a precompensated set-point and the other a feedforward signal which is added directly to the regulating control u(t). Space precludes further discussion; the method has again proved useful though requires significant computation time.

2.6.2 Pole-zero Placement Servos

The generalised minimum-variance algorithm of section 2.5, when used as a model-reference controller, gives an exactly specified closed-loop model $z^{-k}M(z^{-1})$ when all process zeros are in the stability region, and is detuned using Q in difficult cases. Wellstead's algorithm specifies only the closed-loop poles. The servo design of Åström and Wittenmark (1980) differs from these by considering only the noise-free case and by cancelling *only* those process zeros in some 'restricted stability region'.

A general linear regulator of the form:

$$H(z^{-1})w(t) = F(z^{-1})y(t) + G(z^{-1})u(t) \tag{67}$$

is considered, giving a closed-loop transfer-function:

$$y(t) = \frac{z^{-k}HB}{AG + z^{-k}BF} w(t) \tag{68}$$

Here H/G is a precompensator and F/G a feedback element. This transfer-function is to be made equal to a prescribed model:

$$y(t) = z^{-k}M(z^{-1})w(t) = z^{-k}\frac{B_m}{A_m} w(t) \tag{69}$$

by appropriate choice of F, G and H. The control signal is

$$u(t) = \frac{A(z^{-1})}{B(z^{-1})} \cdot \frac{B_m(z^{-1})}{A_m(z^{-1})} w(t) \tag{70}$$

and there are undesirable modes in this control if any root of the denominator BA_m of (70) is outside some region. This is clearly the case if B has unstable roots, but it is also the case when a root of B, though stable, is 'near' the unit circle. Hence factorize B as:

$$B(z^{-1}) = B^{+}(z^{-1})\, B^{-}(z^{-1}) \tag{71}$$

where all the roots of B^{+} are within a specified region (well-damped) and all the roots of B^{-} are outside (poorly damped or unstable). As is known from standard z-transform controller design, or seen from (70), B_m must contain a factor B^{-} so is of the form $B_m = B_m^1 B^{-}$.

Now the degree of A_m is normally less than the denominator of (68), so that there are factors which cancel when comparing (68) and (69). It can be shown that this cancelling factor $A_o(z^{-1})$ is an observer polynomial in state-space parlance, and (63) indicates that an optimum observer in the noisy case is in fact $C(z^{-1})$. However, in the noise-free case here A_o can be chosen as any polynomial with roots in the restricted stability region. The off-line design method is:

(i) choose A_m, B_m (containing a factor B^{-}) and A_o

(ii) Solve for G^1 and F the equation

$$AG^1 + z^{-k} B^{-} F = A_m A_o; \quad g_0^1 \neq 0, \; f_o \neq 0. \tag{72}$$

(iii) Use a control (67) with $G = G^1 B^{+}$ and $H = A_o B_m^1$

The closed-loop transfer-function is:

$$\frac{z^{-k} A_o B_m^1 B^{+} B^{-}}{AG^1 B^{+} + z^{-k} B^{+} B^{-} F} = \frac{z^{-k} A_o B_m^1 B^{+} B^{-}}{A_m A_o B^{+}} = z^{-k} \frac{B_m}{A_m}, \text{ as required.}$$

There are many possible solutions of (72) according to the chosen degrees of G^1 and F; further discussion is in Åström and Wittenmark (1980).

The proposed self-tuner assumes given A_m, A_o, B_m^1 and uses RLS to estimate $\hat{A}$ and $\hat{B}$, for by assumption there is no noise and hence no bias problem (ELS or RML would otherwise be required). The polynomial $\hat{B}$ must be factorised on-line at each step so that the decomposition $B^{+}B^{-}$ can be made and (72) resolved again at each step. The computations can be reduced in cases where all process-zeros are cancelled (as in 'model-reference' self-tuning) or where no zero is cancelled (pole-assignment). Note that the set-point signal w(t) must be persistently exciting (as with all servo methods) in order to get good estimates and hence good control; this is not usually the case with industrial processes so the self-tuner here should be regarded as mainly a design aid for fixed parameter controllers.

Although it has not been suggested in the literature, a combination of the two above pole-placement methods can be achieved for the combined regulation and servo case by using ELS or RML to estimate all A, B, C system parameters, letting $\hat{C}$ be the observer polynomial A_o, and using (63) or (72) to obtain the feedback parameters F and G. A subsidiary calculation would then be used to obtain H, depending on the type and number of open-loop zeros which are desired to appear in the overall closed-loop transfer-function.

2.7 *Some Recent Developments*

One obvious and reasonably straightforward development of the self-tuners described above is to derive their multivariable counterparts; this is, however, outside the scope of this chapter. The other areas of work consist of investigating the behaviour of self-tuners with non-linear plant, and in broadening the scope of self-tuning by considering more general performance criteria and in extending the class of processes that a given algorithm can stabilize. Recall that the control difficulties arise from process zeros outside the stability region, which may result from sampling a minimum phase process at a high rate. The developments described below attempt to overcome these difficulties directly.

2.7.1 State-space Methods

The controller design in an explicit self-tuner needs to be analytically based (which perhaps unfortunately eliminates graphically-aided techniques used in much practical control design); one broad class of design procedures that has been extensively studied in the known-parameter context is that based on linear state-space system models, quadratic cost-functions and Gaussian noise processes - the LQG method (section 1.2.3). An early proposal for a self-tuner employing LQG ideas was that of Peterka and Åström (1973), though their assumed process model excluded coloured noise processes. The method outlined below is based on work by Lam (1980), in which models such as (15) and (60) are discussed.

There are many ways in which an ARMAX model such as (60) can be written in state-space form. One useful structure is the implicit-delay model with an augmented state x of dimension n+k:

$$\left.\begin{aligned} x(t+1) &= A_I x(t) + B_I u(t) + E_I \xi(t) \\ y(t) &= C_I' x(t) + \xi(t) \end{aligned}\right\} \quad (73)$$

where
$$A_I = \begin{bmatrix} -a_1 & \\ \vdots & \\ -a_n & I_{n+k-1} \\ 0 & \\ 0 & 0 \ldots 0 \end{bmatrix} \qquad B_I = \begin{bmatrix} \beta_o \\ \vdots \\ \\ \beta_{n+k-1} \end{bmatrix}$$

$$C_I = \begin{bmatrix} 1 \\ 0 \\ \vdots \\ \\ 0 \end{bmatrix} \qquad E_I = \begin{bmatrix} c_1 - a_1 \\ \vdots \\ c_n - a_n \\ 0 \\ 0 \end{bmatrix}$$

If the system order is n, only n β parameters are non-zero (n-1 if there is no fractional delay); (73) can deal with actual system delays of between 1 and k. To obtain an LQG controller estimates $\hat{x}$ of the states must be generated, and for self-tuning applications a steady-state Kalman filter is sufficient. It can be shown (after some manipulation) that the filter for (73) is of the form:

$$C(z^{-1})\,\hat{x}(t/t) = M(z^{-1})\,\{E_I y(t) + B_I u(t)\} \tag{74}$$

where $M(z^{-1})$ is a (n+k) dimensional square polynomial matrix whose elements depend only on the $C(z^{-1})$ coefficients. In the form of (74) $\hat{x}$ is generated by suitable linear combinations of past input and output data.

The next step in the design is to choose an appropriate cost-function, preferably one which gives fixed control parameters for a constant plant and one which has 'good' closed-loop properties. One such is the N-stage *receding-horizon* function, which minimises a cost of the form:

$$I_N(t) = \Sigma_t^{t+N} \quad (x'Qx + \lambda_N u^2) \tag{75}$$

for a deterministic plant, as we invoke the certainty-equivalence principle (section 1.3.1.3). Typically the leading element of the square weighting matrix Q is 1, the others being zero, so that $x'Qx$ is y^2. The control u(t) is the first of a sequence which would minimise $I_N(t)$, whereas u(t+1) is a control for minimising $I_N(t+1)$ etc: hence the term 'receding-horizon'. The minimum-variance controller of section 2.4 and the 'λ-controller' of Clarke and Hastings-James (1971) can be regarded as finite-stage receding-horizon laws.

The properties of the N-stage law depend on the value of N. In some sense as N increases the control 'improves', and for $N \to \infty$ many important stabilization theorems are available (e.g. Anderson and Moore, 1971) even for zeros of B outside the stability region. Moreover, as N increases, the feedback parameters tend towards a constant set, and the rate of convergence indicates how 'easy' the control problem is. The control u(t) is well known to be of the form:

$$u(t) = K'\hat{x}(t/t) \tag{76}$$

where the Kalman gain vector K is obtained from N iterations of the Riccati equation. For the system (73) and the cost (75) the ith iteration can be written in the form:

$$K(i) = A_I'\,K_1(i), \text{ where } K_1(i) = \frac{P(i+1)B_I}{\lambda_N + B_I'P(i+1)B_I} \tag{77a}$$

and $$P(i) = A_I'\,\{I - K_1(i)B_I'P(i+1)\}A_I + Q \tag{77b}$$

where P(N) = 0. Note the close resemblance of (77) and (31); this can be exploited by using common software for both recursive estimation and

Riccati iteration - an outcome of the duality principle. However, in the above i is an iteration index, and K(i) is interpreted as the Kalman gain used at time t+i for a *fixed*-horizon cost-function. Here we are concerned only with K(0) as it is the feedback vector used to generate the *current* control u(t) and is a *constant* vector if the system parameters are known and constant.

A self-tuning algorithm based on the above is:

(i) use ELS or RML to provide estimates of $\hat{A}$, $\hat{B}$, $\hat{C}$

(ii) use $\hat{C}$ in (74) and compute $\hat{x}$ (t/t)

(iii) iterate the Riccati equation (77) to produce K(0), based on the chosen cost-function

(iv) generate control from (76)

Various useful simplifications are possible, particularly at stage (iii). Rather than iterate an N-stage algorithm at each time step to completion, simply iterate (77) once and store K and P for the next time step. For large t, the resultant value of K would correspond to that given by the *infinite*-stage cost-function, which gives good control. Using this approach, the amount of calculation required at each sample-time is reduced to about twice that of the ELS or RML algorithm and hence is comparable with that required by pole-placement designs.

Further details are in Lam (1980), where it is shown how the state-space method deals with sudden changes of plant gain and delay.

2.7.2 'Hybrid' Self-tuning

As self-tuners are implemented digitally, it has been natural to develop models and control designs in discrete-time; however, section 2.2 has shown that discrete-time zeros outside the stability region can arise for even minimum-phase continuous processes and these may lead to control degradation. Hybrid self-tuners (Gawthrop, 1980; Gawthrop and Clarke, 1980) are algorithms which can overcome these problems as the process models and the control designs are based on continuous-time methods, although the self-tuners themselves are digital. The continuous-time performance criteria can be claimed to be more accessible to practicing control engineers, and the performance should be more consistent as the sample-time is reduced (unlike discrete-time methods which essentially ignore inter-sample behaviour). The method, and its interpretation, follows closely that of section 2.5, but is based on *continuous-time* prediction theory.

The assumed process model is (5), where the delay Δ is taken as known, and a pseudo-process $\psi(t) = P(s)y(t)$ is considered. In the discrete-time case $P = 1/M$, where M is the desired closed-loop model, and hence involves differencings of the sampled output sequence (acceptable); in the continuous-time case P(s) involves differentiating y(t) (usually unacceptable), so ψ does not in general exist. Formally proceeding on the basis of the discrete-time derivation, however, gives rise to a Δ-ahead predictor $\psi^*(t+\Delta)$ which can be shown to have useful properties. Let E_o and F_o be polynomial solutions to:

$$C(s)\,P(s) = E_o(s)A(s) + F_o(s) \tag{78}$$

and let F(s) be the numerator of the *realisable* part of $e^{s\Delta}F_o/A$:

$$\frac{F}{A} = \{e^{s\Delta}\frac{F_o}{A}\}_+ = \int_0^\infty e^{-st}\,h(t-\Delta)dt,\ \text{where } h = L^{-1}\frac{F_o}{A}.$$

and let $E_1(s)$ be the corresponding unrealisable part. The predictor-model of the process ψ is then:

$$C\,\psi^*(t+\Delta/t) = Fy(t) + EBu(t);\ E = E_o + E_1 \tag{79}$$

Roughly speaking, the prediction *error* contains unrealisable terms corresponding to either future noise components (E_1) or derivatives of white-noise (E_o). The predictive control law is:

$$u(t) = \frac{1}{Q(s)}\ [w(t) - \psi^*(t+\Delta/t)] \tag{80}$$

which admits to the continuous-time analogues of the interpretations of section 2.5.1. For example, choosing $P(s) = (1+sT)^3$ with a triple-integrator system gives a self-tuner with a closed-loop model $(1+sT)^{-3}$, with stable control; this is not possible with the discrete-time equivalent as there is a discrete-time zero at about -3.73 (Gawthrop, 1980).

One way of viewing the method is that if P(s) has a degree of deg(A)-deg(B), lemma (2.2.1) shows that the zeros of $B(z^{-1})$ are within the stability region (at least for small h) for minimum phase G(s). Hence the use of P(s) makes the augmented system 'easier' to control in discrete-time, and the prediction-model (79) is a realisable approximation to the desired P(s). P(s) can in principle be generated using analogue components, but it is convenient to approximate to it digitally by using a sample-interval h_1 shorter than the control interval h. In fact, the realisable model

$P_1(s) \equiv P(s)/D(s)$ is simulated, where deg (D) $\geq$ deg (P),

and the roots of D are chosen so that it acts as a filter effective over the shorter interval h_1 but which has negligible phase-lag over the control interval h(i.e. $L^{-1}1/D \approx 0$ when t = h). The model $P_1(s)$ generates an approximation to $\psi(t)$ which can be used in the self-tuner to estimate the parameters of (79).

The 'hybrid' self-tuner is of the same form as its discrete-time counterpart described in section 2.5.2, except that a sub-interval h_1 is used to simulate $P_1(s)$. It is, however, capable of stably controlling systems with a high degree of pole excess at a fast sample rate without the use of Q, though Q itself can be used to reduce excessive control effort and to overcome the remaining problem of controlling non-minimum phase *continuous* systems G(s). The method does, however, require knowledge of the system delay Δ. If Δ is known exactly, there is no problem with fractional delay as the prediction can be based on the sub-interval h_1, but if Δ varies Q has to be used to ensure closed-loop stability. Alternatively, large variations in Δ can be overcome by using an explicit counterpart, as discussed in sections 2.6 and 2.7.1.

2.8 *Conclusions*

Self-tuning is a rapidly advancing field, stimulated by practical interest and by the possibility of economical implementation. It is in some ways an immature subject, as new algorithms are continually being proposed (or reinvented), but there is no clear procedure by which methods can be evaluated. A 'good' adaptive controller should perhaps be characterised by:

(i) closed-loop performance criteria easily understood by control engineers,

(ii) simplicity of coding,

(iii) robustness when applied to a broad a class of processes as possible.

The self-tuners described in this chapter have some of these features, and much of current research is devoted to generating algorithms which satisfy (iii). It is hoped that this work will produce (and has produced) methods which are effective when applied to the non-linear, time-varying, stochastic processes found in industry. Only practical experience will tell.

References

Anderson, B D O and Moore, J B (1971). 'Linear optimal control', Prentice-Hall.

Åström, K J (1970). 'Introduction to stochastic control theory', Academic Press.

Åström, K J (1980). 'Design principles for self-tuning regulators', Symposium on Adaptive Systems, Bochum, FRG.

Åström, K J, Hagander, P and Sternby, J (1980). 'Zeros of sampled systems', IEEE Transactions on CDC (to be published).

Åström, K J and Wittenmark, B (1973). 'On self-tuning regulators', Automatica, Vol 9, pp 185-199.

Åström, K J and Wittenmark, B (1980). 'Self-tuning controllers based on pole-zero placement', Proceedings IEE, Vol 127, Pt D, No 3, pp 120-130.

Clarke, D W (1980). 'Some implementation considerations of self-tuning controllers', in 'Numerical techniques for stochastic systems', ed. F Archetti and M Cugiani, North-Holland.

Clarke, D W and Gawthrop, P J (1975). 'Self-tuning controller', Proceedings IEE, Vol 12, No 9, pp 929-934.

Clarke,D W and Gawthrop, P J (1979). 'Self-tuning control', Proceedings IEE, Vol 126, No 6, pp 633-640.

Clarke, D W and Hastings-James, R (1971). 'Design of digital controllers for randomly disturbed systems', Proceedings IEE, Vol 118, pp 1503-1506.

Edmunds, J M (1976). 'Digital adaptive pole-shifting regulators', PhD thesis, UMIST.

Eykhoff, P (1974). 'System identification', Wiley.

Franklin, G F and Powell, T D (1980). 'Digital control of dynamic systems', Addison-Wesley.

Gawthrop, P J (1977). 'Some interpretations of the self-tuning controller', Proceedings IEE, Vol 124, No 10, pp 889-894.

Gawthrop, P J (1980). 'Hybrid self-tuning control', Proceedings IEE, Vol 127, Pt D, No 5, pp 973-998.

Gawthrop, P J and Clarke,D W (1980). 'Hybrid self-tuning control and its interpretation', Report 1131/80, Department of Engineering Science, Oxford University.

Gustavsson, I, Ljung, L and Soderström, T (1977). 'Identification of processes in closed-loop - identifiability and accuracy aspects', Automatica, Vol 13, pp 59-75.

Kalman, R E (1958). 'Design of a self-optimizing control system', Transactions ASME, pp 468-478.

Kučera, V (1979). 'Discrete linear control', Wiley.

Lam, K P (1980). 'Implicit and explicit self-tuning controllers', D Phil Thesis, Oxford University.

Ljung, L (1978). 'Convergence analysis of parametric identification methods', IEEE Transactions on Aut Control Vol AC-23, No 5, pp 770-783.

Ljung, L (1979). 'Convergence of recursive estimators', IFAC Symposium on 'Identification and System Parameter Estimation', Darmstadt, FRG.

Panuska, V (1969). 'An adaptive recursive least squares identification algorithm', IEEE Symposium on 'Adaptive processes, decision and control'.

Peterka, V (1970). 'Adaptive digital regulation of noisy systems'. IFAC Sympsoium on 'Identification and process parameter estimation', Prague, Czechoslovakia.

Peterka, V and Åström, K J (1973). 'Control of mutivariable systems with unknown but constant parameters', IFAC Symposium on 'Identification and system parameter estimation', The Hague, The Netherlands.

Plackett, R L (1960). 'Principles of regression analysis', Oxford University Press.

Smith, O J M (1959). 'A controller to overcome dead-time', Instrument Society of America Journal.

Wellstead, P E, Edmunds, J M, Prager, D and Zanker, P (1979a). 'Self-tuning pole/zero assignment regulators', Int J of Control, Vol 30, No 1, pp 1-26.

Wellstead, P E and Zanker, P (1979b). 'Servo self-tuners', Int J of Control, Vol 30, No 1, pp 27-36.

Yaglom, A M (1973). 'An introduction to the theory of stationary random functions'. (translated by R A Silverman) Dover.

Chapter Three

Self-tuning multivariable regulators

P E WELLSTEAD and D L PRAGER

3.1 *Introduction*

The early developments of self-tuning methods for single-input/single output systems were based upon the premise that significant computing power would, within a reasonable time span, be available for local loop control. The astounding thing is that in making this prediction some ten years ago, the control community significantly underestimated the rate at which raw computing power would develop. In this spirit, the self-tuning of multivariable systems, which seemed at first glance a daunting computational task, now appears within the range of current sixteen bit processors. Moreover, we hope to illustrate in the body of the paper, that multivariable stochastic regulation poses certain difficulties which are eased by the inherently stochastic setting of self-tuning control. In particular, in the self-tuning case, it is no longer necessary to correctly parametrize the dynamics of the disturbance filter. More will be said of the parametrization problem in section 3.2.

For the moment we turn, by way of introduction, to the more general context of self-tuning and multivariable control. The material presented here is concerned with the closed loop regulation of multivariable systems which are corrupted by an unobservable vector random process. In this sense the approach is essentially different from the mainstream of multivariable theory (Rosenbrock, 1974; MacFarlane, 1979), which is primarily concerned with servo control. It follows from the servo-tracking requirement that the closed-loop system outputs should be "decoupled" in some sense. In the case of stochastic regulation, however, this is not necessarily the case, and as a result the multivariable regulator work which is described here is essentially different from that of, say, Rosenbrock (1974). In particular, (i) Rosenbrock was motivated to find the simplest possible compensator (in the dynamical sense) which would allow stable loop closure. Since self-tuners have a direct digital implementation there is no such constraint on compensator order. (ii) An initial stage in multivariable design is to reduce interaction between input-output channels by an *off-line*, computer aided *design* method. In self-tuning the technique is inherently an *on-line*, computer aided *synthesis* method. In fact, the traditional designer's role is shifted from detail compensator specification (which is now a

†Dr Wellstead is with the Control Systems Centre, UMIST and

*Dr Prager is with Ultra Electronic Controls Ltd.

sythetic procedure) to the more general qualitative design considerations which are more closely related to implementation. One could argue therefore that self-tuning releases the control designer from the tedium of the Bode and Nichol's chart but, because of the on-line nature of the approach, requires intuitive engineering judgements concerning sample rate, slew limiting, and the many other factors which are part and parcel of direct digital controller implementation.

To summarise, the ideas outlined here are motivated in a manner which is characteristically different from off-line multivariable control. We do, however, rely upon input/output descriptions of systems (see section 3.2) and in this sense the theory is supported by the same ideas as those developed by Rosenbrock (1970) for his work.

The paper is laid-out as follows. Section 3.2 discusses certain mathematical preliminaries related to the representation and properties of sampled multivariable systems. Section 3.3 discusses the off-line forms of the synthesis rules used in self-tuning and motivates their various forms. Section 3.4 outlines the self-tuning versions of these algorithms and discusses their relative merits in terms of simulation examples (section 3.5).

It is assumed throughout this chapter that the readers are familiar with the concepts of self-tuning as described in (Wellstead, 1980; Wellstead and Zanker, 1979), and in chapters one and two by Dr O L R Jacobs and Dr D W Clarke. By the same token, no mention is made here of the convergence properties of self-tuners which are studied in detail by Dr Gawthrop in chapter five. The practical aspects of implementing self-tuning have been neglected in the literature, although such problems are treated in Zanker (1980), Oates (1980), and in chapters six and eleven through fourteen.

3.2 *Mathematical Preliminaries*

3.2.1 System Representation

It is assumed that the system to be regulated is controllable and observable and can be described in input/output form by the matrix transfer function (Rosenbrock, 1970; Kailath, 1980)

$$\{I+A(z^{-1})\}y_t = z^{-k}B(z^{-1})u_t + \{I+C(z^{-1})\}e_t \tag{1}$$

where y_t and u_t are respectively a p-vector of outputs and a p-vector of inputs, and e_t is a p-vector white noise process with statistics

$$\left.\begin{aligned} &E\{e_t\} = 0 \\ &E\{e_t\, e_t^T\} = Q \end{aligned}\right\} \tag{2}$$

where, as usual, E{ } is the expectation operator, the suffix T denotes vector/matrix transposition and Q is a positive definite covariance matrix. The terms $A(z^{-1})$, $B(z^{-1})$ and $C(z^{-1})$ are matrix polynomials in the backward shift operator of the form:

$$X(z^{-1}) = X_1 z^{-1} + \ldots + X_{n_x} z^{-n_x} \tag{3}$$

and with orders $n_x = n_a$, n_b and n_c, respectively. It is assumed that the noise corrupting the system is such that $I+C(z^{-1})$ is inverse stable (i.e. the zeros of $\det(I+C(z^{-1}))$ lie in the z-plane unit disc). Furthermore, $I+A(z^{-1})$ and $B(z^{-1})$ are relatively (left) prime, and the structural indices of the system are equal. This assumption of an underlying generic structure incurs little loss of generality because, as shown by Cook (1978), any continuous system when sampled to form a model (equation (1)) will tend to have a generic structure. It is also worth noting that the change from a forward shift to backward shift representation only has a clean minimal form if the discrete model is itself generic (Cook, 1978).

3.2.2 Comments on Sampled Systems

The matrix transfer function model (equation (1)) implies that each of the p transmission channels has associated with it a transport delay of at least k time steps. If certain channels have a larger time delay then this is accommodated by corresponding zero entries in the leading coefficient matrices of $B(z^{-1})$. In practice, most systems have such unequal transport delays and as a result one cannot generally rely upon the leading matrix B_1 in

$$B(z^{-1}) = B_1 z^{-1} + \ldots B_{n_b} z^{-n_b} \tag{4}$$

being non-singular as is (currently) required for the *optimally* derived self-tuning multivariable regulators (see Borrison, 1979; Koivo, 1980).

In addition, as in the single input/single output case, some of the zeros of a sampled multivariable system can, and probably will, be *non minimum phase* (see chapter two, section 2.2). In multivariable terms this means that the zeros of $\det(B(z^{-1}))$ are outside the unit disc in the z-plane. The reasons for this are essentially the same as for the single input/single output case (see chapter two, section 2.2) (Wellstead, Prager, Zanker, 1978), with the additional factor of multiple input channels, actually increasing the likelihood of $B(z^{-1})$ being non-minimum phase. Nevertheless, in the optimal (minimum variable self-tuners described in section 3.3.1 and 3.4.1, the assumption that $B(z^{-1})$ is minimum phase is required.

The classical pole-assignment method (sections 3.3.2, 3.4.2) does not require a minimum phase structure and makes no constraints upon the channel time delays. However, in return for this advantage, we are involved in further computational effort, and of course the optimality of the closed-loop system is lost. However, on this last issue we would ask what performance do we want from a controller, "Good - Bad or Optimal"? (after Rosenbrock and McMorran, 1971).

3.2.3 Commutivity Problems (Wolovich, 1974)

The key obstacle to transposing single variable self-tuners into a multivariable framework is that polynomial matrices do not commute. In other words, we must take account of the *order* in which matrices appear in an expression. A key result in this respect is the "pseudo-commutivity" relation given below,

$$\tilde{G}(z^{-1})F(z^{-1}) = \tilde{F}(z^{-1})G(z^{-1}) \tag{5}$$

which allows us to replace $\tilde{F}(z^{-1})G(z^{-1})$ with the "commuted" pair $\tilde{G}(z^{-1})F(z^{-1})$. The polynomial matrices always exist but (Wolovich, 1974) are not necessarily unique. For appropriately dimensioned (see sections 3.3 and 3.4) matrices equation (5) can be used to synthesize a commuted pair *on-line* in a self-tuning context.

3.3 *Off-line Control Strategies*

A number of intuitive procedures exist for the stochastic regulation of multivariable systems (see chapter eleven). In most cases these consist of inter-channel feedforward terms which are introduced in such a manner that cross-coupling within the underlying system is in some sense mitigated by feed-forward of actuation signals. Such notions have much in common with the well-known off-line procedures of multi-variable pre-compensation mentioned in the introduction. However, this reasoning is not explicitly pursued here, rather we use procedures which treat the multivariable system description directly.

3.3.1 Optimal Regulator Synthesis

Here we assume that in general *each* transmission channel has the same time delay, k, and $\det(B(z^{-1}))$ has all its zeros inside the z-plane unit disc.

3.3.1.1 Minimum Variance Regulation. The aim here is to minimize the following cost function (Borisson, 1974).

$$V = E\{y^T(t+k+1)y(t+k+1)\} \tag{6}$$

The solution to this optimization problem (which is an extension of Åström's (1970) SISO theory) is fully documented in Borisson (1975); it also arises as a special case of the detuned minimum variance algorithm to be explained in the sequel. At this stage therefore we merely quote the results. In particular, the control law required to minimize V is

$$\{I+M(z^{-1})\}\{z\, B(z^{-1})\}u_t = G(z^{-1})y_t \tag{7}$$

where

$$G(z^{-1}) = G_o + G_1 z^{-1} + \dots + G_{n_g} z^{-n_g} \tag{8}$$

and
$$\{I+C(z^{-1})\} = \{I+A(z^{-1})\}\{I+M(z^{-1})\} - z^{-k-1}\,\tilde{G}(z^{-1}) \tag{9}$$

The pseudo-commutative relationship is used

$$\{I+\tilde{M}(z^{-1})\}\tilde{G}(z^{-1}) = G(z^{-1})\{I+M(z^{-1})\} \tag{10}$$

and $$\det\{I+\tilde{M}(z^{-1})\} = \det\{I+M(z^{-1})\} \tag{11}$$

The matrix polynomials $M(z^{-1})$, $\tilde{M}(z^{-1})$ are of order k, and $G(z^{-1})$, $\tilde{G}(z^{-1})$ are of order n_a-1. With this constraint on the orders of the matrix polynomials, equation (10) can be written (by equating coefficients of like powers of z^{-1}) as a set of linear equations in the $(k+n_a-1)p$ unknown coefficients of $G(z^{-1})$ and $\tilde{M}(z^{-1})$.

The off-line minimum variance algorithm is therefore obtained as follows:

(i) Solve equation (9) for $M(z^{-1})$ and $\tilde{G}(z^{-1})$

(ii) Solve equation (10) for $\tilde{M}(z^{-1})$ and $G(z^{-1})$

(iii) Use $\tilde{M}(z^{-1})$, $B(z^{-1})$ and $G(z^{-1})$ to formulate the control law (equation 7).

This results in a closed-loop system defined by:

$$y_t = \{I + M(z^{-1})\}\, e_t \tag{12}$$

3.3.1.2 Detuned Minimum Variance Regulation. The minimum variance regulator can often introduce large loop feedback gains such that the control signal excursions are unacceptably large from an engineering viewpoint. In this case the detuned minimum variance approach (Prager, 1980) is useful since it introduces "moderating poles" into the closed-loop description (equation 12); thus,

$$\{I + z^{-k}\, T(z^{-1})\}y_t = \{I + M(z^{-1})\}e_t \tag{13}$$

where $T(z^{-1}) = T_1 z^{-1} + \ldots + T_{n_t} z^{-n_t}$ specified a set of poles which moderate the control action to meet engineering constraints, albeit at the expense of making the control sub-optimal. To see how this control is arrived at, we define

$$I+\tilde{C}(z^{-1}) = \{I+\tilde{M}(z^{-1})\}\{I+A(z^{-1})\} - z^{-k-1}\, G(z^{-1}) \tag{14}$$

Comparing equations 14, 12 and 11, it is clear that

$$\{I+\tilde{M}(z^{-1})\}\{I+C(z^{-1})\} = \{I+\tilde{C}(z^{-1})\}\{I+M(z^{-1})\} \tag{15}$$

also $\det(I+\tilde{M}(z^{-1})) = \det(I+M(z^{-1})$

implies $\det(I+C(z^{-1})) = \det(I+\tilde{C}(z^{-1}))$

Postmultiplying equation (14) by $I+z^{-k}T(z^{-1})$ gives:

$$\{I+\tilde{C}(z^{-1})\}\{I+z^{-k}T(z^{-1})\} = \{I+\tilde{M}(z^{-1})\}\{I+A(z^{-1})\} - z^{-k-1}L(z^{-1}) \quad (16)$$

$$\left.\begin{aligned} \text{where} \quad L &= G(z^{-1}) - z\{I+\tilde{C}(z^{-1})\}T(z^{-1}) \\ &= L_o + L_1 z^{-1} + \ldots + L_{n_\ell} z^{-n_\ell} \end{aligned}\right\} \quad (17)$$

Now by premultiplying the system description (equation 1) by $I+\tilde{M}(z^{-1})$, substituting from equation (16) and using the relationship (15) it follows that:

$$\begin{aligned} &\{I+\tilde{C}(z^{-1})\}[\{I+z^{-k}T(z^{-1})\}y_t - \{I+M(z^{-1})\}e_t] \\ &\qquad = z^{-k-1}[\{I+\tilde{M}(z^{-1})\}\{z\,B(z^{-1})\}u_t - L(z^{-1})y_t] \end{aligned} \quad (18)$$

It is clear from equation (18) that detuned minimum variance is achieved by the control law

$$\{I+\tilde{M}(z^{-1})\}\{zB(z^{-1})\}u_t = L(z^{-1})y_t \quad (19)$$

Notice that minimum variance regulation follows as a special case of detuning by putting $T(z^{-1}) = 0$. It then follows that $L(z^{-1}) = G(z^{-1})$ and the control law becomes that of equation (7).

3.3.2 Classical Regulator Synthesis

The assumption that the delays in all input channels are equal is highly restrictive. Likewise, the requirement that det $B(z^{-1})$ has all its zeros inside the z-plane unit disk is unlikely to be met in many practical systems (see section 2.2). To this end one is led to seek a formulation which overcomes these restrictions (Prager, 1980). One approach is that of Clarke and Gawthrop (1971) which extends to the multivariable case in a straightforward manner (Koivo (1980) and in chapter eleven) which is analogous to the minimum variance treatment of Section 3.1. A sub-optimal alternative which is inherently insensitive to time delay variations and non-minimum phase behaviour is the multivariable pole-assignment method outlined below.

3.3.2.1 Pole-Assignment. The aim here is to avoid the previously mentioned restrictions on $B(z^{-1})$ by assigning *only* the closed loop poles to prescribed locations using a control law:

$$u_t = G(z^{-1})[I+F(z^{-1})]^{-1}\, y_t \quad (20)$$

$$\left.\begin{aligned} \text{where} \quad G(z^{-1}) &= G_o + G_1 z^{-1} + \ldots + G_{n_g} z^{-n_g} \\ F(z^{-1}) &= F_1 z^{-1} + \ldots + F_{n_f} z^{-n_f} \end{aligned}\right\} \quad (21)$$

The coefficient matrices F_j, G_j are (pxp), and the polynomial matrix orders are given by

$$n_g = n_a - 1, \; n_f = n_b + k - 1 \tag{22}$$

Now, if the controller coefficients are chosen such that

$$\{I+A(z^{-1})\}\{I+F(z^{-1})\} - z^{-k}B(z^{-1})G(z^{-1}) = \{I+C(z^{-1})\}\{I+T(z^{-1})\} \tag{23}$$

then it can easily be shown that the closed loop equation is

$$y_t = \{I+F(z^{-1})\}\{I+T(z^{-1})\}^{-1} \, e_t \tag{24}$$

The poles of the closed loop system are therefore specified by selecting the polynomial matrix $I+T(z^{-1})$ with the required determinant and with order n_t governed by the inequality

$$n_t \le n_a + n_b + k - 1 - n_c \tag{25}$$

The controller parameters are obtained by solving the set of linear equations formed by equating the coefficients of like powers of z^{-1} in equation (23). For these equations to have a solution a fundamental requirement is that the n_a+n_b+k-1 matrix R given below is non-singular.

$$\underbrace{\left[\begin{array}{cccc}
I & & \bigcirc & \\
A_1 & I_1 & & \\
\vdots & A_1 & \ddots & \\
\vdots & \vdots & & I \\
A_{n_a} & \vdots & & A_1 \\
 & A_{n_a} & & \vdots \\
\bigcirc & & & A_{n_a}
\end{array}\right.}_{(n_b+k-1)p}
\underbrace{\left.\begin{array}{cccc}
k\left\{\begin{array}{c} 0 \\ 0 \end{array}\right. & & \bigcirc & \\
\vdots & & & \\
\vdots & -B_1 & \ddots & \\
\vdots & \vdots & & \\
\vdots & \vdots & & -B_1 \\
\vdots & -B_{n_b} & & \vdots \\
\bigcirc & & & -B_{n_b}
\end{array}\right]}_{(n_a)p} = R \tag{26}$$

In the cases considered here, i.e. a correctly parametrized system with $I+A(z^{-1})$, $B(z^{-1})$ relatively (left) prime and observable and controllable, this will almost always be true. In the self-tuning version (section 4) the entries in R become estimated quantities so that singularity will almost never occur.

In a sense, however, this practical viewpoint begs the theoretical question as what would happen if the target convergence point were to be such that R is an ill-posed matrix? In defence we can only remark that no such practical difficulties have been encountered with this scheme.

In the same spirit, the algorithms work well even when R is incorrectly parametrized and hence (one would expect) poorly conditioned.

Having synthesized the controller parameters in equation (20), a pseudo commutation is applied to obtain the control law

$$\{I+\tilde{F}(z^{-1})\}u_t = \tilde{G}(z^{-1})y_t \tag{27}$$

via the relationship

$$\{I+\tilde{F}(z^{-1})\}G(z^{-1}) = \tilde{G}(z^{-1})\{I+F(z^{-1})\} \tag{28}$$

Again, this relationship implies a transformation which in turn must be non-singular and is realised as a set of $(n_f+n_g)p$ simultaneous equations, in the coefficients of $\tilde{G}(z^{-1})$, $\tilde{F}(z^{-1})$.

3.4 *Self-tuning Algorithms*

The "self-tuning concept" in multivariable regulation, is as in the SISO case a means of designing, on-line, a stochastic regulator without explicitly determining the matrix polynomial $I+C(z^{-1})$. In this section we discuss the principal algorithms which have this self-tuning property.

However, as a preliminary to this we should perhaps indicate, albeit in a cursory manner, the operation of multivariable self-tuners. With reference to Fig 3.1, the algorithm consists of, at each sample interval,

(i) A multivariable least squares estimator (Eykhoff, 1974) operating on the p-vector of inputs u_t and p-vector of outputs y_t to form a multivariable system model (M in Fig 3.1).

(ii) An optional synthesis stage which (as per relationship (23)) generates the estimated polynomial matrices which define the required control law (C in Fig 3.1).

(iii) An optional pseudo-commutation (as per relationship (28)) which transforms the control into an implementable matrix difference equation (MDE in Fig 3.1).

Note that the stages (ii) and (iii) are optional in the sense that they only occur in the pole-assignment self-tuner. Minimum variance self-tuning is a so-called implicit method whereby the off-line synthesis rule (e.g. relation (9)) is in some sense hidden in the on-line solution*.

*N.B. See (Wellstead & Sanoff, 1981) for an implicit pole-assignment self-tuner, also see the discussion in (Wellstead, Edmunds, Prager & Zanker (1980) on the ambiguity in the implicit/explicit nomenclature.

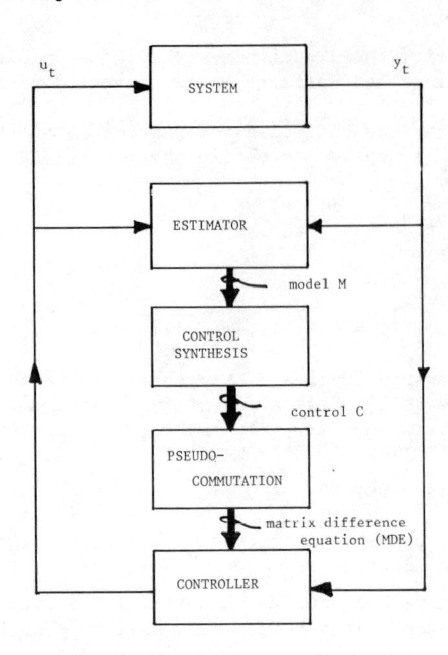

Fig 3.1 Scheme for Multivariable Self-Tuning.

3.4.1 Optimal Self-Tuning Regulators

As in section 3.1, we assume that for optimal (minimum variance related) self-tuners the det $(B(z^{-1}))$ has its zeros inside the z-plane unit disk and that B_1 is non-singular.

3.4.1.1 Minimum Variance Self-Tuning Regulation (Borisson, 1979, Åström and Wittenmark, 1973). This regulator requires that the plant model M be

$$\{I+z^{-k}\hat{A}(z^{-1})\}y_t = z^{-k}\hat{B}(z^{-1})u_t + \varepsilon_t \tag{29}$$

where $\hat{A}(z^{-1})$, $\hat{B}(z^{-1})$ are polynomial matrices whose coefficients are estimated by recursive least squares, and ε_t is the p-vector residual. The orders of the polynomials are defined in terms of the orders of $G(z^{-1})$ and $M(z^{-1})$ (see section 3), thus:

$$n_{\hat{a}} \geq n_g+1$$

$$n_{\hat{b}} \geq n_{\tilde{m}}+n \tag{30}$$

and $\quad n = n_a = n_b$

The regulator is then chosen directly as the matrix difference equation (MDE in Fig 3.1),

$$\hat{B}(z^{-1})u_{t+1} = \hat{A}(z^{-1})y_{t+1} \tag{31}$$

where $\hat{B}_1$ is assumed non-singular. Note that equation (31) may not be the minimum variance strategy. In fact, subject to further assumptions, Boris that the minimum variance strategy is a possible outcome. Only with $C(z^{-1}) = 0$ in equation (1) is the minimum variance strategy the only possible result. Actually, by placing restrictions upon the system, (1975) shows that the minimum variance is obtained, provided

(i) The controlled process has a minimum variance regulator with all observability and controllability indices equal to n+k-1.

(ii) After convergence, the regulator has $\hat{A}^*(z)$, $\hat{B}^*(z)$ relatively left prime with all observability and controllability indices equal to n+k-1.

(iii) The closed-loop system has the maximum observability index not higher than 2n+k-1.

where
$$\left.\begin{aligned} \hat{A}^*(z) &= z^{n_{\hat{a}}-1}(z\hat{A}(z^{-1})) \\ \hat{B}^*(z) &= z^{n_{\hat{b}}-1}(z\hat{B}(z^{-1})) \end{aligned}\right\} \tag{32}$$

(iv) and

$$\begin{aligned} n_{\hat{a}} &= n = n_a \\ n_{\hat{b}} &= n + k \\ n_{\tilde{m}} &= n_m = k \end{aligned} \tag{33}$$

With these constraints the self-tuning system converges to the minimum variance regulator. This in turn implies that the least squares residual ε_t converges to a moving average of the driving noise

$$\varepsilon_t = \{I + M(z^{-1})\}e_t \tag{34}$$

As in the SISO case (section 2.4, chapter 2) the minimum variance self-tuner has the great practical merit of computational simplicity. Again we recall that this advantage is won at the expense of the limitations on $B(z^{-1})$.

<u>3.4.1.2 Detuned Minimum Variance Self-Tuning Regulation</u>. As outlined in Section 3.1b, the aim of detuning a minimum variance regulator is to moderate the frequently excessive control signals associated with minimum variance strategies (Prager, 1980; Edmunds, 1976; Wellstead, Edmunds, Prager and Zanker, 1979). The model M used is defined by

$$\{I + z^{-k}\hat{A}(z^{-1})\}y_t = z^{-k}\hat{B}(z^{-1})u_t + \varepsilon_t \tag{35}$$

and the regulator is chosen as the matrix difference equation (MDE in Fig 3.1),

$$\hat{B}(z^{-1})u_{t+1} = \{\hat{A}(z^{-1}) - T(z^{-1})\}y_{t+1} \tag{36}$$

Under fairly general conditions (Prager, 1980) it is possible to show that for $C(z^{-1}) \neq 0$, the above strategy converges to the detuned minimum variance rule with closed-loop description,

$$[I+z^{-k}T(z^{-1})]y_t = [I+M(z^{-1})]e_t \tag{37}$$

where det $(I+z^{-k}T(z^{-1}))$ defines a desired set of detuning poles and $I+M(z^{-1})$ is the moving average filter associated with the minimum variance solution. Conditions on the system and $\hat{A}(z^{-1})$, $\hat{B}(z^{-1})$ are those which apply for minimum variance self-tuning (in particular the restrictions (i) through (iv) in section 3.4.1.2). The additional constraint upon the order of $T(z^{-1})$ is also required,

$$n_t \leq n_a - n_c \tag{38}$$

3.4.2 Pole Assignment Self-Tuning Regulator

The object of this form of regulator (Prager, 1980; Prager and Wellstead, 1979) is, by abandoning the aim of closed-loop optimality, to overcome the aforementioned restrictions on $B(z^{-1})$. As before, it is assumed that the system to be controlled is described by equation (1). The system model M, however, is taken to be:

$$[I+\hat{A}(z^{-1})]y_t = \hat{B}(z^{-1})u_t + \varepsilon_t \tag{39}$$

in which the transport delay k has been absorbed into the estimated polynomial matrix $\hat{B}(z^{-1})$ and the order of $\hat{B}(z^{-1})$ extended accordingly, thus

$$n_{\hat{b}} = n_b + k \tag{40}$$

The controller (C in Fig 3.1) is synthesized from

$$\{I+\hat{A}(z^{-1})\}\{I+\hat{F}(z^{-1})\} - \hat{B}(z^{-1})\hat{G}(z^{-1}) = I+T(z^{-1}) \tag{41}$$

The matrix difference equation (MDE in Fig 3.1) is

$$\{I+\hat{\tilde{F}}(z^{-1})\}u_t = \hat{\tilde{G}}(z^{-1})y_t \tag{42}$$

by use of the pseudo-commutivity relation discussed in section 3.3.2, vis.

$$\{I+\hat{\tilde{F}}(z^{-1})\}\hat{G}(z^{-1}) = \hat{\tilde{G}}(z^{-1})\{I+\hat{F}(z^{-1})\} \tag{28 bis}$$

Notice that the explicit knowledge of $C(z^{-1})$ is not required in equation (41) (c.f. the off-line relationship (23)). Nevertheless, the algorithm can converge to the closed-loop configuration defined by the off-line design rule, e.g.

$$y_t = \{I+F(z^{-1})\}\{I+T(z^{-1})\}^{-1}e_t \tag{43}$$

The detailed proof of this is contained in the general self-tuning lemma (see Prager and Wellstead, 1979). The conditions under which the algorithm applies are broadly those which occur in minimum variance self-tuning with regard to relative primeness and structural indices. The important assumptions are:

(i) On system order

$$\begin{aligned} n_{\hat{a}} &= n_a \\ n_{\hat{b}} &= n_b + k \\ n_b &= n_a \\ n_g &= n_{\hat{a}} - 1 \\ n_f &= n_{\hat{b}} - 1 \\ n_t &\le n_a + n_b + k - n_c - 1 \end{aligned} \tag{44}$$

(ii) A fixed term pole-assignment regulator for the system exists and has all observability and controllability indices equal to n_f.

(iii) Most important of all, from a practical viewpoint, the system of equations (equation 45 below), which is derived from relation (41) and determines $\hat{F}(z^{-1})$ and $\hat{G}(z^{-1})$, must have a unique solution.

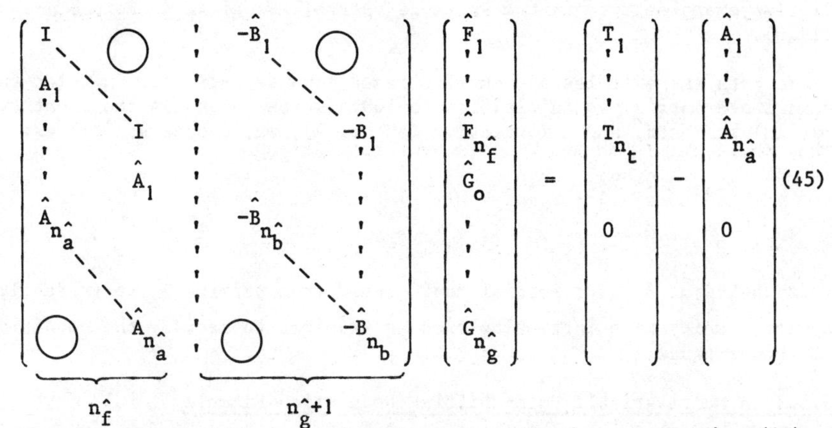

Clearly the existence, or otherwise, of a solution to equation (45) depends upon the non-singularity of the left-hand $(n_f+n_g+1)p$ matrix made up of the estimated model coefficients. Thus, although in theory the matrix may be singular due to parametrization problems, in practice it is most unlikely. However, should singularity occur, the pseudo-inverse methods of numerical analysis may be deployed.

The pole-assignment technique avoids the problems of open-loop non-minimum phase zeros and differing loop time delays. Moreover, the closed-loop poles can be determined by the designer, by specifying the zeros of $\det[I+T(z^{-1})]$. In return for this, optimality of regulation is lost. Although we have only discussed stochastic regulation, pole-assignment servo-following can be incorporated using an extended self-tuning algorithm (Wellstead & Sanoff, 1981). Despite this the algorithm given here is a relatively simple, reliable, robust self-tuning algorithm for the regulator problem. If non-zero reference signals are involved, then incremental control is applied to each input and the signal y_t in the control equation (42) is replaced by the difference y_t-y_r, where y_r is the p-vector of nominally constant reference signals. Thus, for constant reference demands the outputs will, in the steady state tend to track the reference inputs. In transient conditions, however, and because of our inability to specify the closed-loop zeros, the tracking error and interaction levels may be high. The most effective manner of dealing with this problem is by pre-filtering of the reference command vector, in an appropriate manner. In this context it is useful to note that the commonplace d.d.c. (direct digital control) technique of rate-limiting performs a rudimentary (admittedly non-linear) low pass filtering operation which reduces transient interaction in systems under incremental control.

3.5 *Examples*

Two examples are given which illustrate the behaviour of the self-tuning pole-shifting algorithm. Both examples described here were obtained using discrete-time digital simulation. A more practical application, namely the control of a simple hydraulic system comprising two coupled water tanks, is discussed by Prager and Wellstead (1979) Similar examples for minimum variance control are given in Borisson (1976).

In both the examples a recursive least squares estimator in which the covariance matrix was initialized to 100I (where I denotes the identity matrix) was used, and a forgetting factor, λ_t, was chosen as follows:

$$\lambda_{t+1} = 0.99\lambda_t + 0.01$$

$$\lambda_o = 0.96$$

This choice of λ_t aids initial tuning, and then permits λ_t to gradually approach unity with increasing time as required to satisfy the conditions for convergence.

3.5.1 A Multivariable Pole-Shifting Regulation Example

Consider a system described by the discrete time difference equation

$$(I+A_1z^{-1}+A_2z^{-2})y_t = (B_1z^{-1}+B_2z^{-2})u_t + (I+C_1z^{-1})e_t \tag{46}$$

where $A_1 = \begin{pmatrix} 0.8 & -2.9 \\ 0.4 & -1.4 \end{pmatrix}$ $\quad A_2 = \begin{pmatrix} 0.2 & -1.75 \\ 0.2 & -0.95 \end{pmatrix}$

$B_1 = I$ $\quad B_2 = \begin{pmatrix} -0.4 & 0.6 \\ 0.4 & 0.6 \end{pmatrix}$

$C_1 = \begin{pmatrix} -0.25 & 0 \\ 0 & -0.45 \end{pmatrix}$

and where the statistics of the white noise process e_t are given by:

$$E(e_t) = 0$$
$$E(e_te_t^T) = 0.1I$$

$E(\cdot)$ denotes the expectation operator, u_t represents the system input, and $y(t)$ the system output.

The system was simulated for 4000 steps, using an estimation model of the form

$$(I+\hat{A}_1z^{-1}+\hat{A}_2z^{-2})y_t = (\hat{B}_1z^{-1}+\hat{B}_2z^{-2})u_t + \varepsilon_t$$

where $\hat{A}_1$, $\hat{A}_2$, $\hat{B}_1$ and $\hat{B}_2$ were estimated using recursive least squares, and regulator law had the structure

$$(I+\hat{F}_1z^{-1})u_t = (\hat{G}_o+\hat{G}_1z^{-1})y_t$$

The regulator was designed to set $I+T(z^{-1})$ to

$$(I+T(z^{-1})) = I + \begin{pmatrix} -0.1 & 0 \\ 0 & -0.3 \end{pmatrix} z^{-1}$$

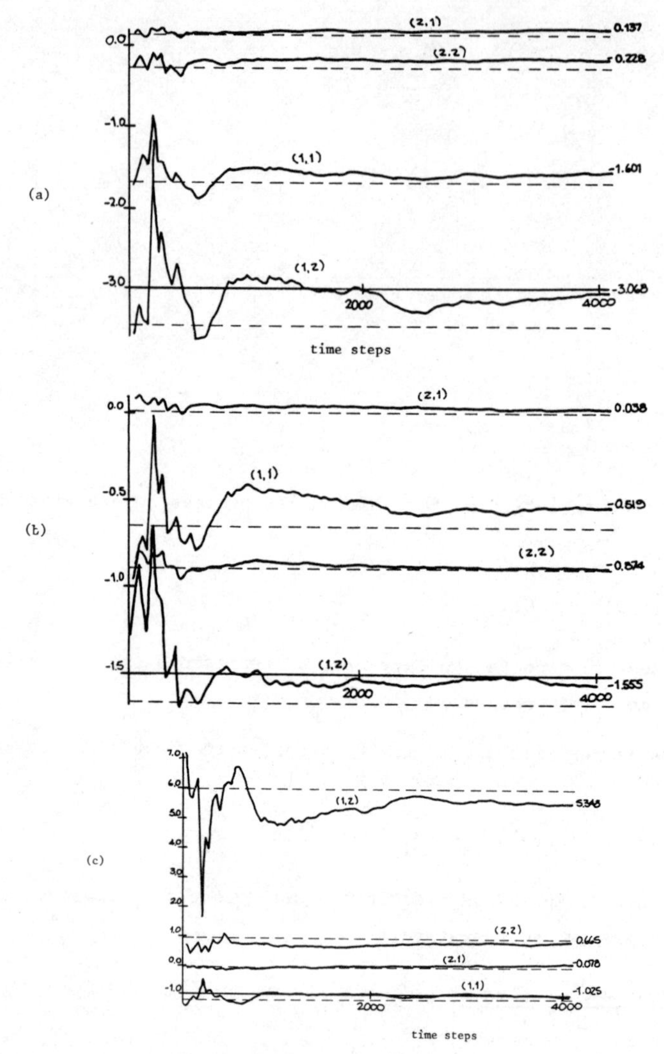

Fig 3.2 Evolution of Controller Coefficients for example 5.1
(a) coefficients of F_1 matrix.
(b) coefficients of G_o matrix.
(c) coefficients of G_1 matrix.

The time evolution of the regulator parameters is shown in Fig 3.2. The regulator parameters at the 4000th step are compared with the true regulator law below:

$$\hat{F}_1 = \begin{pmatrix} -1.601 & -3.068 \\ 0.137 & -0.228 \end{pmatrix} \qquad F_1 = \begin{pmatrix} -1.687 & -3.398 \\ 0.09881 & -0.3307 \end{pmatrix}$$

$$\hat{G}_o = \begin{bmatrix} -0.519 & -1.555 \\ 0.038 & -0.874 \end{bmatrix} \qquad G_o = \begin{bmatrix} -0.6427 & -1.6438 \\ -0.00972 & -0.875 \end{bmatrix}$$

$$\hat{G}_1 = \begin{bmatrix} -1.025 & 5.348 \\ -0.078 & 0.665 \end{bmatrix} \qquad G_1 = \begin{bmatrix} -1.1328 & 5.959 \\ -0.11023 & 0.8428 \end{bmatrix}$$

A useful test for the convergence of the self-tuned system to the desired closed-loop system is to compare the residual sequence ε_t with the system noise e_t. These sequences should ideally be identical.

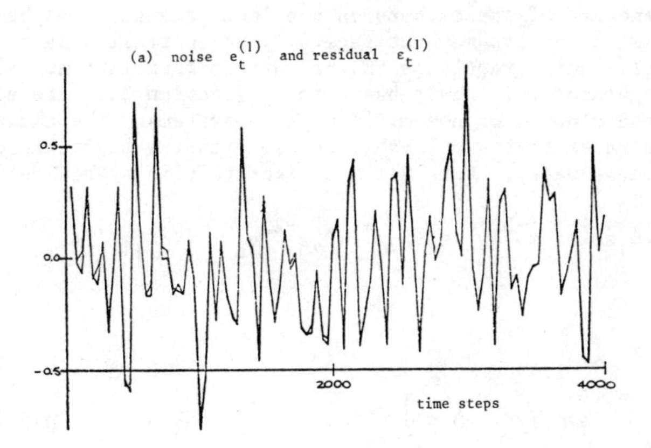

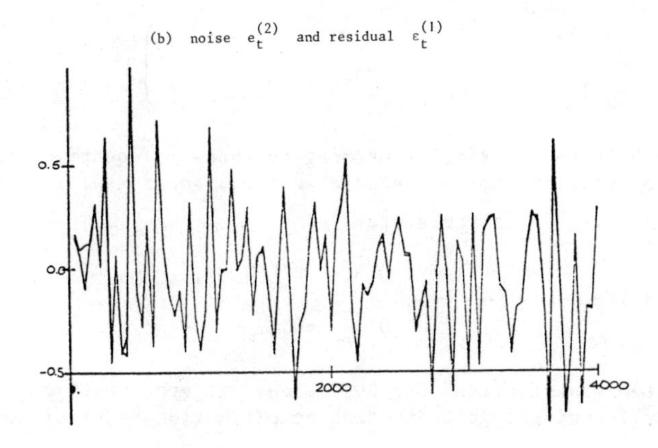

Fig 3.3 Estimation residuals superimposed on the system noise for Example 5.2.

Figure 3.3 shows the elements of these two sequences superimposed, i.e. $e_t^{(1)}$ and $\varepsilon_t^{(1)}$ are superimposed, and $e_t^{(2)}$ and $\varepsilon_t^{(2)}$ are superimposed, where

$$e_t = (e_t^{(1)}, e_t^{(2)})^T$$

The convergence of the algorithm is demonstrated by the convergence of the ε_t and e_t traces as time proceeds.

3.5.2 Multivariable Pole-Shifting Self-Tuning for Systems in which Loop Time Delays Differ.

In many practical systems, the coefficient matrix B_1 of polynomial $B(z^{-1})$ is singular. This may be caused by a different pure time delay in the dependence of the outputs on the input signals. As has been discussed, such systems may not generally be controlled by the Minimum Variance self-tuning regulator in the current formulation. Minimum Variance regulators, in their basic form, (section 2.4) are also not suited to the control of non-minimum phase systems. The example given here demonstrates that a pole-shifting self-tuning algorithm can cope with both these cases. Consider the discrete time system defined by:

$$(I+A_1z^{-1}+A_2z^{-2})y_t = (B_1z^{-1}+B_2z^{-2})u_t + (I+C_1z^{-1})e_t$$

where

$$A_1 = \begin{bmatrix} -1.4 & -0.2 \\ -0.1 & -0.9 \end{bmatrix} \qquad A_2 = \begin{bmatrix} 0.48 & 0.1 \\ 0 & 0.2 \end{bmatrix}$$

$$B_1 = \begin{bmatrix} 1 & 0 \\ 0 & 0 \end{bmatrix} \quad B_2 = \begin{bmatrix} 1.5 & 1 \\ 0 & 1 \end{bmatrix} \quad C_1 = \begin{bmatrix} -0.5 & 0 \\ 0.1 & -0.3 \end{bmatrix}$$

and the symbols have a similar meaning to those in equation (46). The closed loop poles are to be placed at $z = 0.5$ and $z = 0.4$. A suitable choice of $I + T(z^{-1})$ is therefore

$$I + T(z^{-1}) = I + \begin{bmatrix} -0.5 & 0 \\ 0 & -0.4 \end{bmatrix} z^{-1}$$

The system was simulated for 3000 steps, together with a pole shifting self-tuning regulator using an estimation model of the form

$$(I+\hat{A}_1a^{-1}+\hat{A}_2z^{-2})y_t = (\hat{B}_1z^{-1}\hat{B}_2z^{-1})u_t + \varepsilon_t$$ leading to a control law structure:

$$(I+\hat{F}_1z^{-1})u_t = (\hat{G}_o+\hat{G}_1z^{-1})y_t$$

The time evolution of the control parameters if plotted in Figure 3.4 and the values at the 3000th step are compared below with the theoretical values to which they should converge:

$$\hat{F}_1 = \begin{bmatrix} 0.275 & 0.213 \\ 0.139 & 0.292 \end{bmatrix} \qquad F_1 = \begin{bmatrix} 0.213 & 0.203 \\ 0.155 & 0.286 \end{bmatrix}$$

$$\hat{G}_o = \begin{bmatrix} -0.107 & -0.0582 \\ -0.153 & -0.106 \end{bmatrix} \qquad G_o = \begin{bmatrix} -0.101 & -0.0483 \\ -0.16 & -0.115 \end{bmatrix}$$

$$\hat{G}_1 = \begin{bmatrix} 0.0707 & 0.0382 \\ 0.0459 & 0.0332 \end{bmatrix} \qquad G_1 \begin{bmatrix} 0.0682 & 0.0265 \\ 0.0495 & 0.0469 \end{bmatrix}$$

Clearly, the pole shifting regulator works where the basic self-tuning minimum variance regulator would fail.

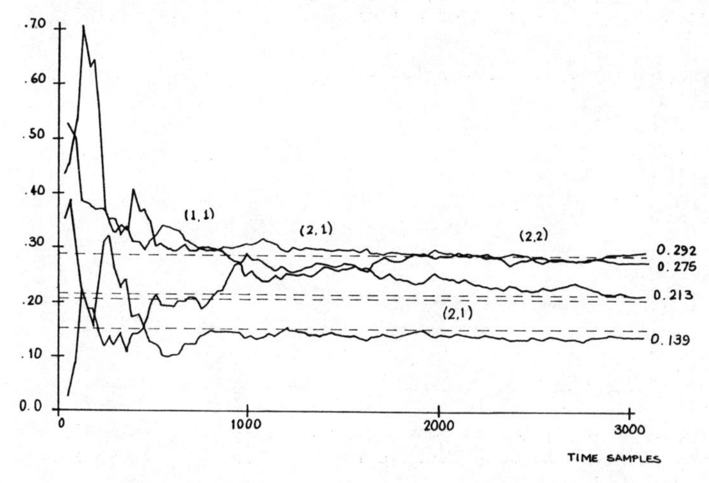

Fig 4a. Time Evolution of regulator parameter matrix F_1 in example 5.2.

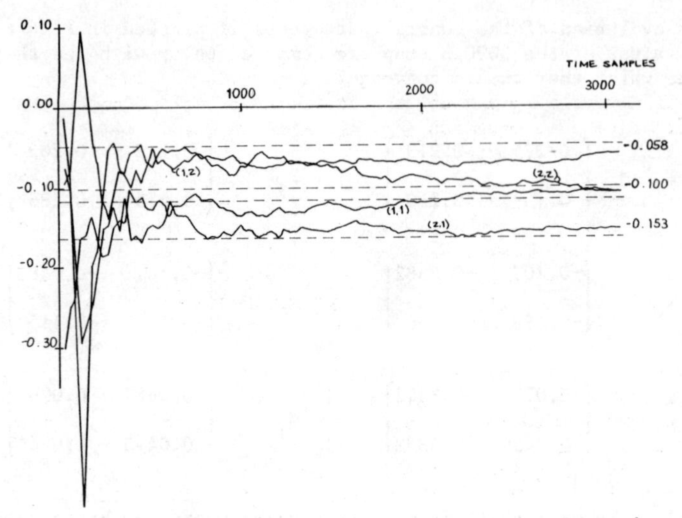

Fig 4b. Time Evolution of Regulator Parameter matrix G_0 in example 5.2.

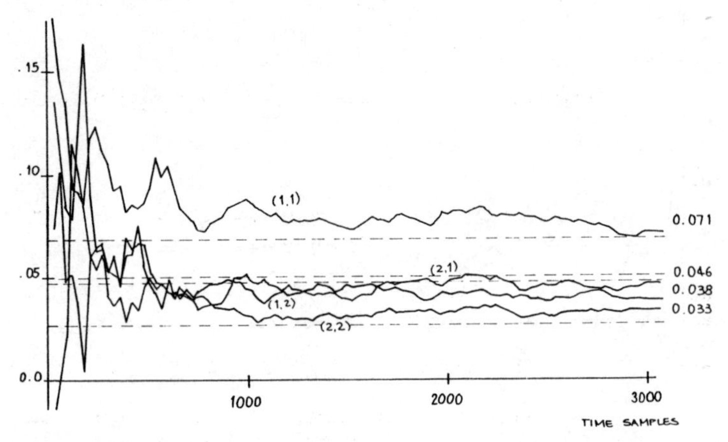

Fig 4.c. Time Evolution of Regulator Parameter Matrix G_1 in example 5.2.

3.6 *Conclusion*

A brief review of the methods available for self-tuning design of multivariable digital regulators has been presented. The extension of the generalised self-tuning regulator of Clarke (chapter 2) to essentially similar (Koivo, 1980) ideas presented above have been omitted for brevity. Multivariable model reference systems are discussed in the next chapter by Parks.

The field of adaptive digital regulation is relatively new, so that the treatment given here is necessarily preliminary in nature. However, certain key issues are well established and will, without doubt, remain in one form or another.

For example, and from a functional viewpoint, the simplicity of least squares in determining a model and its natural link to controller synthesis rules like equation (23) is so elegant and simple to implement that it will prove durable.

In a related vein the philosophical split between optimal control based self-tuners and classically derived algorithm seems also to form a natural division of application areas between those which naturally lead to a well posed optimization problem and those where a robust digital controller is required which can be set-up to achieve a certain closed-loop transient response (Wellstead, 1980).

Finally, it is appropriate to note that this presentation has dealt only with situations in which the basic form of self-tuning property applies. That is to say, the self-tuning assumption is made that $C(z^{-1})$ in equation (1) can be set to zero. In fact as hinted earlier, an extended self-tuning property (Wellstead & Sanoff, 1981) exists which allows $C(z^{-1})$ to take arbitrary form. This leads to numerous possibilities, including self-tuning pole assignment control by extended least squares, implicit pole-assignment and accelerated convergence algorithms; all of which are the subject of intense research.

Acknowledgment

This work was supported by the Science Research Council under grant number GR/A/80358.

References

Rosenbrock, H H. 'Computer Aided Control System Design', Academic Press (1974).

MacFarlane, A G J (ed). 'Frequency Response Methods in Control Systems', IEEE Press (1979).

Rosenbrock, H H. 'State Space and Multivariable Theory', Nelson (1970).

Wellstead, P E. 'Self-Tuning Digital Control Systems: The Pole-Zero Assignment Approach', Lecture at SRC Vacation School on Computer Control Systems (Available as C.S.C. Report 490), (1980).

Wellstead, P E and Zanker, P M. 'Techniques of Self-Tuning', C.S.C. Report 432, UMIST (1979).

Kailath, T B. 'Linear Systems', Prentice Hall (1980).

Cook, P C. 'Some Questions Concerning Controllability and Observability Indices', Proc 7th IFAC World Congress, Published by Pergamon Press (Ed. A Niemi) (1978).

Wolovich, W A. 'Linear Multivariable Systems', Springer Verlag, New York, (1974).

Borisson, U. 'Self-Tuning Regulators for a Class of Multivariable Systems, Automatica, 15 (1979).

Koivo, H N. 'A Multivariable Self-Tuning Controller', Automatica, 16, (1980).

Wellstead, P E, Prager, D L and Zanker, P M. 'A Pole-Assignment Self-Tuning Regulator', Proc IEE, 126, 8 (1979), also Control Systems Centre Report 434 (1978).

Prager, D L. PhD Thesis, Control Systems Centre, UMIST (1980).

Borisson, U. 'Self-Tuning Regulators - Application and Multivariable Theory', Report 7513. Dept of Automatic Control, Lund Institute of Technology, Lund, Sweden (1975).

Clarke, D W and Gawthrop, P J. 'Self-Tuning Controller', Proc IEE, 122 (1971).

Åström, K J. 'Introduction to Stochastic Control', Academic Press, London, (1970).

Åström, K J and Wittenmark, B. 'On Self Tuning Regulators', Automatica, 9, (1973).

Edmunds, J M. PhD Thesis, Control Systems Centre, UMIST (1976).

Zanker, P M. PhD Thesis, Control Systems Centre, UMIST (1980).

Prager, D L and Wellstead, P E. 'Multivariable Pole-Assignment Regulator', Control Systems Centre Report 452. Control Systems Centre, UMIST (1979). (Also in Proc IEE Pt D, 128 9 - 18 (1981)).

Rosenbrock, H H and McMorran, P D. 'Good, Bad or Optimal', IEEE Trans AC-16, 552-554 (1971).

Wellstead, P E, Edmunds, J M, Prager, D L and Zanker, P M. 'Classical and Optimal Self-Tuning Control', Int J Control, 31, 5, (1980).

Wellstead, P E and Sanoff, S P. 'Extended Self-Tuning Algorithm'. Control Systems Centre Report 506, Control Systems Centre, UMIST (1981). (Also to appear in Int J Control, 1981).

Chapter Four

Stability and convergence of adaptive controllers — continuous systems

P C PARKS

4.1 *Introduction*

The first self-adaptive control systems were proposed some 25 years ago in the field of aeronautical engineering. In 1959 an important pioneering symposium was held at the Wright Air Development Center, Dayton, Ohio, USA on "Adaptive Flight Control" (Gregory, 1959). In this symposium some important ideas were put forward, for example "ideal" models followed by high-gain servo-mechanisms "in series", or adaptive systems "in parallel" with the model (the model-reference concept). Various schemes involving system identification followed by adaptive control actions were also suggested. This symposium thus planted the seeds from which most recent work on model-reference adaptive control (MRAC) and self-tuning regulators (STR) has grown.

The early schemes were difficult to implement with the hardware then available, involving, as they did, multiplication and other non-linear processing of signals. Moreover, many of the schemes proposed were not fully understood from a theoretical point of view, leaving doubts about important properties, such as adaptive loop stability or performance in the presence of noise.

Later in the mid-1960s new versions of adaptive control systems were proposed and actually synthesised using new stability concepts being studied at this time, for example Liapunov functions and, later, hyperstability theory. Only quite recently have some long-standing problems such as parameter convergence been resolved (Narendra, 1980).

These developments will be reviewed in subsequent sections of this chapter.

4.2 *Early developments*

The adaptive autopilots proposed in the late 1950s were designed to compensate for changes in aircraft dynamics caused by flight envelopes ranging over wide variations of aircraft speed and altitude. The uncontrolled natural frequency of pitching oscillations of a high performance aircraft can change by a factor of four and the damping ratio by a factor of one-half from their low-speed sea level values. Feedback control can of course modify the response of the aircraft to

Professor Parks is with the Royal Military College of Science, Shrivenham.

pilot inputs, and if this feedback itself is adjusted the response can in theory be made independent of external aerodynamic influences - the "environment".

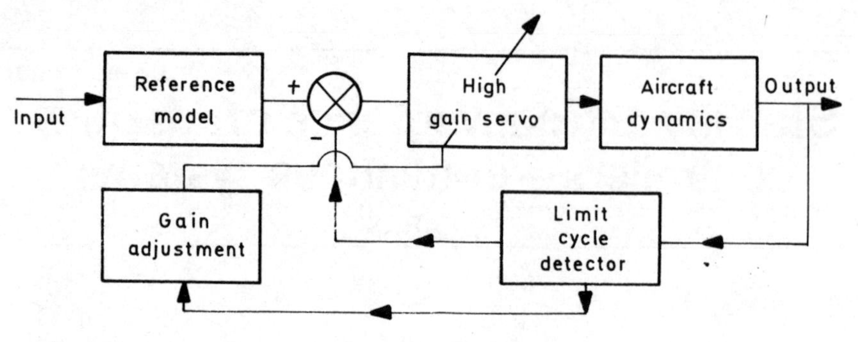

Fig 4.1a

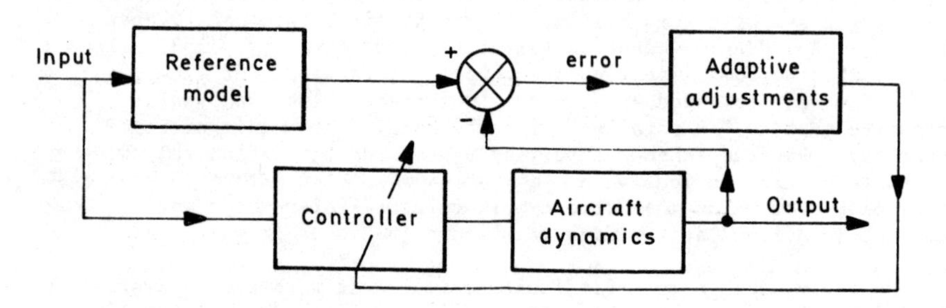

Fig 4.1b

The two schemes using reference models proposed in the 1959 symposium are shown in Fig 4.1. In the first "series" scheme shown in Fig 4.1a the ideal response coming from the reference model is tracked closely by the high gain feedback system. The open loop gain of this system is raised as far as possible without causing instability. In practice a limit cycle is detected at the onset of instability and this is used to prevent further increases in the gain. In the "parallel" scheme shown in Fig 4.1b the actual response is compared with the ideal model response to form an error signal which is then used to make adjustments to the controller, thus compensating for the environmental changes. This is the important model-reference adaptive controller (MRAC) first proposed by Osburn, Whitaker and Kezer (1961). This scheme and its later developments will be the main theme of this chapter.

4.3 *Stability considerations*

The scheme for the adaptive adjustment proposed by Osburn et al (1961) became known as the "MIT rule" and enjoyed considerable popularity in the early 1960s as the adaptive loops were easily synthesised by use of what would be called today "sensitivity functions". A simple example of this technique is shown in Fig 4.2, in which the objective is to drive the gain K_c so that the changes in the plant gain K_v are

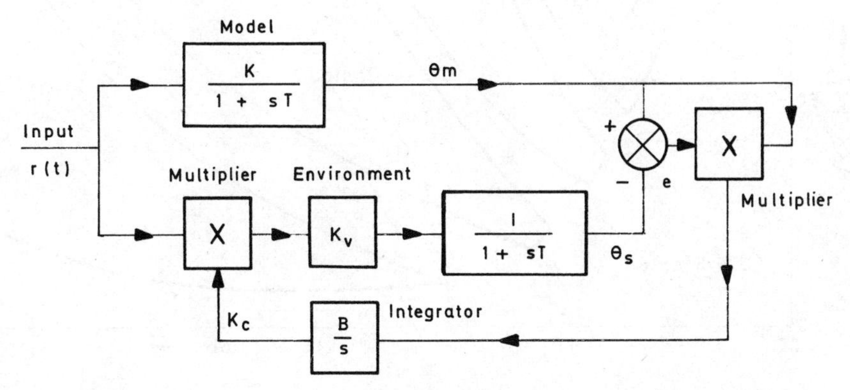

Fig 4.2

compensated and $K_cK_v \to K$, the reference model gain. By a heuristic argument based on an integral-of-error-squared criterion the so-called "MIT rule" is deduced in which

$$\dot{K}_c \propto - e \frac{\partial e}{\partial K_c} \tag{1}$$

the signal $\frac{\partial e}{\partial K_c}$ being found by differentiating the dynamic relationship between e and the input r partially with respect to K_c. In the example of Fig 4.2 the signal $\frac{\partial e}{\partial K_c}$ is effectively $- \theta_m$ leading to the circuit realisation shown there.

If we put $K_v = 1$ and $x = K - K_c$ we can write down a vector differential equation for e and x in the form

$$\begin{bmatrix} \dot{e} \\ \dot{x} \end{bmatrix} = \begin{bmatrix} -\frac{1}{T} & \frac{r(t)}{T} \\ -B\theta_m(t) & 0 \end{bmatrix} \begin{bmatrix} e \\ x \end{bmatrix} \tag{2}$$

For an ideal adaptive control system we should like $e \to 0$, $x \to 0$ as $t \to \infty$ for any r(t) and $B > 0$. Unfortunately this system can be unstable even for quite simple inputs for example $r = R \sin \omega t$ for which the stability diagram shown in Fig 4.3 may be found by numerical techniques (James (1971)).

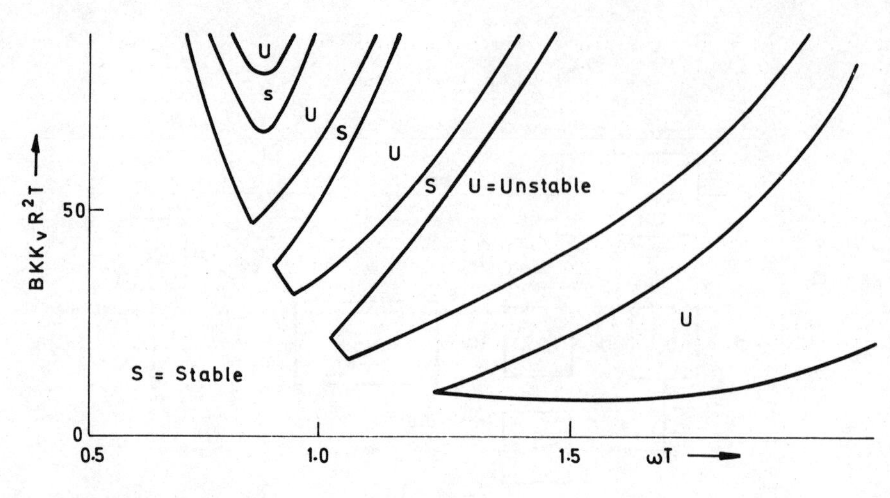

Fig 4.3

4.4 *Liapunov redesign*

The stability problem led E R Rang (Rang 1962), P C Parks (Parks 1966) and others to consider synthesising adaptive controllers using stability theory and in particular Liapunov's second method which was becoming widely known at the time (LaSalle and Lefschetz, 1961).

Considering again the system of Fig 4.2 with $K_v = 1$ let us set up a Liapunov function

$$V = e^2 + \lambda x^2 \quad (\lambda > 0) \tag{3}$$

then $$\frac{dV}{dt} = 2e\dot{e} + 2\lambda x\dot{x}$$

$$= 2e\left(-\frac{1}{T}e + \frac{1}{T}xr\right) + 2\lambda x\dot{x}$$

$$= -\frac{2}{T}e^2 \tag{4}$$

if $$\lambda\dot{x} = -\frac{1}{T}er \qquad \text{or} \qquad \dot{K}_c \propto er \tag{5}$$

instead of the "MIT rule" given by $\dot{K}_c \propto e\theta m$. (6)

The Liapunov redesign leads to the circuit diagram shown in Fig 4.4.

Clearly the point $e = 0$, $x = 0$ in the (e, x) plane is stable, but is it asymptotically stable? In general the answer to this question is "no". For example if $r(t) \equiv 0$ then $e \to 0$ but x is constant and not generally zero. If $r(t) = \exp(-\mu t)$, again $x \to b_o \neq 0$ in general. Thus $r(t)$ must be "sufficiently exciting" in some sense for both e and x to tend to zero as $t \to \infty$. This problem has been only quite recently resolved by Morgan and Narendra (1978).

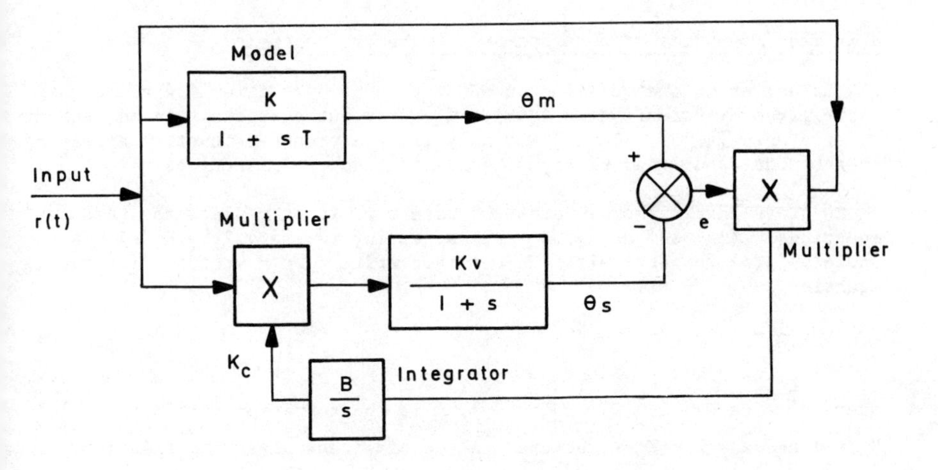

Fig 4.4 Liapunov Redesign of Fig 4.2.

If we now try to extend the Liapunov synthesis given above to the second order system formed by replacing the denominator $1 + sT$ of the first order transfer functions appearing twice in Fig 4.2 by $1 + b_1 s + b_2 s^2$ we find that the Liapunov function of equation (3) can be extended to be

$$V = \frac{b_1}{b_2^2} e^2 + \frac{b_1}{b_2} \dot{e}^2 + \lambda x^2 \tag{7}$$

for which

$$\frac{dV}{dt} = - \frac{2b_1^2}{b_2^2} \dot{e}^2 + \frac{2b_1}{b_2^2} \dot{e}xr + 2\lambda x\dot{x}$$

$$= - \frac{2b_1^2}{b_2^2} \dot{e}^2 \tag{8}$$

$$\text{if } \lambda\dot{x} = - \frac{b_1}{b_2^2} \dot{e}r \text{ or } \dot{K}_c = B\dot{e}r \text{ where } B = b_1/\lambda b_2^2 \tag{9}$$

However in this case we have had to employ the signal $\dot{e}$ instead of e. Differentiation of e however will produce unwanted noise and is not a popular procedure. The question arises: "What are the conditions under which differentiation of e is not necessary in forming the adaptive feedback loop?"

4.5 *The importance of positive-real transfer functions*

The answer to the question posed at the end of section 4.4 above was first given by Parks (Parks 1966). It turns out that the gain adjustment using e only is possible if, and only if, the transfer function appearing in the plant and model of Fig 4.2 are positive real functions.

To prove this rather interesting result we first replace the transfer functions $K/(1 + sT)$ and $1/(1 + sT)$ appearing in Fig 4.2 by $Kq(s)/p(s)$ and $q(s)/p(s)$ respectively. We may then write down a vector differential equation

$$\dot{\underset{\sim}{e}} = \underset{\sim}{A}\underset{\sim}{e} + \underset{\sim}{c}xr \tag{10}$$

where $\underset{\sim}{e}^T = (e, \dot{e}, \ddot{e}, \ldots, e^{(n-1)})$, $\underset{\sim}{A}$ is the companion matrix of $p(s)$, $K_v = 1$ and $x = K - K_c$ as before. A set of linear algebraic relations connect the elements of the vector $\underset{\sim}{c}$ with the coefficients of the polynomials $p(s)$ and $q(s)$.

We consider now the Liapunov function

$$V = \underset{\sim}{e}^T\underset{\sim}{P}\underset{\sim}{e} + \lambda x^2 \tag{11}$$

so
$$\frac{dV}{dt} = \underset{\sim}{e}^T(\underset{\sim}{P}\underset{\sim}{A} + \underset{\sim}{A}^T\underset{\sim}{P})\underset{\sim}{e} + 2\underset{\sim}{e}^T\underset{\sim}{P}\underset{\sim}{c}xr + 2\lambda x\dot{x} \tag{12}$$

If $\underset{\sim}{e}^T\underset{\sim}{P}\underset{\sim}{c} = e$ so that no derivatives of e are involved in gain adjustment rule deduced by writing

$$\lambda\dot{x} = -\underset{\sim}{e}^T\underset{\sim}{P}\underset{\sim}{c}r \tag{13}$$

then we must have

$$\underset{\sim}{P}\underset{\sim}{c} = [100 \ldots 0]^T \tag{14}$$

and also, for V to be a Liapunov function, we must have $\underset{\sim}{P}$ positive definite with

$$-\underset{\sim}{Q} = (\underset{\sim}{P}\underset{\sim}{A} + \underset{\sim}{A}^T\underset{\sim}{P}) \tag{15}$$

positive definite, or semi-definite.

A necessary and sufficient condition for the simultaneous satisfaction of equations (14) and (15) were given in the now famous Meyer-Kalman-Yacubovitch lemma (Kalman, 1963).

"Given a real number γ, two real n vectors $\underset{\sim}{g}$, $\underset{\sim}{k}$ and a real $n \times n$ matrix $\underset{\sim}{F}$, let γ be > 0 and $\underset{\sim}{F}$ stable and completely controllable, then a real vector $\underset{\sim}{q}$ satisfying

(i) $\quad \underset{\sim}{F}^T\underset{\sim}{P} + \underset{\sim}{P}\underset{\sim}{F} = -\underset{\sim}{q}\underset{\sim}{q}^T$

(ii) $\quad \underset{\sim}{P}\underset{\sim}{g} - \underset{\sim}{k} = \sqrt{\gamma}\underset{\sim}{q}$

exists, if and only if

$$\tfrac{1}{2}\gamma + \mathrm{Re}\{\underset{\sim}{k}^T(i\omega\underset{\sim}{I} - \underset{\sim}{F})^{-1}\underset{\sim}{g}\} > 0 \text{ for all real } \omega."$$

To solve equations (14) and (15) we put $\gamma = 0$, $\underset{\sim}{k} = (100 \ldots 0)^T$, $\underset{\sim}{g} = \underset{\sim}{c}$ and $\underset{\sim}{F} = \underset{\sim}{A}$ when the frequency dependent condition above becomes the condition for the transfer function $\frac{q(s)}{p(s)}$ to be a positive real transfer function.

Positive-real transfer functions have played an important rôle in adaptive control system theory in many subsequent developments of MRAC, especially in more recent applications of hyperstability theory.

4.6 *Application of Hyperstability theory to the synthesis of adaptive controls*

Hyperstability theory was developed by V M Popov (Popov, 1973) as a generalisation of the Letov-Lur'e problem which considered a single linear transfer function with a feedback path through a scalar non-linearity. The vector loop considered by Popov is shown in Fig 4.5.

It is supposed that the input vector $\underset{\sim}{v}$ and the output vector $\underset{\sim}{w}$ of the non-linear box B satisfies the "Popov integral inequality"

$$\int_0^{t_1} \underset{\sim}{v}^T\underset{\sim}{w}\, dt \geq -\gamma_o^2 \text{ for all } t_1 > 0 \tag{16}$$

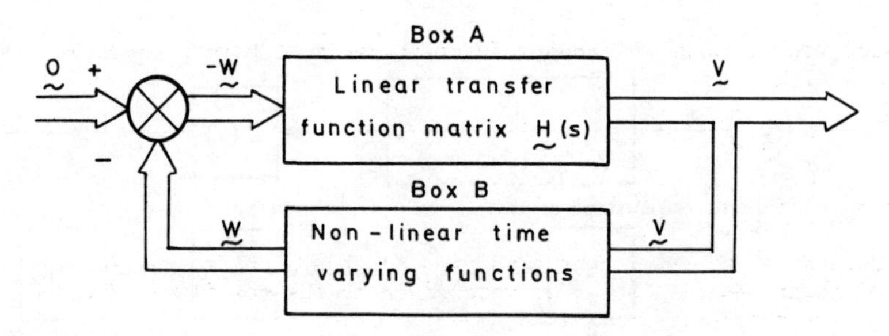

Fig 4.5 Vector Loop Model for Popov's Hyperstability Theory.

If the feedback system is globally (asymptotically) stable for all feedback blocks satisfying the Popov integral inequality, then the system and linear transfer function matrix is said to be (asymptotically) hyperstable. The transfer function matrix must be positive real for hyperstability (and strictly positive-real for asymptotic hyperstability), that is

No poles of $\underset{\sim}{H}(s)$ in Re s > 0 (Re s > 0),

and $\underset{\sim}{H}(j\omega) + \underset{\sim}{H}^T(-j\omega)$ positive semi-definite (positive definite) for all real ω.

Let us now consider how to reshape the adaptive control loop into the form of Fig 4.5. The loop in Fig 4.6a may be redrawn as Fig 4.6b, which is in the required form.

We have the error equation

$$\ddot{e} + a_1\dot{e} + a_2 e = (K - K_c)r, \tag{17}$$

the adjustment equation

$$K_c = \int_0^t \psi_1(\nu, t, \tau)d\tau + \psi_2(\nu, t) + K_o \tag{18}$$

and the filtered error equation

$$\bar{\nu}(s) = D(s)\ \bar{e}(s). \tag{19}$$

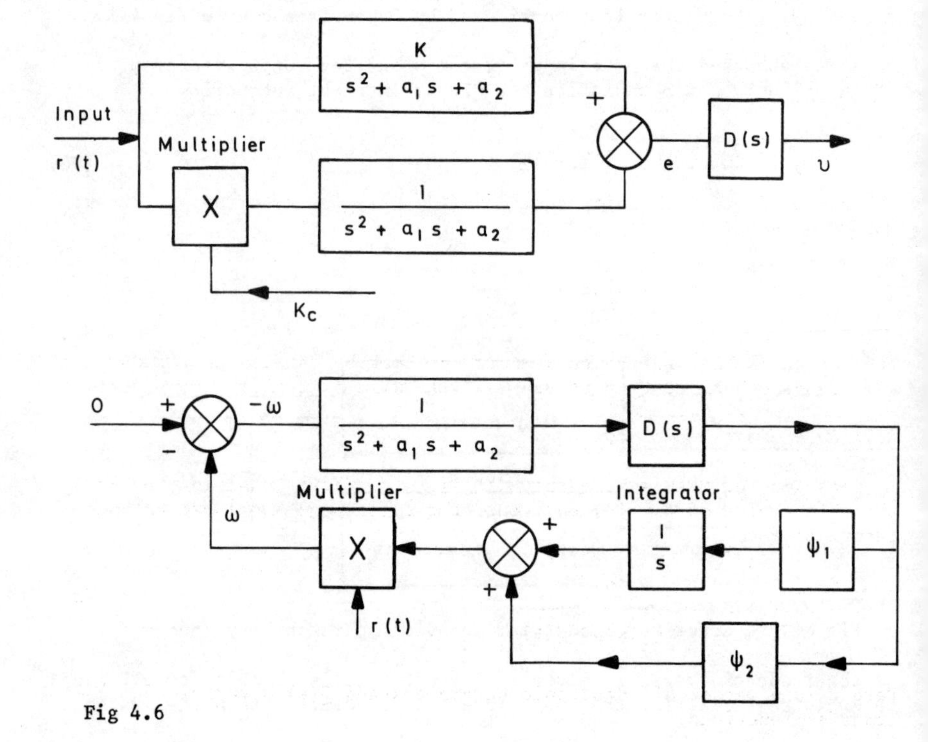

Fig 4.6

To satisfy the two requirements of the Popov hyperstability theory mentioned above we require

(i) $$\int_0^{t_1} \omega\nu \, dt = \int_0^{t_1} \nu r(t) \, [\int_0^t \psi_1 d\tau + \psi_2 + K_o - K]dt \geq -\gamma_o^2 \qquad (20)$$

and

(ii) $$\frac{D(s)}{s^2 + a_1 s + a_2} \text{ a positive real function} \qquad (21)$$

Various solutions of equation (20) are possible, for example

$$\psi_2 = B\nu r \qquad \text{or } \psi_2 = B \text{ sign } \nu r$$

with $\psi_1 = C\nu r$, B and C positive (22)

For then we may split the left-hand side of equation (20) into two parts to form

$$\int_0^{t_1} \nu \, [\int_0^t \psi_1 d\tau + K_o - K]dt + \int_0^t \nu r \, \psi_2 \, dt$$

$$= \tfrac{1}{2}\left|\frac{f^2}{C}\right|_0^{t_1} + \int_0^{t_1} \nu r \, \psi_2 \, dt \geq - \tfrac{1}{2} \frac{f^2(0)}{C} = - \gamma_o{}^2$$

where $f(t) = \int_0^t \psi_1 d\tau + K_o - K$ and $\dot{f}(t) = C\nu r$ if $\psi_1 = C\nu r$

and ψ_2 takes one of the two forms given in equation (22).

To satisfy the condition equation (21) D(s) can take various forms, for example

$$D(s) = 1 + d_1 s \text{ with } d_1 > \frac{1}{a_1}.$$

The design of MRAC is the subject of the book by Y D Landau (Landau, 1979), where both continuous and discrete time systems are considered.

4.7 *Narendra's error models*

Consider the MRAC scheme shown in Fig 4.7. It is assumed that by a suitable choice of $\underset{\sim}{\theta} = \underset{\sim}{\theta}^*$, say, that the system and model can be made identical i.e.

$$W_m(s) \equiv \frac{W_s(s)}{1 - W_s(s)\underset{\sim}{\theta}^{*T}\underset{\sim}{F}(s)} \qquad (23)$$

Here $\underset{\sim}{\theta}$ is a vector of adjustable parameters and $\underset{\sim}{F}(s)$ a vector of transfer functions.

For example if this technique is applied to the system of Fig 4.2 with

$$W_m(s) = \frac{K}{1 + sT} \tag{24}$$

and
$$W_s(s) = \frac{1}{1 + sT} \tag{25}$$

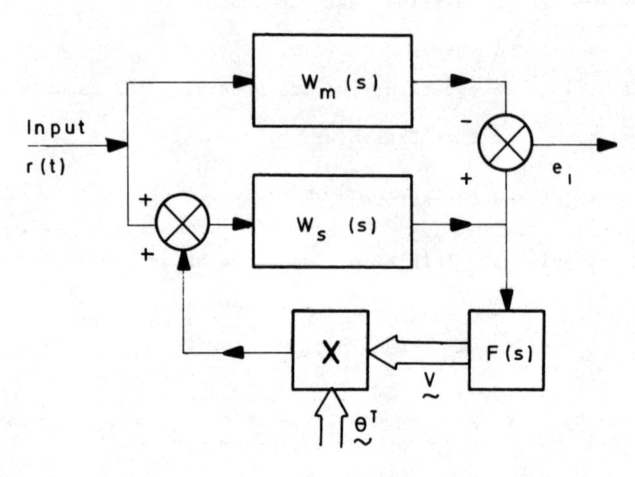

Fig 4.7

then we find that $\underset{\sim}{\theta}^{*T}\underset{\sim}{F}(s) = (K - 1)(1 + sT)$, so that one possibility is for

$$\underset{\sim}{\theta}^{*T} = (K - 1) \tag{26}$$

and
$$\underset{\sim}{F}(s) = (1 + sT)$$

In equation (26) $\underset{\sim}{\theta}^{*T}$ and $\underset{\sim}{F}(s)$ are both scalar rather than vectors.

Going back to Fig 4.7 now consider the error e_1. Working in terms of Laplace transforms this is given by

$$\begin{aligned} e_1 &= y_s - y_m \\ &= W_s(r + \underset{\sim}{\theta}^T\underset{\sim}{v}) - W_m r \\ &= \left(\frac{W_m}{1 + W_m\underset{\sim}{\theta}^{*T}\underset{\sim}{F}}\right)(r + \underset{\sim}{\theta}^T\underset{\sim}{v}) - W_m r \\ &= \frac{W_m\underset{\sim}{\theta}^T\underset{\sim}{v} - W_m\underset{\sim}{\theta}^{*T}\underset{\sim}{F}W_m r}{1 + W_m\underset{\sim}{\theta}^{*T}\underset{\sim}{F}} \end{aligned}$$

Now add and subtract $W_m\theta^{*T}FW_m\theta^Tv$ in the numerator to give

$$e_1 = W_m\theta^T v - \frac{W_m\theta^{*T}FW_m(\theta^T v + r)}{1 + W_m\theta^* F}$$

$$= W_m\theta^T v - W_m\theta^{*T}FW_s(\theta^T v + r)$$

$$= W_m\theta^T v - W_m\theta^{*T}Fy_s$$

$$= W_m(\theta^T - \theta^{*T})v = W_m\phi^T v \qquad (27)$$

where $\phi = \theta - \theta^*$, the "parameter error vector". This gives rise to a new "error model" (Narendra and Peterson (1980) shown in Fig 4.8.

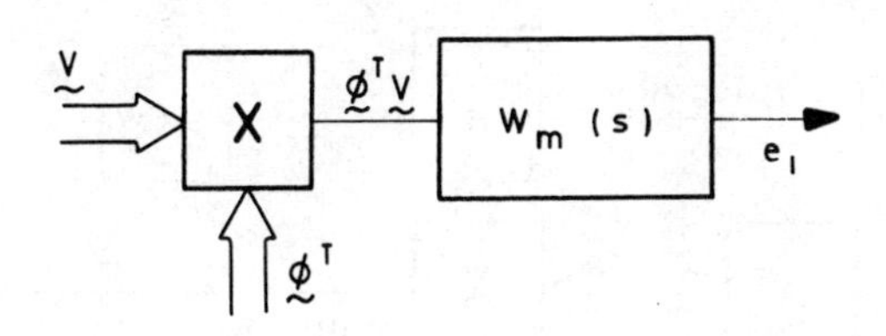

Fig 4.8

We now have to devise the feedback arrangement using e_1 to adjust the parameter error vector ϕ.

A state-space model may be set up to describe $W_m(s)$ of the form

$$\dot{e} = Ae + c(\phi^T v)$$

$$e_1 = (10 \ldots 0)e \qquad (28)$$

Consider now the Liapunov function

$$V = e^T Pe + \lambda\phi^T\Gamma^{-1}\phi$$

for which equation (23) gives $\dot{V} = e^T (PA + A^T P)e + 2e^T Pc(\phi^T v) + 2\lambda\phi^T\Gamma^{-1}\dot{\phi}$

(29)

If $W_m(s)$ is a positive real function we can find a positive definite P so that

$$PA + A^T P = - Q \qquad (30)$$

and $\underset{\sim}{P}\underset{\sim}{c} = (10 \ldots 0)^T$

by use of the Meyer-Kalman-Yacubovitch lemma as in section 4.5 above.

The adaptive law may now be chosen as

$$\dot{\underset{\sim}{\phi}} = -\frac{1}{\lambda}\underset{\sim}{\Gamma}\underset{\sim}{v}e_1 \tag{31}$$

where $\underset{\sim}{\Gamma}^{-1}$ (and $\underset{\sim}{\Gamma}$) is a positive definite real symmetric matrix.

e_1 will tend to zero as $t \to \infty$ and $\underset{\sim}{\phi} \to \underset{\sim}{0}$ also if r(t) is "sufficiently exciting". The feedback arrangement is shown in Fig 4.9.

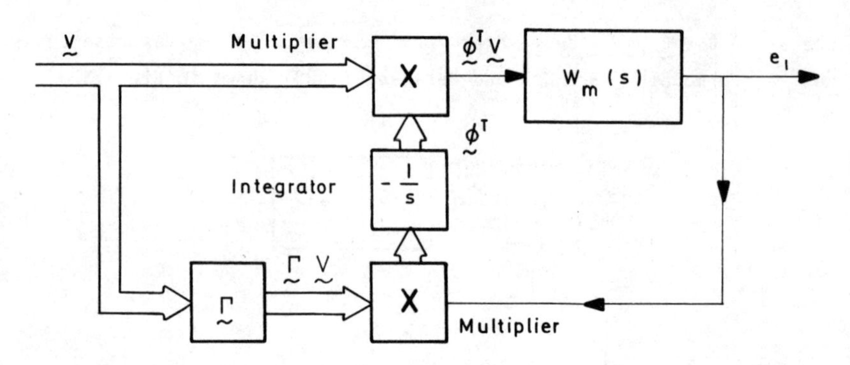

Fig 4.9

If W_m is not a positive real function (which is more likely to be the case), then an augmented error signal has to be used. This important concept was originally due to Monopoli (1974) but is presented here in a rather different way as shown in Fig 4.10.

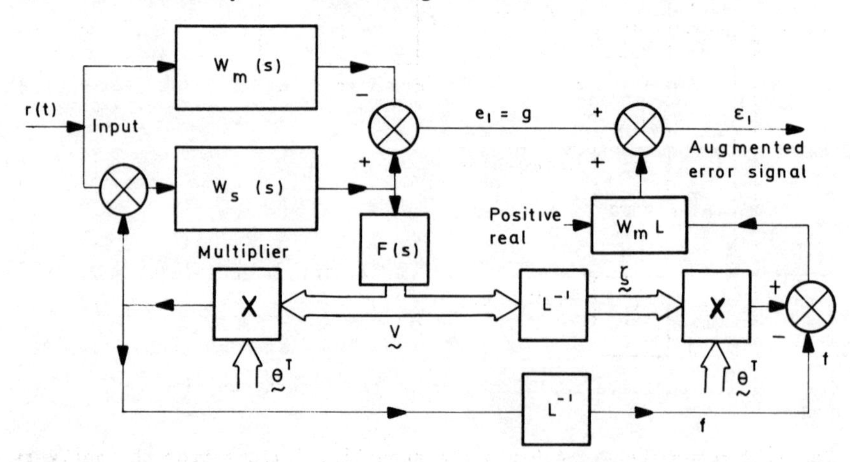

Fig 4.10

This may be redrawn successively as shown in Figures 4.11 and 4.12.

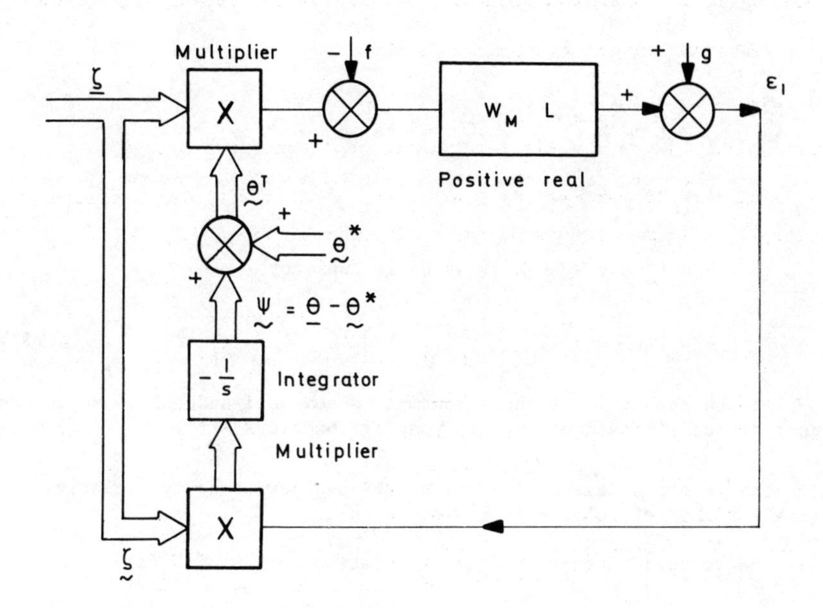

Fig 4.11

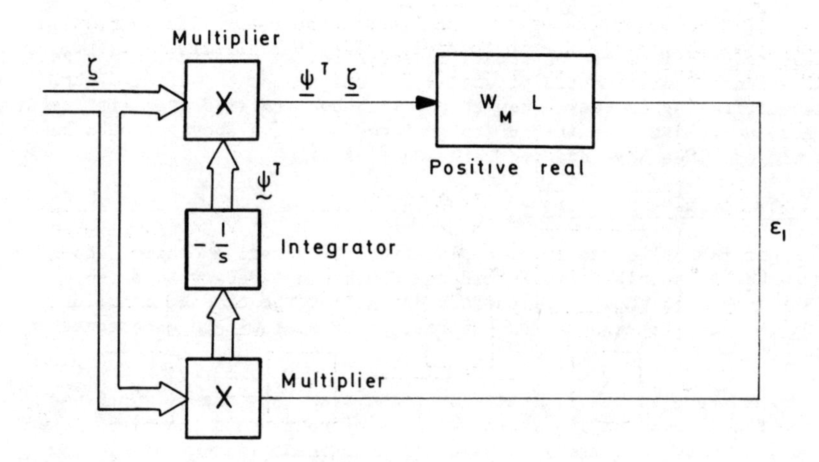

Fig 4.12

Fig 4.12 resembles Fig 4.9 and the stability follows from the analysis leading to Fig 4.9. The signal ε_1 appearing in Fig 4.10 is the "augmented error signal".

In Fig 4.11 we have

$$\varepsilon_1 = W_m L(\underset{\sim}{\theta}^T \underset{\sim}{\zeta} - f) + g$$

$$= W_m L\ (\underset{\sim}{\theta}^T \underset{\sim}{\zeta} - L^{-1}(\underset{\sim}{\theta}^T \underset{\sim}{v})) + W_m(\underset{\sim}{\theta}^T - \underset{\sim}{\theta}^{*T})\underset{\sim}{v}$$

$$= W_m L \underset{\sim}{\theta}^T \underset{\sim}{\zeta} - W_m \underset{\sim}{\theta}^{*T} \underset{\sim}{v}$$

$$= W_m L \underset{\sim}{\theta}^T \underset{\sim}{\zeta} - W_m L \underset{\sim}{\theta}^* L^{-1} \underset{\sim}{v} \text{ as } \underset{\sim}{\theta}^* \text{ is constant}$$

$$= W_m L(\underset{\sim}{\theta}^T - \underset{\sim}{\theta}^{*T})\underset{\sim}{\zeta} \tag{32}$$

While the stability of these schemes is now well understood there are some distinct limitations on the transfer functions $W_m(s)$ and $W_s(s)$ e.g.

(i) the relative degree of $W_s(s)$ should be known exactly (relative degree = number of poles - number of zeros),

(ii) the relative degree of $W_m(s)$ > relative degree of $W_s(s)$,

(iii) there is an upper bound on the number of poles of $W_s(s)$.

(iv) If $W_s(s) = \frac{kq(s)}{p(s)}$ where p(s) and q(s) are monic polynomials then the sign of k must be known,

(v) $W_s(s)$ must be a minimum phase transfer function.

These restrictions cannot always be satisfied in practice, and clearly extensions to distributed parameter systems appear to be ruled out. The present feeling is that a new theory with both relaxed stability requirements and relaxed conditions (to replace (i) - (v) above) should be developed. See Narendra and Peterson (1980).

4.8 *Practical Applications*

A comprehensive review of applications of adaptive control was given in Parks, Schaufelberger, Schmid and Unbehauen (1980) with a table listing some 58 known applications drawn from the cement, metallurgy and chemical industries, process control, power systems and electromechanical systems.

Perhaps due to the limitations of the stability theory mentioned above the most significant applications of continuous adaptive system theory were to be found in the latter two fields listed above.

In the field of power systems adaptive controls based on Landau's hyperstability theory are being developed to improve network stability of the French power transmission system, including extensive nuclear units (Irving et al 1979, Irving 1980).

In the field of electromechanical systems a notable early application of the Liapunov design was to an optical telescope in which the effective inertia changed due to gimbal movements (Gilbart and Winston, 1974). A more recent application of MRAC design techniques has been to the steering of ships whose dynamics vary with speed and loading (Van Amerongen and Udink ten Cate (1975), Van Amerongen (1980)). The effects of noise were examined in these papers where it is found that some of the advantages of the Liapunov or hyperstable designs are lost - in particular the speed of adaption deteriorates.

The total of continuous system designs is outnumbered by other designs such as self-tuning regulator systems, however.

4.9 *Future Developments*

A remarkable feature of recent theoretical developments has been the coming together of various techniques. First, the Liapunov and hyperstability design techniques have been shown to be identical (Narendra and Valavani, 1978). Then the self-tuning regulator and the model reference schemes have been shown to have some similar features (Ljung and Landau, 1978), and in particular the stability of self-tuning algorithms are shown to depend also on positive-real transfer functions (Ljung, 1977).

However, there is a need for a new theory which relaxes the requirement of asymptotic stability and exact model/system matching to a bounded model/system error coupled with less stringent requirements on the model and system transfer functions. This could then admit new classes of systems such as distributed parameter systems which are at present excluded by the present theory. Of course, with the advent of microprocessors much of this theory should be developed directly in discrete time form, or be a discretised form of continuous time results.

References

Gilbart, J W and Winston, G C (1974). 'Adaptive compensation for an optical tracking telescope', Automatica 10, 125-131.

Gregory, P C (ed), (1959). 'Proceedings of the self-adaptive flight control systems symposium', Wright Air Development Center, USA.

Irving, E (1980). 'Implicit reference model and optimal aim strategy for electrical generator adaptive control', in "Methods and Applications in Adaptive Control" (ed H Unbehauen), Springer-Verlag, 219-241.

Irving, E, Barret, J, Charcossey, C, and Monville, J P (1979). 'Improving power network stability and unit stress with adaptive generator control', Automatica, 15, 31-46.

James, D J G (1971). 'Stability of a model reference control system', AIAA Journal 9, 950-952.

Kalman, R E (1963). 'Liapunov functions for the problem of Lur'e in automatic control', Proc Nat Acad Sci, 49, 201-205.

Landau, Y D (1979). 'Adaptive control - the model reference approach', Marcel Dekker.

LaSalle, J P and Lafschetz, S (1961). 'Stability by Liapunov's direct method with applications', Academic Press.

Ljung, L (1977). 'On positive real transfer functions and the convergence of some recursive schemes', IEEE Trans Auto Control AC-22, 539-551.

Ljung, L and Landau, I D (1978). 'Model reference adaptive systems and self tuning regulators - some comparisons', Proc 7th IFAC Congress, 3 1973-1980.

Monopoli, R V (1974). 'Model reference adaptive control with an augmented error signal', IEEE Trans AC-19, 474-484.

Morgan, A P and Narendra, K S (1978). 'On the uniform asymptotic stability of certain linear time-varying differential equations with unbounded coefficients'. Yale University S and IS report 7807.

Narendra, K and Peterson, B B (1980). 'Recent developments in adaptive control', in "Methods and Applications in Adaptive Control", (ed. H Unbehauen), Springer Verlag, 84-101.

Narendra, K S and Valavani, L S (1978). 'A comparison of Lyapunov and Hyperstability approaches to adaptive control', Yale University S and SI Report 7804.

Osburn, P V, Whitaker, H P, and Kezer, A (1961). 'New developments in the design of adaptive control systems', Inst Aeronautical Sciences, Paper 61-39.

Parks, P C (1966). 'Lyapunov redesign of model-reference adaptive control systems', IEEE Trans AC-11, 362-367.

Parks, P C, Schaufelberger, W, Schmid, C and Unbehauen, H (1980). 'Applications of adaptive control systems', in "Methods and Applications in Adaptive Control" (ed. H Unbehauen), Springer-Verlag, 161-198.

Popov, V M (1973). 'Hyperstability of control systems', Springer Verlag.

Rang, E R (1962). Minneapolis-Honeywell Regulator Co Report MR-7905.

Van Amerongen, J and Udink Ten Cate, A J (1975). 'Model reference adaptive autopilots for ships', Automatica, 11, 441-449.

Chapter Five

Some properties of discrete adaptive controllers

P J GAWTHROP

5.1 *Introduction*

Classical control theory came of age when Nyquist derived his well-known theorem: this was not so much due to its ability to analyse known methods, but rather it gave clues to the synthesis of new feedback controllers. Adaptive control seems to be in a state of transition in this respect: analytical tools for stability analysis are becoming available, and such methods are beginning to suggest new adaptive algorithms. This chapter does not attempt to cover all available analytical methods but rather uses a few methods to analyse and compare a number of the simpler algorithms suggested in the literature.

There have been two main streams of development of adaptive algorithms: model-reference adaptive control and self-tuning control. Although these approaches seem quite different at first sight, they are closely related. Egardt (1979) has presented a comprehensive study of the algorithmic relations between a number of methods. In this chapter, the terms 'model-reference' and 'self-tuning' are used more as labels indicating the literature where the algorithms arose rather than implying any fundamental differences.

Every adaptive algorithm has an underlying design method which would be used if the system dynamics were known. The method considered here is the model-reference strategy of making both setpoint response and disturbance response correspond to some pre-defined model in the form of a transfer function. This method relies on the cancellation of system zeros; discrete time zeros, corresponding to most interesting continuous-time systems, are in unsatisfactory or unstable regions of the complex plane. Hence this method is virtually useless. It is considered here as it is simple, and can be modified to give better results (chapter 2).

Adaptive control methods can be divided into two classes: those which directly estimate control law parameters, and those which involve some non-trivial intermediate computation. The former are sometimes called direct or implicit methods, the latter indirect or explicit methods. Direct methods are simpler computationally and are easier to analyse but are restricted to certain simple underlying design methods. Indirect methods may use any underlying design method, but are more difficult to analyse. Because of their simplicity, and because their analysis is a

Dr Gawthrop is with the Department of Engineering Science, Oxford University.

starting point of analysis of indirect methods, a direct method is considered in this chapter.

There are two desirable properties of an adaptive algorithm: it should stabilize any system in the sense of yielding bounded deviation of the output from the set point, and should eventually give a control corresponding to the underlying design method. The former property will be called stability, and the latter convergence.

There are a number of assumptions made when analysing these properties: the controlled system is linear and of known structure, the disturbances are bounded, and the disturbances are stationary with a rational spectral density. The first assumption is for simplicity; the consequences of not using it are discussed by Gawthrop and Lim (1981). The second assumption is used in deriving the stability result, it is weaker than the zero disturbance assumption of much model-reference literature. The final assumption is necessary to give a suitable framework for stochastic convergence.

There are few complete stability results available at the moment, the results presented here are also incomplete and this will be discussed at the appropriate places.

The details of, and motivation for, the algorithms discussed in this chapter are given in chapter two, which should be read first.

The underlying model-reference design method is presented in section 5.2 and some adaptive algorithms in section 5.3. The choice of adaptive gain sequence is discussed in section 5.4 and the properties of the resultant estimators are derived in section 5.5. In section 5.6, block diagram representations of the error evolution equations are noted, and these are used in section 5.7 to give stability results, and in section 8 to give convergence results. Section 5.9 concludes the chapter.

The system representation used here is similar to that used in chapter 2, and is given by:

$$A(q^{-1})y(t) = q^{-k}B(q^{-1})u(t) + C(q^{-1})z(t) \qquad (1)$$

where

$y(t)$ is the measured system output

$u(t)$ is the control input

$z(t)$ a disturbance sequence

k is the (integer) system time delay

Sometimes, $z(t)$ will be taken to be uncorrelated, zero-mean random sequence or white noise.

5.2 *Control objectives*

Any adaptive control algorithm is based on some controller design method which could be applied if the system dynamics were known. The adaptive controller attempts to realise this controller using measured system inputs and outputs rather than a prior knowledge of system dynamics. It is probably true to say that an algebraically simple design method tends to be associated with an adaptive algorithm which is both algorithmically and algebraically simple (Åström and Wittenmark, 1980). Thus the minimum variance regulator is associated with the self-tuning regulator, the model reference controller is associated with the model-reference adaptive controller; both are simple "direct" adaptive algorithms. More complicated design methods, such as the infinite-horizon linear quadratic method, lead to algorithms (Åström, 1974; Lam, 1980) which are algorithmically and conceptually more complicated.

As intimated in the introduction, this chapter will concentrate on the simple direct methods. This section discussed the underlying design method.

5.2.1 Model-reference Control

The approach taken here is based on the notion of a predictor (chapter 2). In particular, a model-reference adaptive control will be defined by the equation:

$$\psi^* (t+k) = w(t) \tag{2}$$

where ψ^* is the prediction of the process:

$$\psi(t) = Py(t) \tag{3}$$

and P is a discrete-time transfer function and w(t) is the desired-value or set point of the process output. As shown in chapter 2, the closed-loop system resulting from applying equation (2) is given by:

$$y(t) = P^{-1}w(t-k) + P^{-1}e(t) \tag{4}$$

where e(t) is the prediction error:

$$e(t) = \psi(t) - \psi^*(t|t-k) = Ez(t) = \sum_{j=0}^{k-1} \gamma_j z(t-j) \tag{5}$$

Thus y(t) responds as the *reference model* P^{-1} both with respect to the delayed set-point and with respect to the prediction error e(t). In the particular case of unit delay (k=1) and a white-noise disturbance process z(t), the reference model P^{-1} defines the spectral density of the closed-loop system error.

An important, though trivial, point is that the control law (2) may be modified, to give a different reference model P_w^{-1} for the set point, as follows:

$$\psi^* (t+k|t) = PP_w^{-1} w(t) \tag{6}$$

The model-reference adaptive control literature has tended to ignore disturbances, thus the e(t) term is omitted from equation (4). Such an omission tends to blur the different roles of P and P_w: indeed if there were no disturbance they could be used interchangably. However, choosing P = 1, giving minimum-variance regulation, often leads to excessive control signals in the presence of disturbances; hence the use of P is essential in practice.

The major drawback of the model reference design policy is the following condition for stability.

Lemma 5.2

The model-reference control strategy gives a stable closed-loop system if:

B and C are stable polynomials

and the model P^{-1} is stable.

The crucial condition is that on B. Usually this discrete-time control strategy is applied to a discrete model of a continuous-system. As continuous-time systems with relative order greater than 2 give an unstable B polynomial in discrete-time when sampled sufficiently rapidly, the method is effectively limited to continuous-time systems with low relative order. There are two partial solutions to this problem: hybrid self-tuning methods and the use of a detuning factor. These points are discussed in chapter 2, they are not pursued further here.

5.2.2 Representations of the control law

Equation (2) is a simple way of expressing the control strategy, but it is not directly suitable for computation of the control law. Some alternative representations are now considered.

Using results from chapter 2:

$$\psi^*(t+k|t) = \frac{F}{C} y'(t) + \frac{G}{C} u(t) \tag{7}$$

where $y'(t) = P_D^{-1} y(t)$

The polynomial C may be viewed in various ways: in model-reference literature it forms a state-variable filter; Åström et al have connected it to the poles of a state observer which in the stochastic case $z(t)$ = white noise, corresponds to a steady-state Kalman filter.

As $c_o = 1$, the first element of the weighting sequence expansion of G/C is g_o. Hence equation (7) may be written as

$$\psi^* (t+k|t) = \psi' (t+k|t) + g_o u(t) \tag{8}$$

where ψ' is the value of ψ^* when $u(t) = 0$.

Thus the control law (2) becomes:

$$u(t) = \frac{1}{g_o} [\psi'(t+k|t) - w(t)] \tag{9}$$

This equation may be parameterised in a number of ways. The self-tuning approach (chapter 2) is to rewrite equation (7) as:

$$\psi^*(t+k|t) = \frac{F}{\hat{C}} y'(t) + \frac{G}{\hat{C}} u(t) + \frac{\hat{C}-C}{\hat{C}} \psi^*(t+k|t) \tag{10}$$

where $\hat{C}$ is a stable polynomial with $\hat{c}_o = c_o = 1$.

This may be written as:

$$\psi^* (+k|t) = X^T(t)\ \theta \tag{11}$$

where $$\theta = \{g_o,\ g_1,\ \ldots;\ f_o,\ f_1,\ \ldots;\ \hat{c}_1 - c_1,\ \ldots\}^T \tag{12}$$

$$X(t) = \frac{1}{\hat{C}} \{u(t),\ u(t-1),\ \ldots;\ y'(t),\ y'(t-1),\ \ldots;\ \psi^*(t*k-1|t-1)\ldots\}^T \tag{13}$$

The feature of this form is that the prediction is now *linear in the parameter vector* θ. Such a representation is well-known in the field of system identification.

Clearly, as this design method is based upon known C, this representation is redundant; it is mentioned here as a link with the adaptive design. Indeed, if the disturbance is zero, or does not have a meaningful stochastic representation, the term $\hat{C}-C$ may be omitted in the adaptive case as well.

Thus equations (12-13) would be replaced by:

$$\theta = \{g_o,\ g_1,\ \ldots;\ f_o,\ f_1,\ \ldots\}^T \tag{14}$$

$$X(t) = \frac{1}{\hat{C}} \{u(t),\ u(t-1),\ \ldots;\ y(t),\ y(t-1),\ \ldots\}^T \tag{15}$$

The model-reference literature usually uses a different representation. Firstly, no disturbances are assumed and thus C may be taken to equal $\hat{C}$ in the adaptive case. Secondly, a controller orientated form based on equation (9) is used:

$$u(t) = X_M^T\ \theta_M \tag{16}$$

where:

$$\theta_M^T = \{1/g_o;\ g_1/g_o\ \ldots;\ f_o/g_o\ \ldots\} \tag{17}$$

$$X_M^T = \{w(t);\ u(t-1)/C\ \ldots;\ y(t)/C\ \ldots\} \tag{18}$$

This form is based on that used in continuous-time algorithms (chapters 2 and 4) where it is well suited to analogue realisation; in a digital context it has much less motivation. It will not be considered further here; see Egardt (1979) for more details.

5.3 *Adaptive algorithms*

The model reference control law of the previous section requires a knowledge of the system parameter vector θ (equation (12)) to generate the prediction ψ^* required in the control law (equation (11)). In the absence of such prior knowledge, an adaptive algorithm may be used to generate an estimate of θ and hence an estimate of ψ^*. A particular parameter estimator (adaption mechanism) is described and motivated in chapter 2. In this chapter a rather more general estimator is used, but the basic motivation remains the same:

$$\hat{\theta}(t) = \hat{\theta}(t-1) + K(t-k)\,\hat{X}(t-k)\,e_1^I(t) \tag{19}$$

The individual elements are considered in detail below: here it is noted that:

$\hat{\theta}(t)$ is the current estimate of θ

$K(t)$ is the estimator gain matrix

$\hat{X}(t-k)$ is an estimate of $X(t-k)$

$e_1^I(t)$ is some measure of estimation error

5.3.1 Prediction estimates

The quantity ψ^* of equation (11) cannot be generated if the parameter vector θ is unknown. If the form of X given in equation (13) is used, X is also unknown. Denoting an estimate of θ by $\hat{\theta}$ and an estimate of X by $\hat{X}$ (this will be discussed in the sequel) a class of estimates of $\psi^*(t|t-k)$ may be defined as:

$$\hat{\psi}_i(t) = \hat{X}^T(t-k)\hat{\theta}(t-i) \qquad 0 < i < k \tag{20}$$

Note that $\hat{\psi}_i$ is realisable if information is available up to time t-k. A corresponding set of prediction error estimates may be defined as:

$$\hat{e}_i(t) = \psi(t) - \hat{\psi}_i(t) \tag{21}$$

Using equation (19), the following useful result is obtained:

$$\hat{e}_0(t) = \hat{e}_1(t) - \hat{X}(t-k)[\hat{\theta}(t)-\hat{\theta}(t-1)]e_1^I(t)$$

$$= \hat{e}_1(t) - \rho(t-k)\,e_1^I(t) \tag{22}$$

where $\rho(t) \triangleq \hat{X}^T(t)K(t)\hat{X}(t)$

5.3.2 Control Law

An obvious way to approximate the control law (2) is to replace ψ^* by $\hat{\psi}_i$. As such a control law must be causal, the only possibility is given by i = k:

$$\hat{\psi}_K(t+k) = w(t) \tag{23}$$

As pointed out by Egardt (1979) certain stability problems would be solved if it were possible to replace $\hat{\psi}_K$ by $\hat{\psi}_o$.

The equation, corresponding to equation (4), for the closed loop system obtained by applying equation (24) is:

$$y(t) = P^{-1}[\hat{e}_K(t) + w(t)] \tag{24}$$

5.3.3 X-vector estimates

If the stochastic properties of the process Cz(t) are not taken into account, the form of X given in equation (14) is used. As the sequences y(t) and u(t) are measured, trivially:

$$\hat{X}(t) = X(t) \tag{25}$$

If, however, equation (13) is used, the elements corresponding to ψ^* are not known; it is natural to replace such elements by $\hat{\psi}$:

$$\hat{X}_i(t) = \hat{C}^{-1}\ \{u(t),\ u(t-1),\ \ldots;\ y'(t),\ y'(t-1),\ \ldots;\ \hat{\psi}_i(t+k-1),\ \ldots\}^T$$

Usually, the index i will be suppressed to avoid excessive notation.

It is convenient to define a scalar error:

$$x_i(t) = [\hat{X}_i(t-k) - X(t-k)]^T\ \theta \tag{26}$$

$$= \tilde{C}(q^{-1})\tilde{e}_i(t)$$

where:

$$\tilde{C}(q^{-1}) \triangleq 1 - C(q^{-1})/\hat{C}(q^{-1})\ ;\ \tilde{e}_i \triangleq \hat{e}_i - e \tag{27}$$

As described in chapter 2, $\hat{X}(t-k)$ is required at time t to update the estimator. From equation (25) it follows that $\hat{\psi}_i(t-1)$ is the most recent estimate of ψ needed here, thus *the estimator is realisable for all $0 \leq i \leq k$.*

In the original self-tuning regulator of Åström and Wittenmark, the control policy is to set w = 0 thus

$$\hat{\psi}_K(t+k) = 0$$

Use of $\hat{X}_K$ then means that the $\hat{\psi}$ elements disappear and thus $\hat{C}$-C is not estimated. In the extension of this method (Clarke and Gawthrop, 1975) to the case with $w \neq 0$, it seemed natural to continue to use $\hat{X}_K$; this leads to difficulties in the analysis. However, as noted by Solo (1979, 1980) in the adaptive context, and Landau(1979) in the model-reference context, use of $\hat{X}_{K-1}$ improves stability properties.

As first sight, it would seem that use of $\hat{X}_i$ $0 \leq i < k-1$ would give even better results. However consideration of the quantities involved shows that:

Iff $i = k-1$ or k

$$E\{\hat{X}(t-k)\ e(t)\} = 0 \tag{29}$$

As equation (29) is crucial in convergence analysis the only feasible possibilities are $i = k$ or $k-1$. Solo (1979), in considering systems with $k = 1$ has labelled the former choice recursive maximum likelihood (RML) and the latter choice approximate maximum likelihood (AML).

5.3.4 Estimator output error

It is convenient to define the following estimator output errors:

$$v_1(t) = \hat{X}^T(t-k)[\hat{\theta}(t-1) - \theta] \tag{30}$$

$$v_o(t) = \hat{X}^T(t-k)[\hat{\theta}(t) - \theta] \tag{31}$$

Using the same argument as that used to derive equation (29) gives the following *key result:*

$$v_o(t) = v_1(t) + \rho(t-k)\ e_1^I(t) \tag{32}$$

Landau (1979) has termed v_1 an a-priori quantity and v_o an a-posteriori quantity.

5.3.5 Estimator input error-stochastic algorithm

This section considers an algorithm taking explicit account of the properties of the process $Cz(t)$. As discussed in chapter 2, the estimator input is given by:

$$e_1^I(t) = \hat{e}_1(t) \triangleq \phi(t) - \hat{X}_i(t-k)\hat{\theta}(t-1) \tag{33}$$

Using equations (5) and (27) gives

$$\hat{e}_1(t) = e(t) + \tilde{e}_i(t) \tag{34}$$

and $$\tilde{e}_1(t) = - [v_1(t) + x_i(t)] \tag{35}$$

Using equation (27):

$$\tilde{e}_1(t) = -v_1(t) - \tilde{C}\,\tilde{e}_i(t) \tag{36}$$

Where the subscript i takes the value k or k-1 depending on the choice of $\hat{X}_i$. If the adaptive algorithm converges then v(t) = 0 and so:

$$\hat{e}_i(t) = e(t) \tag{37}$$

As discussed in chapter 2, e(t) is uncorrelated with $\hat{X}_i(t-k)$, this fact is crucial in proving convergence.

5.3.6 Estimator input error - non-stochastic case

As $\psi(t)$ is generated by passing y through an inverse model, the continuous-time analogue of (33) involves pure derivatives and is thus unrealisable. In the continuous-time case it is thus necessary to prefilter the quantity $\hat{e}(t)$; the discrete-time analogue of this is to replace (33) by:

$$e_1^I(t) = \hat{e}_1^F = L^{-1}\,\hat{e}_1(t) \tag{38}$$

Choosing $L^{-1} = P^{-1}$ = desired model and considering the simple case when the system delay is minimal gives the traditional model reference controller where:

$$\hat{e}_1^I(t) = y(t) - P^{-1}\,w(t-1) \tag{39}$$

As equation (37), the convergence point of this algorithm is characterised by:

$$\hat{e}^F(t) = L^{-1}\,e(t) \tag{40}$$

This is *not* uncorrelated with $\hat{X}(t-k)$ and thus the algorithm cannot converge to the desired control law unless e(t) = 0. Hence if $L \neq 1$, it is fruitless to impose a stochastic structure on the problem. Hence the simple form of X given in equation (15) is used, and so:

$$x(t) = 0 \tag{41}$$

Defining:

$$\tilde{L} = L - 1 \tag{42}$$

then equation (20) may be written as:

$$\hat{e}_1^F(t) = -\,\hat{e}_1(t) - \tilde{L}\hat{e}_1^F(t)$$

$$= -(e(t) - v_1(t)) - \tilde{L}\hat{e}_1^F(t) \tag{43}$$

Equation (43) is similar to equation (36) with i = 1, this suggests the *modified algorithm* corresponding to equation (36) with i = 0

$$\hat{e}_1^F(t) = -[e(t) - v_1(t) - \tilde{L}\hat{e}_o^F(t)] \tag{44}$$

Where by analogy with equation (22), $\hat{e}_o^F$ is generated as

$$\hat{e}_o^F(t) = [1-\rho(t-k)]\ \hat{e}_1^F(t) \tag{45}$$

The extra term $\rho(t-k)\hat{e}_1^F$ is well known in the model reference literature. For a fuller discussion into the relations between stochastic and non-stochastic algorithms see Egardt (1979).

Some adaptive algorithms have been presented which differ in the choice of the data vector approximation $\hat{X}_i$ and the error filter $L^{-1}(t)$. Two particular classes will be analysed corresponding to: L = 1 (the stochastic design case), and $\hat{X} = X$ (the deterministic design case). An important observation is that these two methods give rise to very similar error equations: (36), (43) and (44).

5.4 *The estimator gain*

In the previous section, the choice of estimator gain K(t) (equation (19)) was unspecified. Here a number of possibilities are considered. Before going into details, the intuitive analysis of a simple case will be given which suggest a suitable choice of gain matrix.

5.4.1 A simple example

Consider the simple case of estimating a single parameter $\hat{\theta}$ where both $\tilde{C}$ and $\tilde{L}$ are zero. Defining

$$\tilde{\theta}(t) = \hat{\theta}(t) - \theta(t) \tag{46}$$

equations (19, (34) and (36) or (43)

$$\tilde{\theta}(t) = [1-K(t-k)\hat{X}^2(t-k)]\ \tilde{\theta}(t-1) + K(t-k)\hat{X}(t-k)e(t) \tag{47}$$

This is a difference equation with time varying gain. Roughly speaking, however, to avoid unstable response:

$$K > 0 \text{ and } KX^2 < 2 \tag{48}$$

To avoid oscillatory response, it is further required that:

$$KX^2 < 1 \tag{49}$$

This intuitive analysis makes the following (matrix) form of K(t) plausible:

$$K(t) = (\beta+\hat{X}^T(t)S^{-1}(t-1)\hat{X}(t))^{-1}S^{-1}(t-1) \qquad (50)$$

where S(t) is some positive definite matrix and β a positive scalar. If X and S are scalars, equation (50) satisfies expressions (48) and (49).

Another possibility is the use of the scalar matrix:

$$K(t) = s^{-1}(t)\ I \qquad (51)$$

where I is the unit matrix of relevant dimension and s positive scalar. If X were a scalar, this would satisfy expressions (43) if:

$$s^{-1}(t)\ X^2(t) < 2$$

5.4.2 Matrix Gains

The class of matrix gains considered is described recursively by:

$$S(t) = \alpha\hat{X}(t)\hat{X}^T(t) + \beta S(t-1) \qquad (52)$$

$$S(0) = S_o > 0$$

where the scalars α and β are constrained by:

$$0 \le \alpha < 2;\ 0 < \beta \le 1 \qquad (53)$$

Using the matrix inversion lemma (chapter 2):

$$S^{-1}(t)X(t) = \frac{1}{\beta}\left[1 - \frac{\alpha\sigma'(t)}{1+\alpha\sigma'(t)}\right] S^{-1}(t-1)X(t)$$

$$= \frac{1}{\beta(1+\alpha\sigma'(t))} S^{-1}(t-1)X(t) \qquad (54)$$

where $\sigma'(t) = \beta^{-1}X^T(t)S^{-1}(t-1)X(t)$

Also define the quantity:

$$\sigma(t) = X^T(t)S^{-1}(t)X(t) = \sigma'(t)/(1+\alpha\sigma'(t)) < 1/\alpha \qquad (55)$$

This leads to the following alternative expression for the gain matrix:

$$K(t)\hat{X}(t) = \frac{1+\alpha\sigma'(t)}{1+\sigma'(t)} S^{-1}(t)X(t) \qquad (56)$$

Using equation (50) or equation (51), the scalar ρ(t) of equation (23) is given by:

$$\rho(t) = \hat{X}(t)^T K(t)\hat{X}(t)$$

$$= \sigma'(t)/(1+\sigma'(t)) < 1 \qquad (57)$$

Particular choices of α and β correspond to a variety of algorithms appearing in the literature:

$\alpha = 0, \beta = 1$

The constant gain matrix much used in the model-reference literature (Narendra and Lin, 1980). Also used with $S_o = 1$ by Goodwin, Ramadge and Caines (1980).

$\alpha = 1$

Recursive least squares with exponential forgetting much used in the self-tuning literature. See chapter 2 for a detailed discussion and numerical algorithms.

$\beta = 1$

The integral adaption algorithm of Landau (1979) p. 191. In this case α corresponds to $1/\lambda$ in Landau's notation.

Note that a necessary condition for convergence (though not sufficient) is that:

$$\underset{t\to\infty}{\mathrm{Lt}}\ K(t) = 0 \tag{58}$$

This is implied by:

$$\beta = 1,\ \alpha > 0 \tag{59}$$

5.4.3 Scalar Gains

The properties of scalar gain algorithms are a rather different to those of matrix gain algorithms. By analogy with the matrix gain case define:

$$s(t) = \alpha\hat{X}^T(t)\hat{X}(t) + \beta\, s(t-1) \tag{60}$$

$$s(0) = s_o > 0$$

The scalar version of equation (50) does not seem to give useful properties; equation (51) is used instead.

In connection with this algorithm it is useful to define a scalar σ_s analogous to σ of equation (55):

$$\sigma_s(t) = \hat{X}^T(t)\, s^{-1}(t)\hat{X}(t) < \alpha^{-1} \tag{61}$$

and a scalar $\sigma'_s(t)$:

$$\sigma'_s(t) = \beta^{-1}\hat{X}^T(t)\, s^{-1}(t-1)\hat{X}(t)$$

$$= \sigma_s(t)/(1-\alpha\sigma_s(t)) \tag{62}$$

Particular choices of α and β lead to a number of algorithms appearing in the literature.

$\alpha = 1$

The so called stochastic approximation with exponential forgetting much used in self-tuning literature.

$\beta = 1$

A scalar decreasing gain algorithm used, for example, by Goodwin, Sin and Saluja (1980). In this case α corresponds to $1/\bar{a}$ in their notatioı

5.5 *Properties of the Estimator*

In this section certain properties of the estimator of equation (19) will be derived. A useful viewpoint is to regard the estimator equation (19), combined with estimator output error equation (31) or (32) as a single-input single-output system. Some different representations of the same system will be considered.

5.5.1 Representation E_L - matrix case

The time varying system E_L is defined by the output equation (30)

$$v_1(t) = \hat{X}^T(t-k)\ \tilde{\theta}\ (t-1) \tag{63}$$

The input scalar is defined by analogy with the least-square ($\alpha = 1$) method as:

$$e_L(t) = \frac{(1+\alpha\sigma')}{(1+\sigma')}\ e_1^I(t) \tag{64}$$

then, in the matrix case, (64) may be rewritten as:

$$\tilde{\theta}(t) = \tilde{\theta}(t-1) + S^{-1}(t-k)\hat{X}(t-k)\ e_L(t) \tag{65}$$

5.5.2 Representation E_s - scalar case

This is similar to representation E_s except that equation (65) is replaced by

$$\tilde{\theta}(t) = \tilde{\theta}(t-1) + s^{-1}(t-k)\hat{X}(t-k)\ e_1^I(t) \tag{66}$$

5.5.3 Representation E_o - matrix case

As discussed by Landau (1979, chapter 5) useful results are obtained by replacing the representation E_L by the a-posteriori representation E_o. Thus the system input $e_L(t)$ is replaced by:

$$e_o^I = [1-\rho(t-k)]\ e_1^I(t) \tag{67}$$

$$= (1+\sigma'(t-k))^{-1}\, e_1^I(t)$$

$$= (1+\alpha\sigma'(t-k))^{-1}\, e_L(t) \tag{68}$$

This corresponds to equation (22) in the stochastic case or equation (45) in the non-stochastic case.

The system output v_1 is replaced by equations (31) and (32) to give

$$v_o(t) = \hat{X}(t-k)\tilde{\theta}(t)$$

$$= v_1(t) + \rho(t-k)\; e_1^I(t)$$

$$= v_1(t) + \sigma'(t-k)\; e_o^I(t) \tag{69}$$

It is important to realise that E_o and E_L are two representations of the same algorithm.

5.5.4 Quadratic forms

Stability analysis of time varying systems requires the definition of certain quadratic forms; according to the context these may be used as Liapunov function candidates, energy storage functions or positive super martingales. There are two particular forms which are useful:

In the matrix gain case define:

$$V(t) = \tilde{\theta}^T(t)S(t-k)\tilde{\theta}(t) \tag{70}$$

$$\bar{V}(t) = s^{-1}(t-k)V(t) \tag{71}$$

In the scalar case these become:

$$V(t) = s(t-k)\tilde{\theta}^T(t)\,\tilde{\theta}(t) \tag{72}$$

$$\bar{V}(t) = \tilde{\theta}^T(t)\tilde{\theta}(t) \tag{73}$$

It will be seen that V is the natural choice for matrix gain estimators and $\bar{V}$ for scalar gain estimators

Using equation (71) and (6);

$$\bar{V}(t) - \bar{V}(t-1) = s^{-1}(t-k)[V(t)-V(t-1)] - \left[\frac{1}{s(t-k-1)} - \frac{1}{s(t-k)}\right] V(t-1)$$

$$= s^{-1}(t-k)[V(t)-\beta V(t-1)] - \alpha\sigma_s\bar{V}(t-1) \tag{74}$$

5.5.5 Properties of E_L - matrix case

Stability results are obtained by *differencing* quadratic forms: compare the analogous process of *differentiation* used in continuous time (chapter 2). Hence consider:

$$V(t)-V(t-1) = \tilde{\theta}^T(t-1)[S(t-k) - S(t-k-1)]\tilde{\theta}(t-1)$$
$$+ 2\tilde{\theta}(t-1)S(t-k)[\tilde{\theta}(t) - \tilde{\theta}(t-1)]$$
$$+ [\tilde{\theta}(t) - \tilde{\theta}(t-1)]S(t-k)[\tilde{\theta}(t) - \tilde{\theta}(t-1)]$$
$$= \alpha v_1^2(t)-(1-\beta)V(t-1)$$
$$+ 2\, v_1(t)e_L(t)$$
$$+ \sigma(t-k)e_L^2(t) \qquad (75)$$

The first term follows from equation (52); the second and third terms from equations (66) and (55).

Define:

$$\bar{\sigma} = \sup_{t > t_o} \sigma(t) \qquad (76)$$

Then if $\alpha > 0$; equation (75) may be rearranged to give:

$$\sum_{t=t_o}^{N} (v_1(t) + \frac{1}{\alpha} e_L)^2 > \frac{1}{\alpha^2} (1-\alpha\sigma_o)\sum^{N} e_L^2(t) - \frac{1}{\alpha} V(t_o-1) \qquad (77)$$

Noting from expression (55) that:

$$\alpha\bar{\sigma} \leq 1 \qquad (78)$$

Using the definition of gain given by Desoer and Vidyasagar (1975)

$$\text{gain} \quad \{(1+\alpha E_L)\} > (1-\alpha\bar{\sigma}) \qquad (79)$$

This result has been obtained previously for the case $\alpha = 1$ (Gawthrop, 1980).

5.5.6 Properties of E_o - matrix case

Using the result (equation (75)) of the previous section and equations (68) and (69) (various arguments and superscripts being temporarily omitted):

$$V(t) - V(t-1) = \alpha(v_o-\sigma' e_o) - (1-\beta)V(t-1)$$
$$+ 2(v_o - \sigma' e_o)(1+\alpha\sigma')e_o$$
$$+ \sigma(1+\alpha\sigma')^2 e_o{}^2 \qquad (80)$$

Noting that from equation (55):

$$\sigma(1+\alpha\sigma') = \sigma' \qquad (81)$$

it follows that:

$$V(t) - V(t-1) = \alpha v_o^2(t) - (1-\beta)V(t-1) + 2v_o(t)e_o(t) - \sigma'(t-k)e_o^2(t) \tag{82}$$

Note the significant difference in the third terms of equations (75) and (82).

Proceeding as in the previous section, the following σ - *independent* result is obtained:

$$\sum^N e_o^2 < \sum^N (\alpha v_o + e)^2 + \text{const}$$

$$\text{i.e. gain } \{(1+\alpha E_o)^{-1}\} < 1 \tag{83}$$

Alternatively, equation (82) may be rearranged as:

$$\sum_{t=t_o}^{N} v_o(t)[e_o(t) + \frac{\alpha}{2} v_o(t)] > - v(t_o-1) \tag{84}$$

This gives

$$\text{The system } (\alpha E_o)^{-1} + \tfrac{1}{2} \text{ is passive} \tag{85}$$

Expression (83) and (85) are different ways of describing the same result (equation (82)); expression (83) is particularly useful when $\alpha = 1$: expression (85) is particularly useful when $\alpha = 0$.

5.5.7 Properties of E_1 - scalar case

Following the same arguments as in the matrix case and using the appropriate form of V(t) (equation (72)):

$$V(t) - V(t-1) = \alpha X^2(t-k)\overline{V}(t-1) - (1-\beta)V(t-1) + 2v_1(t)\hat{e}_1(t) + \sigma_s(t-k)\hat{e}_1^2(t) \tag{86}$$

The first term of equation (86) differs from that of equation (75) due to the dyad XX^T being replaced by the scalar X^TX. This term renders equation (86) useless for stability analysis; however use of the quadratic form of equation (73) gives:

$$\overline{V}(t) - \overline{V}(t-1) = s^{-1}(t-k)[2v_1(t)\hat{e}_1(t) + \sigma_s(t-k)\hat{e}_1^2(t)] \tag{87}$$

This equation (87) suggests the use of the system E_1^M with input and output:

$$V_1^M = s^{-\frac{1}{2}} V_1 \tag{88}$$

$$e_1^M = s^{-\frac{1}{2}} \hat{e}_1 \tag{89}$$

The scalars $s^{-\frac{1}{2}}$ are known as *multipliers* in the theory of input-output stability.

Rearranging equation (87) leads to the statement:

$$\text{The system } \left(E_1^M + \frac{\bar{\sigma}_s}{2} \right) \text{ is passive} \tag{90}$$

where $\sigma_s = \sup_{t>t_o} \sigma_s(t)$

As: $\sigma_s(t) < 1/\alpha$, it also follows that:

$$\text{The system } (\alpha E_1^M + \tfrac{1}{2}) \text{ is passive} \tag{91}$$

It is interesting to compare expression (91) with expression (85). The a-posteriori matrix system inverse E_o^{-1} beocmes E_1^M; the scalar α becomes α^{-1}.

Alternatively, the small gain formulation is:

gain $\quad (1 + [\alpha E_1]^{-1})^{-1} < 1$

5.6 *Error feedback systems*

The stability and convergence of adaptive algorithms is usefully analysed by deriving an equivalent feedback system. This conceptual simplification has been much used by Landau (1979) and also by Narendra and his coworkers (Narendra and Lin, 1980). The idea of reducing the problem to the analysis of a simple block diagram is appealing to those with a classical control theory background.

It will be shown that these error models are driven by the prediction error e(t) (equation (5)), which in turn depends on the process disturbance z(t). Depending on the assumption made about the process z(t), various results about the behaviour of the adaptive controller emerge. In model-reference analysis (e.g. Narendra and Lin, 1980) z(t) = 0, and thus deterministic convergence of some error measure to zero is found. In stochastic analysis, it is assumed the z is a white noise process, and thus stochastic convergence results are found. Finally, if z(t) is merely assumed to be bounded in some sense, results relating to the boundedness of some error are derived.

For the purposes of such analysis, two slightly different error models will be derived: one to be used with non-stochastic assumptions on the disturbance z(t) (zero or bounded), and one when z is assumed to be a white noise process.

5.6.1 Non stochastic error system - matrix case

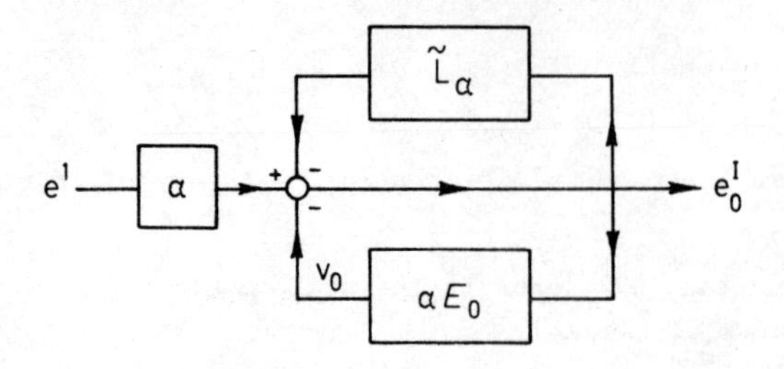

Fig 5.1

The form of error system used is displayed in Fig 5.1. The various components of this system are now described.

As no stochastic structure is placed on z(t), there is no loss in generality in replacing z(t) by:

$$z'(t) = \frac{\hat{C}}{C} z(t) \tag{93}$$

and

$$e'(t) = \frac{\hat{C}}{C} e(t) \tag{94}$$

In effect, the unknown C is replaced by the known $\hat{C}$, and the control law based on equations (11-13) is replaced by one using equations (16) and (17). Hence, for the purposes of stability analysis, use of equations (93) and (94) allows the stochastic algorithm to be analysed in the non-stochastic environment as a special case of non-stochastic algorithm with L = 1.

The system E_o represents the estimator equations with input $\hat{e}_o$ as discussed in section 5.5. The properties of E_o derived in section 5.5 are most easily expressed in terms of αE_o, where α is the scalar appearing in equations (52) and (60) so the error system includes this term.

Similarly to equation (42) define:

$$\tilde{L}_\alpha = \alpha L - 1 \tag{95}$$

equation (43) then becomes:

$$\hat{e}_1^F = (\alpha L)^{-1}(\alpha \hat{e}_1(t))$$

$$= \alpha(v_1(t) + e(t)) - \tilde{L}_\alpha \hat{e}_1^F \tag{96}$$

Combining equations (32) and (45):

$$\hat{e}_o^F(t) + v_o(t) = \hat{e}_1^F(t) + v_1(t) \tag{97}$$

Applying this result to equation (44)

$$\hat{e}_o^F(t) = -(v_o(t) + e(t)) - \tilde{L}\, e_o^F(t) \tag{98}$$

Using equation (95):

$$\hat{e}_o^F(t) = -\alpha(v_o(t) + e(t)) - \tilde{L}_\alpha e_o^F(t) \tag{99}$$

Equation (99) corresponds to the upper part of Fig 5.1; equation (96) is used for the scalar case (Fig 5.2). Using equation (22) in place of equation (45), a similar result holds for the stochastic algorithm but with $\tilde{L} = 0$.

5.6.2 Non stochastic error system - scalar case

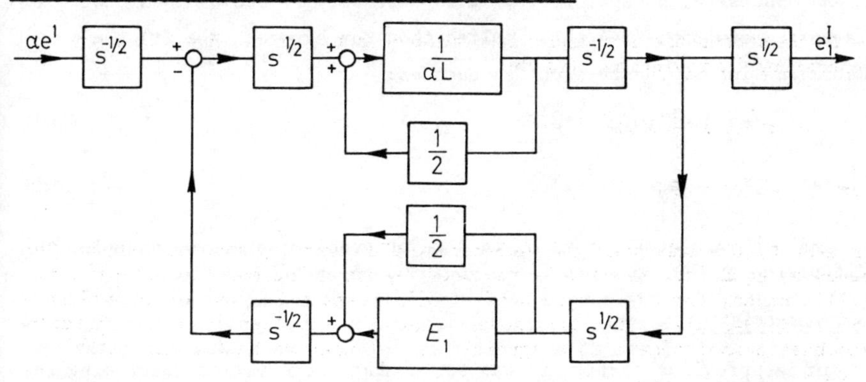

Fig 5.2

The derivation of this diagram is similar to that for the matrix case; the upper part of the diagram follows from equation (96). The two main differences are: the *multipliers* $s^{\frac{1}{2}}$ have been introduced as in equations (88) and (89); and the errors e_1^I and v_1 appear in place of e_o^1 and v_o. The feedback and feedforward terms of magnitude 1/2 do not change the input-output properties of the feedback system but are useful when generating stability results.

5.6.3 Stochastic error systems (k=1)

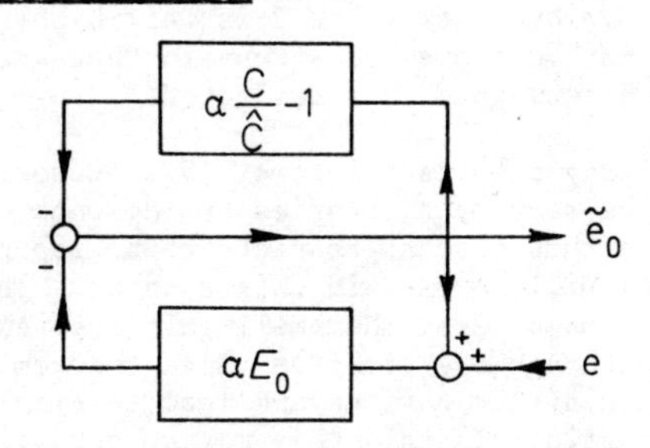

Fig 5.3

These systems are applicable when the input filter L is unity and z is taken to be a white noise process. The block diagram for the matrix case is given in Fig 5.3. $\hat{X}_0$ is used in the matrix case and $\hat{X}_1$ in the scalar case.

The upper part of the diagram follows from equation (36) instead of equations (43) and (44). Note that the noise term e(t) enters at a different place on the diagram; the fact that when $\tilde{e} = 0$, the estimator input is the moving average process e(t) is crucial for convergence analysis. The scalar case gives a diagram bearing the same relation to Fig 5.2 as Fig 5.3 does to Fig 5.1.

5.6.4 Control law error system

As discussed in section 5.3, it is necessary to use a *causal* control law, from equation (24) this implies that the error $\tilde{e}_K \neq e^I(t)$ the estimator input. Hence equation becomes:

$$y(t) = P^{-1}(w(t) + e^I(t) + e_c(t)) \tag{100}$$

where $$e_c(t) = \hat{e}_K(t) - e^I(t) \tag{101}$$

The analysis of the term e_c is a major problem; see, for example, the discussion by Narendra and Lin (1980) and by Egardt (1979).

5.7 *Stability*

In the previous section it was shown that the equation describing the evolution of the various error signals associated with adaptive algorithms could be written as a feedback system see Figs 5.1 and 5.2. There are a number of powerful stability methods available to analyse such systems; the question is: which is the most appropriate method for this problem?

The traditional tool for analysing the stability of model-reference adaptive control systems is the method of Liapunov (Parks, 1965; Narendra and Lin, 1980). The basic idea of Liapunov methods is to examine the unforced motion of the state of some difference equation. In this context, Liapunov methods assume no disturbance and analyse the stability of the parameter estimate $(\hat{\theta})$ trajectories. As successful control requires that the *scalar* quantity $\tilde{e} = \hat{X}^T\hat{\theta}$ is well behaved in the presence of noise, Liapunov's method seems too strong in that states are analysed and too weak in that disturbances are neglected.

Input-output stability methods (Willems, 1970; Desoer, 1975) do not explicitly consider the state trajectories but do consider certain disturbances, they are thus more appropriate than Liapunov methods in this context, and they will be used in this chapter. There are two theorems which will be used here: the small-gain theorem and the passivity theorem. Although the proof of these theorems is rather esoteric, they are intuitively obvious and lead to simple graphical frequency domain criteria. In this they may be compared to Nyquist's theorem. These two theorems will now be applied to the analysis of the feedback system derived in section 5.6.

A problem which remains unsolved in general is the analysis of the signal $e_c(t)$ (equation (101)) due to the need to apply a causal control law. Consideration of this term is deferred to the end of the section.

The feedback system of Fig 5.1 may be looked at in two ways: the system $(1+\alpha E)^{-1}$ in feedback with $\tilde{L}_\alpha$; or the system αE in feedback with $(1+\tilde{L}_\alpha)^{-1} = (\alpha L)^{-1}$. The former representation is used for the matrix case, the latter for the scalar case (Fig 5.2).

5.7.1 Matrix gain algorithms

The small gain theorem (Desoer and Vidyasagar, 1975) states that two systems in a negative feedback arrangement are stable if the loop gain is less than one - an intuitively clear result. Equation (23) says that the gain of $(1+\alpha E_o)^{-1} < 1$; thus stability results if the gain of $L_\alpha < 1$. L_α is a linear time-invariant system and its gain (in the required sense) may be characterised as the radius of the smallest circle enclosing the Nyquist locus of L_α:

$$r \triangleq \sup_{|q|=1} [1-\alpha L(q^{-1})] \tag{102}$$

This argument, combined with the precise statement of the small gain theorem (Desoer and Vidyasagar, 1975) gives the following theorem:

Theorem 5.7.1

Any of the adaptive algorithms described in section 5.3 combined with the matrix gain algorithm of equation (52) have the following stability property:

If $r < 1$:

$$\left[\sum_{t=0}^{N} e_o^{I^2}(t)\right]^{\frac{1}{2}} < \frac{\alpha}{(1-r)} \left[\sum_{t=0}^{N} e'^2(t) + \text{const}\right]^{\frac{1}{2}} \tag{103}$$

This inequality has two interpretations. If the input disturbance is transient with finite *energy* then so is the output error e_o^I. Alternatively, both summations may be divided by N; the result then states that persistent inputs with finite *power* yield an output error e_o^I with finite power.

5.7.2 Scalar gain algorithm

The *passivity* theorem (Desoer and Vidyasagar, 1975) states that two systems in a negative feedback arrangement are stable if both are *passive* in the sense that:

$$\sum^{N} [\text{input} \times \text{output} - \varepsilon\ \text{input}^2] > \text{constant};\ \varepsilon = 0$$

One system must also be strictly passive ($\varepsilon > 0$) (Desoer and Vidyasager, 1975) - a slightly stronger condition. A circuit analogy is a resistor with an input voltage producing an output current: passive systems are analogous to circuits which dissipate energy. Equation (91) says that the system $[\alpha E_1^M + \frac{1}{2}]$ is passive. Apply the simple loop transformation shown in Fig 5.2 to give $\alpha E_1 + \frac{1}{2}$ in feedback with:

$$(\alpha L^M)^{-1}/(1-\tfrac{1}{2}(\alpha L^M)^{-1} = 1/(\alpha L^M - \tfrac{1}{2}) \tag{104}$$

where $$L^M = s^{-\frac{1}{2}} L s^{-\frac{1}{2}} \tag{105}$$

Let $Z = (\text{input to L})\ (\text{output of L}) - \varepsilon\ (\text{input to L})^2$ (106)

and define Z^M similarly then:

$$\sum^N Z^M = \sum^N s^{-1} Z$$
$$= s^{-1}(N+1) \sum_{t=0}^{N} Z(t) - \sum_{j=1}^{N} \left(\sum_{t=0}^{j} Z(t)\right)(s^{-1}(t+1) - s^{-1}(t)) \tag{107}$$

This is Abel's partial summation formula and yields the following Lemma.

Lemma 5.7.1

If:

(i) L is strictly passive

(ii) s^{-1} is a non-increasing sequence (β=1)

Then $s^{-\frac{1}{2}} L s^{\frac{1}{2}}$ is strictly passive.

A frequency domain condition that a linear transfer function $L(q^{-1})$ be strictly passive is that its real part be positive for all $|q| = 1$. These considerations give rise to the following theorem.

Theorem 5.7.2

If the adaptive algorithms described in section 5.3 are combined with the scalar gain algorithm of equation (60) and if:

(i) $\mathrm{Re}\{\alpha L(q^{-1}) - \frac{1}{2}\} > 0\ \ |q| = 1$

(ii) β=1 (no 'forgetting')

Then:

$$\left[\sum_{t=t_o}^{N} s^{-1}(t-k)\ e_1^{I^2}(t)\right]^{\frac{1}{2}} < g_o\left[\sum_{t=t_o}^{N} s^{-1}\ e'^2(t)\right]^{\frac{1}{2}} + g_1 \tag{108}$$

where g_o and g_1 are finite positive constants.

Remark

As $s^{-\frac{1}{2}}Ls^{\frac{1}{2}} = 1$ if L=1, both conditions may be replaced by:

(i*) $L = 1$

(ii*) $\alpha > 1/2$

The appearance of the multiplier $s^{-\frac{1}{2}}$ in expression (108) makes this result rather weaker than the corresponding matrix gain result (equation (103)). For example, if the right-hand side of (108) is uniformly bounded, it follows that:

$$s^{-1}e^{I^2} \rightarrow 0$$

However, as s is unbounded, this does not imply that $e^I \rightarrow 0$.

If conditions (i*) and (ii*) are used, it is not necessary that $\beta = 1$ and thus s may be bounded. This seems to be related to Egardt's method which effectively uses (i*) coupled with the requirement that $\beta < 1$.

5.7.3 Comparison of algorithms

Conditions for the stability of the matrix gain algorithm (Theorem 5.7.1) involve a gain condition on $L\alpha$; the corresponding condition for the scalar gain algorithm involved a passivity condition. However, both involve regions of the complex plane, and by considering the inverse function it is possible to compare the matrix and scalar gain algorithms. It is a simple exercise in complex analysis to show the following equivalence:

Given a complex number z then:

$$|z-1| < 1 \iff \text{Re}\,\{\tfrac{1}{2}-\tfrac{1}{2}\} > 0$$

This leads to the comparison of table 5.1, where L is to be taken as the locus of $L(q^{-1})$, $|q| = 1$.

Table 5.1 Stability conditions

	Gain condition	Passivity condition
Matrix gain	$\left\|L - \frac{1}{\alpha}\right\| < 1/\alpha$	$\text{Re}\{L^{-1} - \frac{\alpha}{2}\} > 0$
Scalar gain	$\|L^{-1}-\alpha\| < \alpha$	$\text{Re}\{L - \frac{1}{2\alpha}\} > 0$

Note that the scalar gain case requires the additional condition that the 'forgetting factor' $\beta = 1$ or that no filter is used ($L = 1$).

An interesting feature of these results is the reciprocal roles that the term α plays in the matrix and scalar gain cases. This is due to the e^{I^2} term not appearing in the matrix representation E_o and the v^2 term not appearing in the scalar representation E_1. However, if it can be shown that the quantity $\sigma_s \to 0$, equation (90) gives that the passivity condition, in the scalar case, may be replaced by Re $\{L\} > 0$. This point is discussed further in section 5.8.

The stochastic algorithm (L = 1) satisfies the conditions of Table 5.1, thus stability of this algorithm does not depend on the transfer function $\hat{C}/C$ although convergence (section 5.8) does.

5.7.4 The control error

To obtain results about the properties of the system output y(t), the stability analysis of this section is not sufficient. From equation (100) the signal e_c representing the difference between the approximation error $\hat{e}_k$ associated with the casual control law and the estimator input e^I also must be analysed. No analysis will be presented here, but some known results in the literature will be described.

5.7.4.1 Scalar case: $\alpha=0$, $\beta<1$

Egardt (1979) has considered algorithms which involve a stochastic approximation estimator. The algorithms are similar to the ones treated here where $\alpha=1$, $\beta<1$ and L=1; they incorporate modifications which ensure that $\hat{\theta}$ is uniformly bounded. These algorithms have the strong property that uniformly bounded disturbances give uniformly bounded error signals.

5.7.4.2 Multiple recursion methods

In a recent paper, Goodwin, Sin and Suluja (1980) introduced a scalar multiple recursion. These algorithms involve k (the system delay) interlaced estimators each of which effectively operates on a unit delay system. In the particular case L=1, $\hat{e}^I=\hat{e}_1$, and thus in effect, e_c is zero for each of the k estimators. Because of the lack of this term, this algorithm seems to be the only method for which stability and convergence has been shown.

5.8 *Stochastic Convergence*

There are three methods of analysing the stochastic convergence of adaptive algorithms which will be treated here: the ordinary differential equation method of Ljung (1977 a,b) and Kushner and Clarke (1979); the generalised martingale convergence method of Solo (1979) as used by Goodwin, Sin and Saluja (1980), and the input-output stability/martingale method of Gawthrop (1980 a,b).

The method of Ljung was the first to be applied in adaptive control theory, it involves a new theory of its own and is applicable to a wide range of algorithms. The more recent martingale methods are based on standard martingale theory (Doob, 1953; Chung 1975). They are restricted to a narrower range of algorithms, but require no explicit conditions on the allowed range of the parameter estimates. See Solo (1979) for a discussion of the relative merits of the methods in an estimation, rather than control, context.

In view of equation (24) the convergence result which would be hoped for is:

$$\underset{t\to\infty}{\text{Lt}} \; \tilde{e}_k(t) = 0 \qquad \text{wp1} \tag{109}$$

In fact the weaker result that the power of $\tilde{e}_k(t)$ tends to zero will be obtained. The sort of conditions necessary to deduce convergence results are now discussed.

Two conditions are essential prerequisites for convergence: The estimator gain K(t) must decrease to zero i.e.:

(A) $\alpha>0,\ \beta=1$

The vector $\hat{X}_i$ must be uncorrelated with $L^{-1}e(t)$
this means that the filter L^{-1} cannot be used and the versions of $\hat{X}_i$ with i=k or k-1 must be used giving

$$\text{(B)} \quad E\{\hat{X}_i(t-k)e(t)\} = 0 \tag{110}$$

All the results presented here require some stability conditions on the adaptive control prior to showing convergence. However, in certain circumstances, Goodwin, Sin and Saluja (1980) have shown that these convergence results can be used to give stability results using arguments by contradiction; this will not be pursued further here. The basic condition required by all algorithms is that the components of the X vector have finite power:

$$\text{(C)} \quad \frac{S(N)}{N} \text{ is uniformly bounded} \tag{111}$$

This condition would hold if stability, as considered in section 5.7, could be shown.

For the matrix gain algorithms, it is also required that:

$$\text{(D)} \quad \sup_{t>t_o} \rho(t) = \hat{X}^T(t)\, K(t)\hat{X}(t) < 1 \tag{112}$$

a condition that has been noted by Egardt (1979); this is critically stronger than expression (57) ($\rho<1$). A much stronger condition which greatly extends the possible results is

$$\text{(E)} \quad \underset{t\to\infty}{\text{Lt}} \; \rho(t) = 0 \tag{113}$$

This result is implied by the conditions imposed by Solo (1979) in an estimation context.

For a detailed discussion of the relation between these conditions and their validity see Gawthrop (1980 b). It is interesting to note that if conditions (A) and (C) are true then condition (E) is equivalent to S(N)/N having a positive definite limit:

$$\text{(E*)} \quad \underset{N\to\infty}{\text{Lt}} \frac{S(N)}{N} = P_{\infty} > 0 \qquad (114)$$

5.8.1 Martingale convergence theorems

Two fundamental martingale theorems are required in the following proofs. An informal discussion of these results is now given; formal results may be found in standard texts (Doob, 1953; Chung 1975). A stochastic process Z(t) is called a *martingale* if:

$$E\{Z(t+1)|t\} = Z(t) \qquad (115)$$

For example the following process is a martingale

$$Z(t) = \sum_{i=0}^{t} \gamma(i)z(i) \qquad (116)$$

where z(t) is a white noise process.

An important class of martingales are those with uniformly bounded variance:

$$E\{Z(t)Z^T(t)\} < \text{constant} < \infty \qquad (117)$$

for example the process of equation (116) obeys (117) if:

$$\sum_{i=0}^{t} \gamma^2 < \text{constant} \qquad (118)$$

Such martingales have the property:

$$\underset{t\to\infty}{\text{Lt}} Z(t) = Z^{\infty} \qquad (119)$$

where Z^{∞} is a realisation dependent finite constant.

A process Y(t) is said to be a *positive super martingale* if:

$$Y(t) > 0; \; E\{Y(t+1)|t\} \le Y(t) \qquad (120)$$

Using Doob's decomposition (Chung, 1975) such processes may be shown to be the sum of a martingale obeying equation (117) and a decreasing sequence. Hence:

$$\underset{t\to\infty}{\text{Lt}} Y(t) = Y^{\infty} \qquad (121)$$

where Y^{∞} is a positive finite constant dependent on the realisation. By analogy with their deterministic counterparts, positive supermartingales are sometimes called stochastic Liapunov functions.

5.8.2 Martingale methods - matrix case

This method has been previously used to analyse the least-squares algorithm ($\beta=1$, $\alpha=1$) with the X vector approximation $\hat{X}_k$. Here it is used for the more general case of expression (109), and using the approximation $\hat{X}_{k-1}$ (equation (26)):

$$\hat{e}_o + v_o = \hat{e}_1 + v_1 \tag{122}$$

The estimator part of the diagram follows from representation E_o; equations (67) and (69), and the fact that the estimator forms a *linear* system.

$$\begin{aligned} v_o(t) &= E_o \hat{e}_o(t) \\ &= E_o \tilde{e}_o(t) + \varepsilon(t) \end{aligned} \tag{123}$$

where $$\varepsilon(t) = E_o e(t) \tag{124}$$

The idea of the method is simple: show that the feedback system is input-output stable, and show that

$$\underset{N\to\infty}{\text{Lt}} \frac{1}{N} \sum_{t=t_o}^{N} \varepsilon^2(t) = 0 \qquad \text{w.p.1} \tag{125}$$

As far as the general reader is concerned w.p.1 (with probability one) may be ignored - it merely implies that the limit holds for each particular sequence of interest of random numbers generating e(t)

Using equations (5), (19), (67) and (69)

$$\begin{aligned} \varepsilon(t) &= E_o e(t) = \sum_{j=0}^{k-1} E_o \gamma_j z(t-j) \\ &= \hat{X}_o^T(t-k) \left| \sum_{j=0}^{k-1} Z_j(t) + Z_o \right| \end{aligned} \tag{126}$$

where $$Z_j(t) = \sum_{i=0}^{t} K(i-k)\hat{X}(i-k)(1+\sigma^1(i-k)z(i) \tag{127}$$

and Z_o is an arbitrary initial condition.

Using condition B, Z_j is a martingale. Following Goodwin and Payne (1977) Appendix D, and Gawthrop (1980 b) it follows that, using equation (50):

$$E\{Z_j(t)Z_j^T(t)\} = E\{\Sigma^t S^{-1}(t-k-1)\hat{X}_o(t-k)\hat{X}_o^T(t-k)S^{-1}(t-k-1)\}E\{z^2\} \qquad (128)$$

Now: $\Sigma S^{-1}(t-k-1)\hat{X}_o(t-k)X_o^T(t-k)\ S^{-1}\ (t-k-1)$

$$= \Sigma(1+\alpha\sigma^1)S^{-1}\ (t-k)\ \frac{S\ (t-k)-S(t-k-1)}{\alpha}\ S^{-1}(t-k-1)$$

$$< \frac{(1+\alpha\ \sup\ \sigma^1)}{\alpha}\ S_o^{-1} \qquad (129)$$

As $\sigma^1 = \rho/(1-\rho)$ the assumption:

$$\sup \rho < 1 \qquad (130)$$

implies that Z is of uniformly bounded variance and thus converges. As discussed elsewhere (Gawthrop, 1980 b), it is possible to choose

$$Z_o = \sum_{j=0}^{k-1} Z_j^\infty \qquad (131)$$

and suitably modify the initial condition on the integrator implicit in the system $E_o\tilde{e}(t)$. Hence repeating for j = 1 to k:

$$\underset{t\to\infty}{\text{Lt}}\ Z(t) = 0 \qquad \text{w.p.1.}$$

Using the boundedness condition C, equation (131) implies:

$$\underset{t\to\infty}{\text{Lt}}\ \frac{1}{N}\sum^{N} \varepsilon^2(t) = 0 \qquad \text{w.p.1.} \qquad (132)$$

In the particular case that k=1, i=0 may be used. In this case ε is the only input to the system of Fig 5.3 and thus if the feedback system is stable, $\hat{e}_o$ obeys an equation similar to equation (132). Noting that:

$$\tilde{e}_1 = (1-\rho)\tilde{e}_o \qquad (133)$$

the following theorem is proved:

Theorem 5.8.1

If:

(i) The system delay k = 1

(ii) $\hat{X}_o$ is used

(iii) conditions A, B, C and D are true

(iv) $|\alpha C(q^{-1})/\hat{C}(q^{-1}) - 1|<1; \qquad |q| = 1$

Then:

$$\underset{N\to\infty}{\text{Lt}} \frac{1}{N} \sum^{N} \tilde{e}_k^2(t) = 0 \qquad \text{w.p.1} \tag{134}$$

Remarks:

1. The result that $\rho(t)<1$ (equation (57)) falls critically short of condition (iii) which requires that ρ be uniformly < 1.

2. The result of equation (134) falls short of the ideal of (109), but seems typical of the results obtained.

It is interesting to examine the consequences of replacing $\hat{X}_o$ by $\hat{X}_1$ theorem 5.8.1. From equation (133), $\tilde{C}$ in the diagram may be replaced by $\frac{\tilde{C}}{1 - \rho}$. Noting that:

$$\text{gain}\left\{\frac{\tilde{C}_\alpha}{1-\rho}\right\} \leq \frac{1}{1-\sup\rho} \;\; \text{gain} \;\; \tilde{C} \tag{135}$$

The following theorem results.

Theorem 5.8.2

If conditions (ii) and (iv) in theorem 5.8.1 are replaced by:

(ii*) $\hat{X}_1$ is used

(iv*) $|\alpha C(q^{-1})/\hat{C}(q^{-1})-1| < (1-\sup\rho)$

Then result (134) is true.

Remark

The essential difference between the two cases $\hat{X}_o$ and $\hat{X}_1$ is that in the former case the term $(1-\sup\rho)^{-1}$ appears outside the feedback loop, and in the latter case inside. In general, (iv*) is more restrictive than (iv). If, however assumption (E)(expression (113)) were made, the two methods give identical results.

It has been shown by Gawthrop (1980), that if condition (E) is true then, given conditions (A) to (D):

$$\underset{N\to\infty}{\text{Lt}} \frac{1}{N} \sum^{N} (\hat{e}_k(t)-\hat{e}_o(t)) = 0$$

Hence the following theorem:

Theorem 5.8.3

If condition (i) of theorem 5.8.1 is replaced by condition (E) then result (134) is true (for all delays k).

5.8.3 Scalar gain algorithms - supermartingale method

Solo (1979) has introduced a convergence method based on a generalisation of the martingale convergence theorem. To avoid introducing more theorems, the better known but essentially similar technique based on Doob's decomposition and the supermartingale convergence theorem will be used here. The diagram of Fig 5.4, which is similar to Fig 5.2, is appropriate here.

Using equation (87), the scalar gain algorithm has the property that:

$$\bar{V}(t) - \bar{V}(t-1) = s^{-1}[2v_1(t)[\tilde{e}_1(t)+e(t)] + \sigma_s(t-k)[\tilde{e}_1(t) + e(t)]^2 \qquad (137)$$

Noting the crucial condition (B) (equation (110)):

$$E\{\bar{V}(t)|t-1\} = s^{-1}(t-k)[2v_1(t)\tilde{e}_1(t)+\sigma_s(t-k)\tilde{e}_1^2(t)] + s^{-1}(t-k)\sigma_s(t-k)E\{e^2(t)\}$$
$$+ \bar{V}(t-1) \qquad (138)$$

The two terms of the right hand side of equation (138) are examined in turn. From Fig 5.3:

$$\tilde{e}_1 = \frac{\hat{C}}{C} v_1$$

Using lemma 5.7.1, using similar arguments to those supporting Theorem 5.7.2, and noting that $\sigma_s < \frac{1}{\alpha}$ it follows that if $C/\hat{C}-1/2\alpha$ is strictly passive then:

$$\Sigma^N s^{-1}(t-k)[2v_1(t)\tilde{e}_1(t)+\sigma_s(t-k)\tilde{e}_1^2(t)] < - \varepsilon\Sigma^N s^{-1}(t-k)\tilde{e}_1^2(t) \qquad (139)$$

for some $\varepsilon > 0$.

The second term if (138) is examined next.

$$s^{-1}(t)\sigma_s(t) = \hat{X}^T(t)x^{-2}(t)\ \hat{X}(t)$$
$$= \text{trace}\ [s^{-1}(t)X(t)X^T(t)x^{-1}(t)] \qquad (140)$$

In a similar fashion to equation 129, it follows that:

$$\sum^N s^{-1}(t-k)\sigma_s(t-k) < s_o^{-1} \qquad (141)$$

Define: $\bar{V}_1(N) = \Sigma^N E\{\bar{V}(t)-\bar{V}(t-1)\} - s_o^{-1}$ (142)

Then from (139) and (141):

$$\bar{V}_1(N) < 0 \qquad (143)$$

The process:

$$\bar{V}_2(N) = \bar{V}(N) - \bar{V}_1(N)$$

is thus positive, and, by definition, a martingale and thus:

$$\underset{N\to\infty}{\text{Lt}} \quad \bar{V}_2(N) = \bar{V}_2^{\infty} \tag{144}$$

It follows that $\bar{V}_1(N)$ is uniformly bounded below; and using (139) and (141) it follows that:

$$\Sigma s^{-1}(t-k)\tilde{e}^2(t) < \infty \tag{145}$$

Using Kroneckers lemma:

$$\underset{N\to\infty}{\text{Lt}} \quad s^{-1}(N-k) \; \Sigma e_1^2(t) = 0 \tag{146}$$

Combining equation (146) with assumption (C) (equation (111)) gives the following theorem.

Theorem 5.8.4

If:

(i) the system delay $k = 1$

(ii) $\hat{X}_1$ is used

(iii) conditions (A), (B), (C) and (D) are true

(iv) Re $\{C(q^{-1})/\hat{C}(q^{-1}) - \frac{1}{2\alpha}\} > 0$; $|q| = 1$

Then:

$$\underset{N\to\infty}{\text{Lt}} \; \frac{1}{N} \sum^{N} \tilde{e}_k^2(t) = 0 \qquad \text{w.p.1} \tag{147}$$

As in derivation of Theorem 5.8.3, condition $E(\sigma_s \to 0)$ allows condition (i) to be removed. In view of equation (138) it has the further consequence that σ_s may be replaced by zero rather than $1/2\alpha$. This gives the following result:

Theorem 5.8.5

If condition (i) in theorem 8.4 is replaced by condition (E), then condition (iv) may be weakened to:

(iv*) Re $\{C(q^{-1})/\hat{C}(q^{-1})\} > 0$; $|q| = 1$ and result (147) holds

5.8.4 The method of Ljung

The first rigorous method applied to the stochastic convergence of adaptive algorithms is due to Ljung (1978 a,b) although similar methods were derived by Kushner and Clarke (1979). Although the results are difficult to derive formally, various informal derivations and discussions are available (Ljung and Wittenmark, 1974; Åström et al, 1977; Solo 1980). The informal discussion here is restricted to the particular algorithms considered in this chapter.

The basic idea used in this method is that as the estimator gain $K(t)$ (equation (46)) becomes small, and assuming conditions (A-E*) the parameter estimate $\hat{\theta}$ will change slowly. Thus a given finite change is due to a large number of random variables which, as the control law is approximately constant, are approximately stationary. Also, assumption (E*) means that:

$$\bar{S}(t) = S(t)/t \tag{148}$$

will also change slowly.

Assumption (E) implies that:

$$K(t) \sim S(t)$$

and so:

$$\tilde{\theta}(t+\Delta t)-\tilde{\theta}(t) \simeq \frac{\Delta t}{t}\,\bar{S}^{-1}(t)\,\frac{1}{\Delta t}\,\Sigma_t^{t+\Delta t}\,\hat{X}\,(i-k)\hat{e}(i) \tag{149}$$

$$\bar{s}(t+\Delta t) - \bar{s}(t) = \frac{\Delta t}{t}\,[\alpha\frac{1}{t}\,\Sigma_t^{t+\Delta t}\,\hat{X}(i)\hat{X}^T(i) - \bar{s}(t)] \tag{150}$$

These equation suggest the introduction of the logarithmic time scale:

$$\tau = \log t \tag{151}$$

so $\quad \Delta\tau \simeq \dfrac{\Delta t}{t}$

Using this transformation, it is possible to have a small $\Delta\tau$ to give an approximate derivative whilst, for large t, having a large Δt to allow the summations to be replaced by expectations. Hence using the key condition (B), the parameter estimate errors $\tilde{\theta}$ are asymptotically given by:

$$\frac{d\bar{\theta}}{d\tau} = -R^{-1}\;E\;\{\hat{X}\,\frac{\hat{C}}{C}\,X^T\}\;\bar{\theta} \tag{152}$$

$$\frac{dR}{d\tau} = \alpha E\;\{\hat{X}(t)\hat{X}^T(t)\} - R \tag{153}$$

Note that the expectations are evaluated as if the control law:

$$\hat{X}^T\hat{\theta} = w$$

were used for a long time.

A similar pair of equations with R replaced by:

$$r(\tau) = \text{trace}\,[R(\tau)] \tag{154}$$

describes the corresponding scalar gain algorithms.

The results of Ljung (1977 a,b) show that asymptotic stability of $\bar{\theta}$ imply w.p.1 convergence of $\tilde{\theta}$ to zero. The Liapunov analysis of these equations give such a result under exactly the same condition of $\hat{C}/C$ as derived using the martingale methods and using assumptions (A-E).

These ordinary differential equations provide a useful alternative to Monte Carlo simulation of the actual algorithm when investigating particular cases. These equations may also be linearised about $\bar{\theta}=0$ to generate examples of $\hat{C}/C$ where the algorithms do not converge.

5.8.5 Comparison of Algorithms

The following table is analogous to table 5.1; $C/\hat{C}$ represents the locus of $C(q^{-1})/\hat{C}(q^{-1})$ for $|q| = 1$.

Table 5.2 Convergence conditions

	Gain condition	Passivity condition
Matrix gain	$\left\lvert \frac{C}{\hat{C}} - \frac{1}{\alpha} \right\rvert < \frac{1}{\alpha}$	$\mathrm{Re}\{\frac{\hat{C}}{C} - \frac{\alpha}{2}\} > 0$
Scalar gain	Not applicable	$\mathrm{Re}\{\frac{\hat{C}}{C}\} > 0$

If assumptions (A-E) are made, convergence is dependent on the frequency domain conditions of table 5.2. Unlike the stability results of section 5.7, the transfer function $C/\hat{C}$ is not known: C is the unknown noise numerator polynomial. The fact that these conditions cannot be verified a-priori would be expected: if C were known, it would be possible to set $\hat{C}=C$ and the conditions would be automatically satisfied.

Ideally, conditions (A-E) would be automatically verified using the stability results of section 5.7; unfortunately this is not so. In particular, condition (C) does not imply condition (E). It is possible that adaptive algorithms could be modified in such a way that conditions (A-E) could be verified. Note, however, that in the estimation or a adaptive prediction context, it is usually easy to verify these conditions. For example Solo (1979) uses condition (E*) which implies condition (E).

5.9 *Conclusion*

A number of the simpler adaptive algorithms have been analysed. Although results concerning a prediction error are readily obtained, the need to use a different (causal) signal in the control law leads to problems which are not completely solved.

Although the more complicated explicit methods, involving an extra computation between estimated and control coefficients, have not been discussed, it is believed that the results displayed here are a starting point for the analysis of such algorithms.

References

Åström, K J (1974). 'A self-tuning regulator for non minimum-phase systems', Report 7411(C), Lund Institute of Technology, Sweden.

Åström, K J and Wittenmark, B (1973). 'On Self Tuning Regulators', Automatica, 9, pp 185-199.

Åström, K J and Wittenmark, B, (1980). 'Self-tuning controllers based on pole-zero placement", Proc IEE, 127 pt D, No 3, pp 128-130.

Clarke, D W and Gawthrop, P J (1979). 'Self-tuning control', Proc IEE, 126, No 6, pp 633-640.

Chung, K L (1975). 'A course in probability theory', 2nd end, Academic Press.

Desoer, C A and Vidyasagar, M (1975). 'Feedback systems: input-output properties', Academic Press.

Doob, J L (1953). 'Stochastic processes'. John Wiley and Sons.

Egardt, B (1979). 'Stability of adaptive controllers', Lecture notes in control and information sciences No 20, Springer-Verlag,

Gawthrop, P J (1977). 'Some interpretations of the self-tuning controller'. Proc IEE, 124, No 10, pp 889-894.

Gawthrop, P J (1980 a). 'On the stability and convergence of self-tuning controllers', in "Analysis and optimisation of stochastic system". (ed Jacobs), Academic Press.

Gawthrop, P J (1980 b). 'On the stability and convergence of a self-tuning controller', Int J Control, 31, No 5, pp 973-998.

Gawthrop, P J and Lim, K W (1981). 'Robustness of self-tuning controllers'. In preparation.

Goodwin, G C and Payne, R L (1977). 'Dynamic system identification: experiment design and data analysis', Academic Press.

Goodwin, G C Ramadge, P J and Caines, P E (1980). 'Discrete-time multivariable adaptive control'. Trans IEEE, AC-25, No 3, pp 449-456.

Goodwin, G C, Sin, K S and Saluja, K K (1980). 'Stochastic adaptive control and prediction - the general delay - coloured noise case'. Trans IEEE, AC-25, No 5, pp 956-950.

Kalman, R E (1958). 'Design of self-optimising control systems', Trans ASME 80, pp 468-478.

Kushner, H J and Clarke, D S (1978). 'Stochastic approximation methods for constrained and unconstrained systems', Springer-Verlag.

Lam, K P (1980). 'Implicit and explicit self-tuning controllers', D Phil thesis, Oxford University.

Landau, I D (1979). 'Adaptive control: The model reference approach', Marcel Dekker.

Ljung, L (1977 a). 'On positive-real transfer functions and the convergence of some recursive schemes', Trans IEEE, AC-23, No 5, pp 539-550.

Ljung, L (1977 b). 'Analysis of recursive stochastic algorithms'. Trans IEEE, AC-23, No 5, pp 551-575.

Ljung, L and Wittenmark, B (1974). 'Analysis of a class of adaptive regulators'. Proceedings IFAC symposium on stochastic control, Budapest.

Narendra, K S and Lin, Y (1980). 'Stable discrete adaptive control'. Trans IEEE, AC-25, No 3, pp 456-461.

Parks, P C (1966). 'Liapunov redesign of model reference adaptive control systems'. Trans IEEE, AC-11, pp 362-367.

Solo, V (1979). 'The convergence of AML', Trans IEEE, AC-24, No 6, pp 958-962.

Solo, V (1980). 'Some aspects of recursive parameter estimation'. Int J Control, 32, No 3, pp 395-410.

Willems, J C (1970). 'The analysis of feedback systems'. M.I.T. Press.

Chapter Six

Implementation of self-tuning controllers

D W CLARKE

6.1 *Introduction*

Kalman's 'self-optimizing control system' (1958) was implemented in hardware that was 'externally digital, internally analogue' - coefficients were normalised and stored on servo-potentionmeters, and the programming was done on a patch-panel. The argument which mitigated the complexity of programming a special-purpose computer was that the system would be applicable to a range of control tasks; this is still valid with modern self-tuners implemented on microprocessors, although the algorithms have become more refined and digital technology has been completely transformed. On the other hand, a special-purpose system is not really suitable for developing a new algorithm, but is best kept until the properties of the method are fully understood. This may be one reason why Kalman's proposal was not pursued further.

By the 'seventies the minicomputer had become predominant for process control applications, and many such systems were connected on-line to laboratory and industrial plant. It was not surprising, therefore, that the first reported applications of self-tuning - readily implemented on minicomputers - came soon after the basic theory had been developed by Åström and Wittenmark (1973). These applications covered a range of industries: ore-crushing (Borisson and Syding, 1976), ship-steering (Kallström et al, 1977), paper-making (Cegrell and Hedqvist, 1975), titanium-dioxide kilns (Dumont and Bélanger, 1978), distillation columns (Sastry, Seborg and Wood, 1977), cement-blending (Keviczky et al, 1978), enthalpy exchangers (Jensen and Hänsel, 1974), and many others. This spate of applications is probably unprecedented in the history of 'modern' control theory, and shows that the method is usable with only a relatively modest computing power. This was confirmed by the development of a microprocessor-based self-tuner (Clarke, Cope and Gawthrop, 1975).

There are many ways in which a self-tuner can be implemented. It can form part of a DDC package in a minicomputer control system, where the self-tuner can act continuously on a few critical loops or can cycle around the loops in turn, 'trimming' the feedback parameters and leaving them fixed until the next cycle. This transient rather than continuous tuning could be interpreted as a *commissioning aid* for a control loop which could be exercised by the control engineer on start-up or on significant set-point or process changes. In this mode, a complex but fixed-parameter controller can be tuned.

Dr Clarke is with the Department of Engineering Science, Oxford University.

A self-tuner can be built as an 'out-station' in a microprocessor-based distributed control system - the current trend in process automation - communicating with an operator's console via the system bus. As it is typically found that the PID algorithm in an out-station takes only a fraction of the computational power available, the extra cost of the hardware and software to implement a self-tuner instead should be quite low. On the other hand, the potential benefit of a self-tuner available for critical loops could be considerable. Indeed, it should be possible to construct dedicated microprocessor-based self-tuners for single-loop (or interactive multiple-loop) control. There are several possibilities in this area. For example, by augmenting the standard self-tuner with appropriate special-purpose software, a control system for dedicated markets such as pH control could be developed. Alternatively, a suitably developed system could be produced for 'non-traditional' control markets where the skill for PID tuning is absent. In all these areas, however, a clearly defined and documented user interface is essential.

This chapter describes various aspects of the implementation of self-tuning controllers, giving both a general over-view and details of a system based on an industry-standard microprocessor (the Intel 8080). Section 6.2 presents listings of FORTRAN modules which can be used to create the simplest minimum-variance self-tuning regulator and shows how other self-tuners can be constructed by generalising the algorithms. Section 6.3 discusses techniques for robust parameter estimation which can be used with the short word-lengths typical of microprocessor systems. Section 6.4 considers various solutions to the problems of constant offsets and load disturbances, whilst section 6.5 is concerned with the vitally important topic of *integrity*, describing the software 'jackets' that can be built into a self-tuner so that it can still perform even if the plant does not satisfy the assumptions of the theory. This discussion includes a partial answer to the question of when a self-tuner should be switched off. Section 6.6 lists typical parameters that need to be chosen by the user before starting a self-tuning run, indicating that the values of most of these are not critical. Finally, section 6.7 describes a self-tuning experiment on an industrial chemical batch reactor which illustrates some of the points made in the rest of the chapter.

6.2 *An elementary self-tuning regulator*

In a self-tuner based on minimum-output-variance regulation theory, a regression model of the form:

$$y(t) = f_o y(t-k) + \ldots + g_o u(t-k) + \ldots + \text{error} \qquad (1)$$

is proposed, so that recursive-least-squares can be used to estimate $\hat{F}(z^{-1})$, $\hat{G}(z^{-1})$ for direct use by the feedback control:

$$\hat{f}_o y(t) + \hat{f}_1 y(t-1) + \ldots + \hat{g}_o u(t) + \hat{g}_1 u(t-1) + \ldots = 0 \qquad (2)$$

If the parameter $\hat{g}_o$ is 'fixed' to avoid possible numerical difficulties, its value $\bar{g}$ should be related to some extent to the first point on the system's pulse response b_o (see Åström and Wittenmark, 1973), and (2) becomes:

$$u(t) = - \{\hat{f}_o y(t) + \hat{f}_1 y(t-1) + \ldots + \hat{g}_1 u(t-1) + \ldots\}/\bar{g} \quad (3)$$

and the corresponding regression model is:

$$y(t) - \bar{g}u(t-k) = f_o y(t-k) + \ldots + g_1 u(t-k-1) + \ldots + \text{error} \quad (4)$$

In practical systems there are constraints on the amplitudes of control signal that are allowed (e.g. 0-100% valve opening), so u must in general be 'clipped':

$$u_{min} \leq u(t) \leq u_{max} \quad (5)$$

If the desired value of u is clipped it is important that the *actual* value be saved in the data vector used in later parameter estimation. One method of overcoming offsets (see section 6.4) is to cascade a digital integrator $1/(1-z^{-1})$ after the self-tuner. 'Wind-up' in this integrator should be avoided, and if the control u(t) causes the integrator output to saturate, the value of u(t) should be recomputed so as to just cause clipping, and this should be the value saved in the data array.

6.2.1 Data structures and basic subroutines

Suppose NU and NY are the numbers of u and y parameters to be estimated (n+k-2 if $\bar{g}$ is fixed, and n), then a parameter vector THETA of dimension NU + NY can be used, with THETA (1) corresponding to $\hat{f}_o$, etc. As is usual with discrete-time controllers, the input/output data vectors require more attention, for in the theory they are indexed by t, but in a program need to be stored in fixed-length arrays. The convention is, therefore, to let the first element of an array correspond to time t (the current data), and to shift all data arrays by one unit at each sample instant. (In languages such as Pascal it is straightforward to use *ring buffers* instead, in which an item such as u(t) is stored using a pointer t *modulo*NU). A subroutine which combines saving of data and shifting the corresponding array is SAVE, given in Table 6.1.

Table 6.1 Routine for saving data

```
      SUBROUTINE SAVE (DATA, ARRAY, LENGTH)
      DIMENSION ARRAY (1)
      DO 1 I = LENGTH, 2, -1
1     ARRAY (I) = ARRAY (I-1)          ! SHIFT OLD DATA ALONG
      ARRAY (1) = DATA                 ! INSERT NEW
      RETURN
      END
```

In order to use the estimation routine, a data vector x(t) must be primed with data from y(t-k) ..., u(t-k-1) ..., which is done using a routine MOVE. The fact that estimation uses data down to times t-k-NU and t-k-NY implies that the USAVE and YSAVE arrays must be appropriately dimensioned. The control given by (3) is conveniently calculated using a function SCAPRO which computes the scalar product of two arrays (data and estimates here); MOVE and SCAPRO are listed in Table 6.2

Table 6.2 MOVE and SCAPRO

```
   SUBROUTINE MOVE (FROM, TO, LENGTH)
   DIMENSION FROM (1), TO (1)
   DO 1 I = 1, LENGTH
1  TO(I) = FROM (I)
   RETURN
   END


   REAL FUNCTION SCAPRO (ARRAY 1, ARRAY 2, LENGTH)
   DIMENSION ARRAY 1 (1), ARRAY 2 (1)
   SCAPRO = 0.0
   DO 1 I = 1, LENGTH
1  SCAPRO = SCAPRO + ARRAY 1 (I) * ARRAY 2(I)
   RETURN
   END
```

With these elementary routines defined, we are now in a position to consider the main self-tuning code. As the recursive estimator is a key part of all self-tuners, involving significant numerical analysis problems, section 6.3 is devoted entirely to its theory and coding, and we simply assume here that a suitable estimation routine is available. We also assume that there is a real-time executive which initiates the self-tuner at intervals h and acquires current plant output data Y via an ADC. Three logical variables EST, DC and INT are involved, having the value .TRUE. if parameter estimation is to be used, if the model is to be augmented by a dc level (see section 6.4), and if a digital integrator is to be cascaded. Table 6.3 shows the basic coding for the minimum-variance self-tuner which generates the new control signal U.

Table 6.3 Basic self-tuner routine

```
C  GET READY FOR ESTIMATION
   PHI = Y - GBAR * USAVE (K)                   ! 'output' with fixed parameter
   CALL MOVE (YSAVE (K), X, NY)                 ! prime x(t) with y(t-k) ...
   CALL MOVE (USAVE (K+1), X(NY+1), NU)         ! and u(t-k-1) ...
C
   NPAR = NY + NU                               ! number of parameters
   IF (DC) NPAR = NPAR + 1                      ! estimate dc?
   IF (DC) X (NPAR) = 1.0                       ! so put in 1 in x vector
C  CALL ESTIMATION ROUTINE
   IF (EST) CALL SQRTES (PHI, X, S, GK, THETA, NPAR, PERR, FORGET)
C  SAVE Y
   CALL SAVE (Y, YSAVE, K + NY)
C  COMPUTE CONTROL
   U1 = SCAPRO (THETA, YSAVE, NY) + SCAPRO (THETA (NY + 1), USAVE, NU)
   IF (DC) U1 = U1 + THETA (NPAR)               ! remove offset effect
   U1 = - U1/GBAR                               ! self-tuner output
C  SEE IF INTEGRATOR REQUIRED
   IF (.NOT. INT) OLDU = 0.0                    ! if not, clear old value
   U = OLDU + U1                                ! integrate
   IF (U.GT. UMAX) U = UMAX                     ! clip the
   IF (U.LT.UMIN) U = UMIN                      ! control
   U1 = U- OLDU                                 ! recompute U1 (desaturation)
   OLDU = U                                     ! save control for next time
   CALL SAVE (U1, USAVE, K + NU)                ! and STR control
C  OUTPUT CONTROL U AND RETURN FOR NEXT INTERRUPT.
```

Just as a practical digital PID algorithm involves considerably more code than that required simply for implementing the mathematical relation, so the listing of Table 6.3 is only a fraction of a practical self-tuner. In addition there must be a 'set-up' phase for initialising variables such as NY, NU, FORGET, etc. and 'diagnostic' or 'jacket' software for checking the validity of the estimation. Details of these practical additions will be given later.

Further implicit self-tuners, such as those which minimise a generalised variance (Clarke and Gawthrop, 1975, 1979) or include feedforward action, can be programmed by adding elements to the data vector in the listing of Table 6.3. In some cases, such as in RML (Recursive Maximum Likelihood) estimation, a *filtered* data vector or variable is required, based on the equation:

$$T(z^{-1})\, d^f(t) = d(t);\ \text{with}\ T_o = 1 \tag{6}$$

which produces filtered data $d^f(t)$ from input data $d(t)$ using the function FILTER of Table 6.4.

Table 6.4 A filtering routine

```
REAL FUNCTION FILTER (DATA, STATE, FILPAR, LENGTH)

DIMENSION STATE (1), FILPAR (1)

D = DATA - SCAPRO (STATE, FILPAR, LENGTH)    ! STATE contains old
                                               filtered data
CALL SAVE (D, STATE, LENGTH)                 ! save for next time

FILTER = D

RETURN

END
```

If a single filtered variable is required, the value of FILTER can be used, whereas a vector is obtained as output of the STATE argument. Note that initial values for STATE need to be specified; usually the elements are set to zero.

With *explicit* self-tuners a control design subroutine is inserted between the parameter estimation and control generation phases. For example, the Diophantine equation;

$$A'(z^{-1})\, G(z^{-1}) + z^{-k}\, B'(z^{-1})F(z^{-1}) = T(z^{-1}) \tag{7}$$

needs to be solved for the F and G parameters at each sample-instant, given estimates $\hat{A}'$ and B' and a user-specified closed-loop characteristic polynomial $T(z^{-1})$. Algorithms for solving such equations are given in Kučera (1979). One possibility, described in Edmunds (1976) is to define:

$$\phi(t) = T(z^{-1})\xi(t),\ \psi_1(t) = \hat{A}'(z^{-1})\xi(t)\ \text{and}\ \psi_2(t) = \hat{B}'(z^{-1})\xi(t-k) \tag{8}$$

where $\xi(t)$ is an uncorrelated random sequence generated internally, so that (7) becomes the regression relation:

$$\phi(t) = G(z^{-1})\psi_1(t) + F(z^{-1})\psi_2(t) + \text{error} \tag{9}$$

The error here is simply due to fitting errors of the parameters, and RLS can be used to estimate $\hat{F}$, $\hat{G}$ using the *same* basic routine as in the estimation of $\hat{A}'$, $\hat{B}'$. This saving of coding is, not surprisingly, at the cost of extra on-line computation compared with a direct algorithm for solving (7).

The self-tuners as implemented above require a fair amount of computation between the acquisition of y(t) and the generation of control u(t), and this may involve a delay which is an appreciable fraction of the sample interval in some cases. This can be avoided by rearranging the algorithms so that the control is based on the *preceding* parameter estimates and hence calculated immediately. This has been found (for slowly varying systems) not to lead to significant reduction in performance. Indeed, self-tuners for slowly varying systems could be implemented with a multi-tasking executive in which the parameter estimation has a low priority and is interruptable by more urgent tasks, provided that it is equipped with a private data vector x(t) which is not changed until the estimation is completed.

6.3 *Numerically Stable Parameter Estimation*

A recursive parameter estimator is at the heart of all self-tuning algorithms, and as a self-tuner may be left in operation for many tens of thousands of iterations using a microprocessor with short word-lengths, it is important that the estimator is numerically stable. This is not the case with the standard RLS (Recursive Least Squares) algorithm, as can be seen by the following single-parameter example:

Consider estimating θ in the model $\phi(t) = \theta x(t) + \varepsilon(t)$, using RLS:

$$\left.\begin{aligned} \hat{\theta}(t) &= \hat{\theta}(t-1) + K(t)\,(\phi(t) - \hat{\theta}(t-1)\,x(t)) \\ K(t) &= P(t-1)x(t)/(1+P(t-1)x(t)^2) \\ P(t) &= (1-K(t)x(t))P(t-1) \end{aligned}\right\}$$

Defining $\tilde{\theta}(t)$ to be the parameter error $\theta - \hat{\theta}(t)$, we have

$$\tilde{\theta}(t) = (1-K(t)x(t))\tilde{\theta}(t-1) + K(t)\varepsilon(t)$$

which is a stochastic difference equation which is unstable if $|1-Kx|>1$, and in particular if $Kx < 0$, that is if P ever becomes negative. This is just possible in practice if $Kx \sim 1$ (e.g. if P is large), and if both K and x are computed and stored using short word-lengths. Note that once P becomes negative the algorithm retains its instability until $|Px^2|$ exceeds 1. In the multiparameter case the problem is more severe, as P must be a symmetrical positive-definite matrix for stability of the algorithm, and estimation fails if P ever becomes negative definite. This tends to happen in simulations on minicomputers (e.g. on a PDP 11 with 32-bit floating point representation) after about 5-10,000 iterations, which is rather more than the 1000 or so generally used in typical simulation runs but less than would be involved in an industrial situation (a possible corollary of Murphy's Law). Similar problems are encountered in Kalman filtering (see Bierman, 1977).

The solution to this form of numerical instability is to replace the updating of P using (1-Kx) P by the updating of a *factor* of P such that the resulting P is guaranteed to be non-negative definite. Peterka (1975) derived such an algorithm, in which P(t) is factorized as:

$$P(t) = S(t)\ S'(t) \tag{10}$$

where S is an upper-triangular matrix called the *square-root* of P, and S(t) is updated at each iteration. As details are in Peterka (1975), we concentrate here on outlining the basic idea with a single-parameter example.

Including the forgetting factor β, the update of $P(t) = S(t)^2$ is:

$$S(t)^2 = (1-S(t-1)^2 x(t)^2/\mu^2).S(t-1)^2/\beta, \text{ where}$$

$$\mu^2 = 1 + S(t-1)^2 x(t)^2.$$

Now if $f \triangleq S(t-1)x(t)$, this can be written as

$$S(t).S(t) = \frac{S(t-1)}{\sqrt{\beta}} \left\{(1\ \ jf/\mu)\begin{pmatrix} 1 \\ jf/\mu \end{pmatrix}\right\} \frac{S(t-1)}{\sqrt{\beta}};\ j \triangleq \sqrt{-1} \tag{11}$$

Introducing a 2-dimensional orthogonal matrix T:

$$T = \begin{bmatrix} u & jv \\ -jv & u \end{bmatrix} \quad \text{where } u^2 - v^2 = 1 \tag{12}$$

then (11) can be written as:

$$S.S = \frac{S}{\sqrt{\beta}}\ (1\ \ jf/\mu)\ TT' \begin{pmatrix} 1 \\ jf/\mu \end{pmatrix} \frac{S}{\sqrt{\beta}} \tag{13}$$

Choosing the elements of T such that:

$$(1\ \ jf/\mu)\ T = (h\ \ 0)$$

means that S itself can be updated as:

$$S(t) = S(t-1).\ h/\sqrt{\beta} \tag{14}$$

Equations (12) and (13) give:

$$u + fv/\mu = h \text{ and } j(fu/\mu + v) = 0, \text{ so using}$$

$$u^2 - v^2 = 1 \text{ gives } u = \mu/\sqrt{\mu^2-f^2} \text{ and } v = -f/\sqrt{\mu^2-f^2}$$

so that h is $(\mu^2 - f^2)/\mu\sqrt{\mu^2-f^2}$ which is correct as $\mu^2 - f^2 = 1$ in this case. Note also that K(t) is given by $S(t-1)^2x(t)/\mu^2 = S(t-1)\ f/\mu^2$ so that a complete update of S and $\hat{\theta}$ is obtained *without* having to evaluate P.

When estimating n parameters the orthogonal matrix T is of dimension n+1 and is the product of n different orthogonal matrices T_i each of which is used to annul an element of the last column of a matrix $(H\ \underset{\sim}{0})$ where H is upper triangular, such that $S(t) = S(t-1)\ H/\sqrt{\beta}$. Peterka (1975) derives the basic theory, and Clarke, Cope and Gawthrop (1975) give coding suitable for an 8080 microprocessor. Table 6.5 gives a FORTRAN listing of a subroutine SQRTES which is called by the self-tuning code of section 6.2.

Table 6.4 The square-root parameter estimation routine

```
      SUBROUTINE SQRTES (PHI, X, S, G THETA, N, E, FORGET)
      DIMENSION X(1), S(1), G(1), THETA (1)
C
C     PHI    - 'output' φ(t)
C     X      - data vector x(t)
C     S      - upper-triangular square-root S(t) of the 'covariance' matrix
               P(t)
C            - elements stored in vector as (1,1),(1,2), (2,2),(1,3), (2,3)
               (3,3), ...
C     G      - Kalman gain vector is G/SIGSQ
C     THETA  - vector θ̂ of parameter estimates
C     N      - number of parameters (dimension of X, B, THETA)
C     E      - prediction error of the 'old' model
C     FORGET- square-root of the 'forgetting-factor' β
C
      E= PHI - SCAPRO (X, THETA, N)          ! e =φ- x'(t)θ̂(t-1)
C
      SIG = FORGET
      SIGSQ = SIG* SIG                       ! β
      IJ = 0
      JI = 0
C
      DO 30 J = 1, N
      FJ = SCAPRO (S(JI), X, J)              ! f = S'x
      JI = JI + I
      A = SIG/FORGET
      B = FJ/SIGSQ
      SIGSQ = SIGSQ + FJ * FJ                ! β + Σfj^2
      SIG = SQRT (SIGSQ)
      A = A/SIG
      G(J) = S(JI) * FJ
      S(JI) = S(JI) * A
      J1 = J-1
      IF(J1.EQ.0) GOTO 20
C
      DO10 I = 1, J1
      IJ = IJ + 1
      C = S(IJ)
```

```
   S(IJ = A * (C- B*G(I))
10 G(I) = G(I) + C * FJ
20 CONTINUE
30 IJ = IJ + 1
C
   A = E/SIGSQ
   DO 40 I = 1,N
40 THETA (I) = THETA (I) + G(I) * A     ! θ^(t) = θ^(t-1) + K(t0)e
   RETURN
   END
```

Although the square-root method requires n extractions of a square-root, it is still remarkably efficient, involving about $2n(n+4)$ multiplications per update. This compares well with even efficient coding of the RLS algorithm (naive coding involving $O(n^3)$ multiplications), so the method is to be recommended even if long word-lengths are involved. An alternative approach which does not require square-root extraction is the UD method of Thornton and Bierman (1978); see Bierman (1977) for general numerically stable methods and Clarke (1980) for FORTRAN coding of the UD method. It has been found that such methods give similar results using *single*-precision arithmetic as the standard RLS (Kalman filter) algorithm using *double*-precision arithmetic.

The great advantage of RLS algorithms is that P (or S) is available to give a guide to the variances of the parameter estimates, provided it is scaled using an estimate of the noise variance obtainable from the mean-square prediction error. P can be used by jacketing software to monitor the performance of the estimator. Sometimes, however, computational speed is vital, and simplifications are required. There are various approaches that can be taken: $K(t)$ can be replaced by a scalar gain-sequence such as α/t (as in Stochastic Approximation) or by some constant value (but which needs to be chosen with care); $P(t)$ can be approximated by a Toeplitz matrix so that computations depend on n rather than n^2; 'fast' methods (Ljung, Morf and Falconer, 1978) can be used. Nevertheless, it is probably better in such cases to retain RLS but relegate estimation to a background task.

6.4 *Offsets and Load Disturbances*

The linearised input-output behaviour of a practical process is of the form:

$$y(t) = G(s)\,u(t) + d(t) \qquad (15)$$

where $d(t)$ is a generalised disturbance term whose mean $\bar{d}$ is unlikely to be zero. An estimate of the mean can, however, be obtained using a *low-pass* filter such as a simple lag:

$$\hat{\bar{d}}(t) = \frac{d(t)}{1 + sT} \qquad (16)$$

where T is chosen to be larger than typical process time-constants. Similarly, estimates of the means $\bar{u}$ and $\bar{y}$ of the signals $u(t)$ and $y(t)$

can be generated, and the variations of u(t) and y(t) *around* their means can be obtained using *high-pass* filters of the form:

$$\tilde{u}(t) = u(t) - \hat{u}(t) \tag{17}$$

which in z-transform terms gives the filtered data $\tilde{u}$ as:

$$\tilde{u}(t) = \left(1 - \frac{(1-q)}{1-qz^{-1}} \right) u(t); \; q = \exp(-h/T) \tag{18}$$

When estimating parameters of a process model, whether in normal or in predictive form, the mean $\bar{d}$ should be taken into account, leaving $C(z^{-1})$ and feedforward terms to account for the variations in d(t). This can be done in two ways:

(i) use high-pass-filtered signals $\tilde{u}$, $\tilde{y}$ in the data-vector x(t) and in the 'output' $\phi(t)$, or

(ii) explicitly include the estimation of $\bar{d}$ by augmenting the data-vector x(t) by 1, as shown in Table 6.3. This implies that the system model (in predictor form) is:

$$Cy^*(t+k) = Fy(t) + Gu(t) + d' \tag{19}$$

Much self-tuning *regulator* theory is based on deviations of y(t) around zero, but in practice the objective is to regulate y(t) around some prespecified set-point w(t), which is often assumed to be a constant $\bar{w}$, such that there is zero (average) 'steady-state error': $\bar{y} = \bar{w}$. There are several approaches to this problem:

(i) cascade a discrete-time integrator $1/(1-z^{-1})$ with the self-tuner, as outlined in Table 6.3 and shown in Fig 6.1, and use the error signal $y(t) - \bar{w}$ in the control law in place of y(t), as in a conventional controller. The corresponding prediction model is:

$$(1-z^{-1})Cy^*(t+kt) = (1-z^{-1})Fy(t) + Gu(t) + (1-z^{-1})d'$$

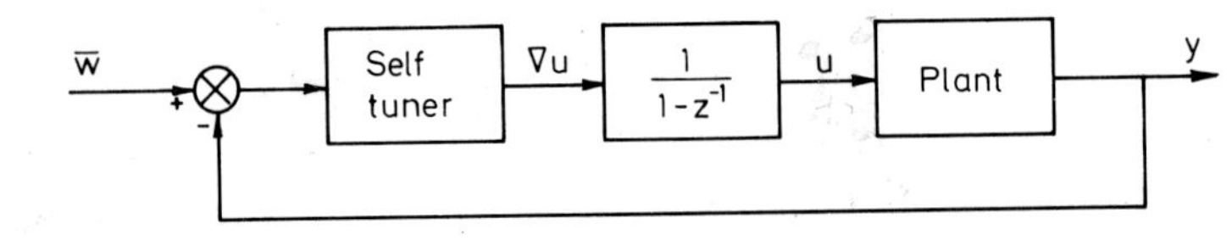

Fig.6.1 Cascade of integrator

Hence one more F parameter needs to be estimated, but as d' is supposedly constant $(1-z^{-1})d'$ is zero and so no constant term needs to be estimated. Note, however, that $C' = (1-z^{-1})C$ has a root on the unit-circle, implying that (in theory, anyway) there may be convergence difficulties. The addition of a simple integrator is in fact rather drastic (particularly as F' attempts to cancel it again!); a better transfer-function to cascade is the P+I term which in the s-domain is:

$$G_c(s) = \frac{s + \varepsilon}{s} \quad ; \quad \varepsilon \text{ small} \tag{20}$$

Here the value of ε is not particularly crucial, but should be such that $G_c(jw)$ is approximately constant over the 'important' frequency range of the plant.

(ii) recall that for the generalized minimum-variance self-tuner, the closed-loop is

$$y(t) = \frac{B}{PB + QA} w(t-k) + \frac{EB + QC}{PB + QA} \xi(t) \tag{21}$$

Now if $s(t) = \bar{w}$ and $\bar{\xi} = 0$, then $\bar{y} = \bar{w}$ if $P(1) = 1$, $Q(1) = 0$. As $1/P = M(z^{-1})$ the desired closed-loop model, choosing $M(1)$ = closed-loop gain = 1 is straightforward. $Q(1) = 0$ may be obtained by choosing $Q = (1-z^{-1})Q'$, but interpreting $L = 1/Q$ as a cascade transfer-function in predictive control, it is seen that the appropriate choice of Q is $1/G_c$, which in z-transforms is of the approximate form:

$$Q = \lambda(1-z^{-1})/(1-\alpha z^{-1}), \text{ where } \alpha \sim 1-\varepsilon \tag{22}$$

In this approach a constant d" still needs to be estimated in the prediction model for ψ^* to eliminate offset due to non-zero $\bar{d}$, giving a model:

$$C\psi^*(t+k/t) = Fy^f(t) + Gu(t) + d'' \tag{23}$$

Note, however, that this self-tuner also works for *time-varying* w(t), whereas in method (i) variations of w(t) affect the estimates of the self-tuner parameters (see Wittenmark, 1973) and so changes in the set-point involve a further period of re-tuning.

(iii) put an integrator into an *outer* loop, as shown in Fig 6.2.

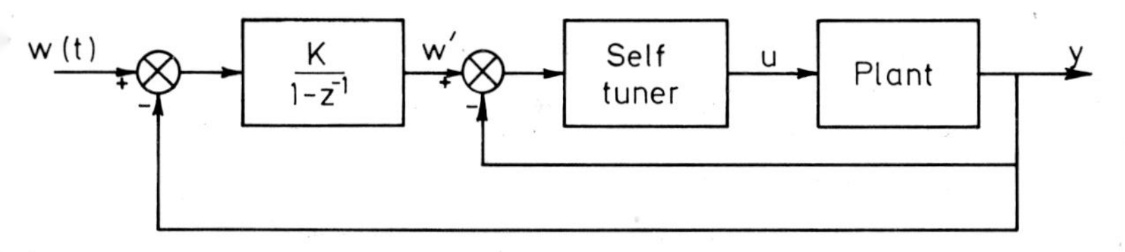

Fig.6.2 Outer-loop integration

The dc level in the inner self-tuning loop needs to be accounted for in the estimation (otherwise the F and G parameters are biased), but $w^1(t)$ is persistently exciting even if w(t) is constant. This implies that a simple regression model of the form:

$$\phi(t) = F'y^{f}(t-k) + G'\ u(t-k) + H'w^{1}(t-k) + d + \varepsilon(t) \tag{24}$$

can be used in the self-tuning in which the parameters of H' are identifiable.

6.4.1 Load Disturbances

In most cases the offset d will be time-varying, not only due to plant-variations but also due to load disturbances which cannot be accounted for by feedforward terms. This implies that the forgetting factor associated with the dc term should be smaller (i.e. faster) than that associated with the process dynamics estimation. In this case it is simpler to estimate d separately, using the prediction error of the estimated model as shown in the following:

$$\hat{d}(t) = \hat{d}(t-1) + (1-\beta^{1})e(t-1);\ \beta^{1} = \text{dc forgetting factor} \tag{25}$$

$$\phi^{1}(t) = \phi(t) - \hat{d}(t) = x'(t)\theta + \varepsilon(t);\ \text{regression model} \tag{26}$$

$$e(t) = \phi^{1}(t) - x'(t)\ \hat{\theta}(t);\ \text{prediction error} \tag{27}$$

6.5 *Integrity*

Just as with PID regulators, there will be occasions when a self-tuner will fail, although the reasons for and the manner in which a self-tuner fails will be in general more complex. Experience has shown how to minimise PID failures (such as Ziegler-Nichol's tuning rules), and practical experiments have also suggested how (to some extent) a self-tuner can be made more reliable. As the self-tuner is intended to give 'superior' control performance its integrity is of great interest (and should be the central question of self-tuning research), for on the other hand a low-gain proportional controller is generally robust against process changes but usually gives poor control. In some loops absolute reliability is crucial, which may mean that a self-tuner can be used only as a commissioning aid and not trusted to operate continually.

When deriving a self-tuner, assumptions have to be made about the process and its environment. Usually these assumptions are that the process is adequately locally linearizable, that the signals are persistently exciting such that the estimator can produce a reasonable model, and that the parameters of the self-tuner have been correctly chosen. When applied to a real process, however, there may be significant non-linearities, signals may remain constant over long periods of time and incorrect parameters may have been given. A well-designed self-tuner is one in which there is a robust control algorithm, jacketing software to ensure reliable estimation, and such that it is relatively insensitive to incorrect parameters. It must, moreover, be easy to use by non-experts and have a designed closed-loop performance which is based on engineering intuition.

6.5.2 Controller Integrity

The control design procedure (assuming a correct model) should be relatively insensitive to the assumptions made about the process, provided it is not at the cost of poor performance. For example, the minimum output-variance design is highly sensitive to process zeros and should not be used. Some explicit methods can, in principle, deal with processes with varying time-delay but involve a design calculation which itself might be numerically unsound. In fact, it is a matter of current research as to which method (if any) is uniformly best.

Nonlinearities, particularly in the actuator, can be a serious problem in both the control and the estimation phases. For example, a valve with a dead-band may induce limit cycles, and difficulties can be experienced with split-range valves (e.g. steam over part of the range and water over another part) as there is often both a dead-band and a change of gain in the different ranges. One solution to this problem is to have a simple PI loop controlling the valve output such as flow whose set-point is the output of the self-tuner; this has the effect of linearizing the valve characteristics. Similarly a highly non-linear process can be (partially) linearized by interposing some approximate static inverse characteristic in the forward path, so that the overall system is more readily controlled by the self-tuner. Provided these precautions are taken, control integrity should not be significantly worse than with fixed-parameter controllers, though this again is a matter of current research.

In some cases, control action reaching prespecified limits is an indication of potential failure. It may be necessary to have a simple back-up controller, and the one used for start-up of the self-tuner can be adopted. The limits themselves can be time-varying, and the controls of two PID regulators with, e.g., high and low gains can be taken in some cases.

6.5.2 Estimation Integrity

There are many causes of estimator failure, and it is important that they are dealt with by appropriate software, as bad parameters naturally lead to bad control. Some remedies modify the estimator while retaining estimation, some simply freeze the updating of parameters so that the control law is fixed, whilst in the worst cases a default control is brought in and the tuning has to be restarted from scratch (the philosophy of 'graceful degradation'). Some of the causes of (and solutions to) failure are:

(i) a signal is not 'persistently exciting', implying that the P matrix (proportional to the parameter covariances) increases. If it is a feedforward signal that is constant the appropriate set of parameters can be frozen and only a subset estimated. If, however, it is the input of the output signal (possibly due to control saturation or transducer failure) the overall estimation should be stopped.

(ii) the signals, though seemingly persistently exciting, are not in fact perturbing the plant. For example, if there is a dead-zone in the actuator or quantization levels in the transducer the estimated model between $u(t)$ and measured $y(t)$ does not reflect the true process dynamics. This can be overcome by choosing appropriate 'gap-action' levels within which the estimation is (temporarily) frozen. For example, if $y(t)$ is 'close' to the set-point estimation can be stopped.

(iii) the process is strongly nonlinear and subjected to load disturbances which move y(t) 'rapidly' away from the set-point (e.g. in pH control), the process model being radically changed. This is again (partially) overcome by having bounds *outside* which estimation is frozen.

(iv) the process is rapidly time-varying. Case (iii) can be interpreted in this light, but often (as in aerospace) variations can be such that no estimator can properly track the parameters. Some form of parameter scheduling, with the estimator 'trimming' the values is called for here.

Tests can be made based on the P matrix and the model's prediction error. The relative size of diagonal terms in P indicate how individual parameters are significant, and some norm of P (such as the trace) can be used as an overall guide to estimation quality. With a forgetting factor β of 1 (i.e. no forgetting) $|P|$ should not increase, but $\beta < 1$ (i.e. the normal case as it allows for tracking of slowly varying parameters) can allow $|P|$ to increase if the data is poor. One possibility (Fortescue et al, 1979) is to have a *variable* β which tends to 1 if the variance of the prediction error is below some bound, but this does not in itself solve all the above problems.

In some cases estimation integrity may not matter, particularly if the associated (short) periods of poor control can be tolerated. For example, in simulation studies step variations in gain or time constant of 10 - 100 have been tried (see, e.g., Lam (1980)) without difficulty. If integrity is important, limits should be chosen such that estimation is *only* performed when all signals are guaranteed to be exciting, and this would be at the cost of infrequent tuning. In the worst case, estimation can be undertaken only on the command of a commissioning engineer and then suitably prescribed test-signals can be generated.

6.6 *Choice of Parameters*

To give its best performance, a self-tuner needs to be 'tuned' by the appropriate choice of parameters. It could be argued, however, that most of these parameters are relatively easy to choose and that the closed-loop control is reasonably insensitive to their values. Some parameters depend on the control design method involved, so the discussion below concentrates on the generalized minimum-variance self-tuner.

6.6.1 System Parameters

(i) the model order n is often arbitrary in practice, and depends on the frequency range of interest and hence on the controller sample-time. The number of parameters to be estimated increases with n, whereas too low a value implies a sub-optimal steady-state controller, and a value of 2 is often adequate. (Too high a value in explicit self-tuners may lead to uncontrollable common modes).

(ii) the integer delay k depends on the sample-time h and the process delay Δ. In fact if there is an appreciable delay (such as greater than $\frac{1}{2}$ the dominant process time-constant), h is itself often chosen so that k is 2 or 3 as the number of parameters (and data storage) to estimate increases with k. To avoid control problems with some self-tuners kh should be chosen to exceed Δ (slightly), but if Δ varies appreciably k can either be estimated (Kurz, 1979) or explicit self-tuner should be considered.

Typically, therefore, 2 y parameters and 2-4 u parameters would be estimated.

6.6.2 Controller parameters

(i) the desired closed-loop model $M(z^{-1}) = 1/P(z^{-1})$. This should have an order approximately that of the system, and dynamics related to those of the system. A closed-loop much 'faster' than the open-loop process implies tight control with relatively large signals, whereas if the closed-loop is approximately of the same speed as the open-loop, control is loose (and presumably 'robust'). $M(z^{-1})$ is best chosen as the discretization of a specified $M(s)$.

(ii) control weighting being the discretization of $Q(s) - \lambda L(s)$, where $L(s) - (s+\varepsilon)/s$ is to ensure zero steady-state error. The value of λ should be set to 0 initially and varies later by the user to modify the closed-loop performance without affecting parameter estimates. If there are large system changes a value of λ in the form $\hat{g}_o\lambda'$ may give more consistent results.

(iii) the sample-time h is a critical parameter; for the control action it should be about 1/10 of the dominant plant time-constant or such that k is not too large to avoid non-minimum phase discrete-time models. This is less important with hybrid self-tuning where h is chosen so that high-order dynamics can be ignored, but too small a value would lead to both numerical problems (all poles migrating to 1 + j0) and to excessive control signals.

6.6.3 Estimator Parameters

A self-tuner is best started by having a (low-gain) initial controller or manual control in operation, running through a preliminary estimation phase with the possible injection of test-signals, and then closing the loop with the self-tuned law.

(i) the initial parameters $\hat{\theta}(0)$ should correspond to the low-gain law (saved as a default law for emergencies) but modified by the initial estimation phase.

(ii) the initial value of the P (or S) matrix is usually of the form αI where α is large (10, say) for rapid tuning. For batch processes the initial estimation phase is important, and both $P(0)$ and $\hat{\theta}(0)$ should best correspond to values obtained in the previous batch. $P(0)$ itself can be used in the testing for later estimation integrity.

(iii) the forgetting-factor β should be chosen on the basis of the expected variations in system parameters, a value near 1 (0.999, say) implying 'slow' variations whereas a smaller value (0.95, say) implying 'fast' variations. Recall that with a non-linear system frequent set-point changes leads to rapid changes in the linearized model. It may be appropriate to use a value of 0.95 initially and a value of 0.999 after tuning is complete to allow for slow parameter changes.

(iv) any parameter can be fixed by making the corresponding entry in P(0) zero. One approach is to use the initial estimation phase to obtain rough estimates to that the fixed parameter may be appropriately chosen. In some self-tuners this may not be necessary.

(v) various dead-zones for the estimator can be chosen based on expected characteristics of actuators, transducers, etc. A safe method which however curtails self-tuning is to have large estimation dead-zones.

6.7 *A Practical Example*

To verify the utility of self-tuning and the robustness of a typical algorithm to the assumptions made about the process, it was felt that practical industrial trials were crucial. To this end a portable control system was constructed, based on the Intel 8080 microprocessor and standard industrial circuit boards, so that experiments could be set up quickly on remote sites. Fig 6.3 shows the basic configuration of the self-tuner, SESAME. Although the first microprocessor-based self-tuner (Clarke, Cope and Gawthrop, 1975) was programmed using a macro-assembly language and hence only required about 2½ K bytes of ROM and ½ K of RAM, a higher-level language (Control Basic - Clarke and Frost, 1979) was used for later experiments with SESAME as on-site modifications to the program could then be made. Similarly, a floppy-disc system was added so that data could be logged for later analysis.

In view of the likely dynamic range of parameters, floating-point working is necessary with self-tuning, but a short word-length (16 bits fractional part) was found to be satisfactory provided that a robust estimator was used. Hence calculations could be made reasonably quickly, involving an overall computation time of 3 down to ½ second per sample-time for about 10 estimated parameters, depending on whether an interpretive language was used or not.

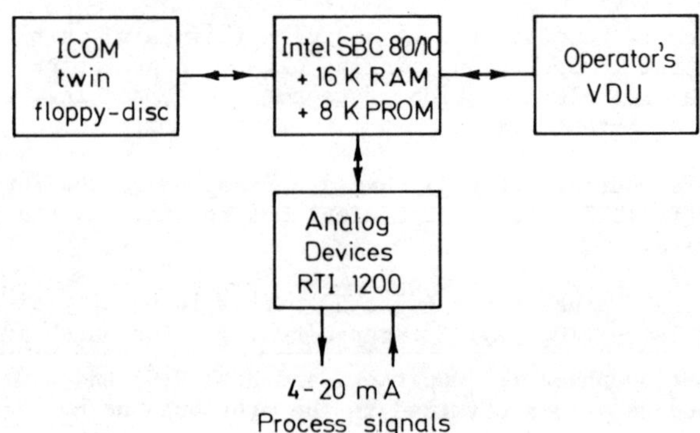

Fig 6.3 SESAME - a portable self-tuner

6.7.1 The Process

One investigated process was the batch chemical reactor shown in Fig 6.4. The stirred-tank reactor is surrounded by a jacket which can either be heated by steam or cooled by water, the control being actuated by a split-range valve. The production cycle is as follows: the reactor is filled with reagents and heated by pressurized steam to boil off any water, the lid is closed and the temperature rises towards the desired operating temperature. A few degrees *below* the set-point a strong exothermic reaction starts so that cooling must be initiated to avoid serious overshoot. As the reaction progresses less heat is generated as there is a decreasing amount of residual reagents, so the cooling effect of the jacket must be reduced in order to maintain the temperature. If the batch temperature is allowed to fall too much, the reaction period must be considerably increased to ensure product quality.

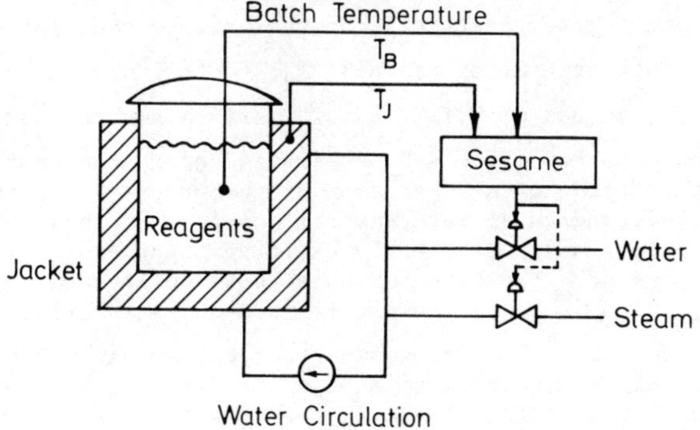

Fig 6.4 Control of a batch reactor

Exothermic (and endothermic) reactions can be difficult to control, as the heat generated (or lost) by the reaction varies exponentially with temperature. Furthermore, heat-transfer characteristic may vary significantly during the cycle. Often a PID regulator can be used, but it may require expensive tuning, either by using a mathematical model or during a series of production runs. If the same reactor is used for a range of products PID may become impracticable, but the problem seems a potential candidate for self-tuning.

6.7.2 The Experiments

The control objective and the self-tuner's parameters were chosen partly on the basis of the existing control scheme. This method involved the use of a pre-computed jacket temperature profile which was designed to remove the 'expected' heat generated by the exotherm, leaving to a PI regulator the task of trimming off variations due to material quality, etc. It was found that this technique was sensitive to the temperature at which cooling was initiated, with changes as small as half a degree causing large changes in the actual profile. It was seen (and is verifiable theoretically) that the exotherm causes a stable system to become increasingly unstable, so a self-tuner control objective was of the model-reference type, in which the desired closed-loop model was that of the open-loop process *without* exotherm. Studying the initial heating response, before the exotherm is active, showed that (not surprisingly) a second-order dynamic model with some dead-time gave a reasonable approximation, so the closed-loop was chosen as:

$$M(s) = \frac{1}{(1 + sT)^2}$$

with the value of T corresponding to the open-loop dynamics. Again, inspection of the response to the initial cooling signal suggested a delay of 2 minutes, so a sample-time of one minute and a k of 2 were chosen.

In the initial trial the self-tuner output directly controlled the split-range valves, and the algorithm was sequenced on the basis of batch temperature T_B. During heat-up, full steam was applied; data was logged after reaching temperature T_1 so that arrays could be filled; estimation was initiated after reaching T_2; control was exercised after reaching T_3. The values of T_1, T_2 and T_3 were chosen so that a 'reasonable' process model could be estimated before the exotherm was active and with regard to the operation of the process. In the event this trial failed, and there were several possible reasons. One was that the lack of a truly persistently exciting input caused an invalid model to be estimated. The cure for this is to apply a test signal from T_1 to T_3, and this was successfully used on later trials on a different test reactor. The second was that the actuator characteristics were grossly non-linear, with major differences between the heating and cooling effects. The cure for this is to apply local feedback around the actuator, and this was done with the reactor described here.

In the second series of trials, therefore, a PI regulator was placed in cascade with the self-tuner whose measured output was jacket temperature, the output of the self-tuner now corresponding to desired jacket temperature. The tuning of this loop was neither difficult nor critical; the integrator required desaturation and the value of the 'desired jacket temperature' was recomputed on actuator saturation and handed back to the self-tuner to avoid estimation errors. As variations in the control signal during heat-up were not allowed for production reasons, the $\bar{g}$ parameter was fixed at a value obtained during preliminary experiments, though later simulations showed that the value was not critical.

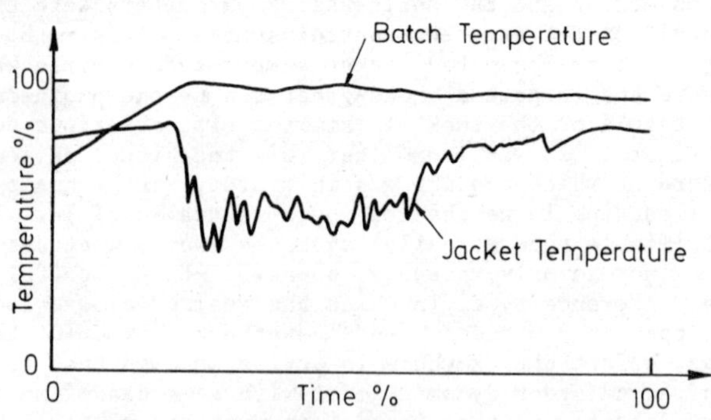

Fig 6.5 A typical self-tuning run

Fig 6.5 shows the temperature profiles for a typical self-tuning run. It is seen that the required batch temperature is maintained using a jacket temperature profile which is of the shape predicted from simple theoretical considerations.

Fig 6.6 shows an atypical run in which the quality of the reagents is poor, possibly due to contamination by water. The heat of reaction is less than usual, so the jacket temperature follows a distinctly different profile yet maintaining good batch temperature control.

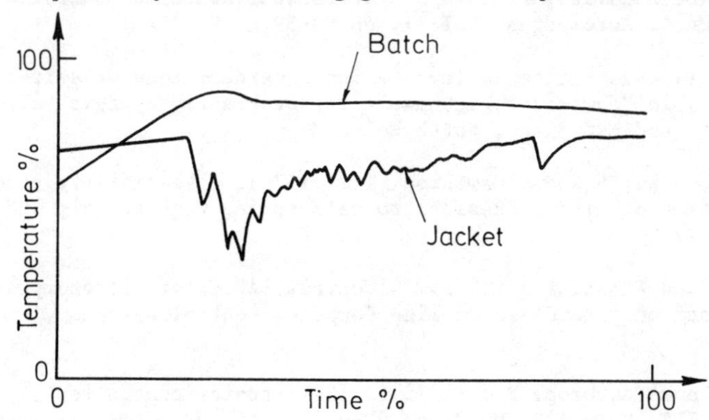

Fig 6.6 An atypical run

Both these cases show that self-tuning can be successfully applied to the control of batch reactors. In the event, as the reactor was designed for only one reaction a fixed parameter controller could be used *provided* that its parameters are carefully chosen. The utility of self-tuners is for reactors with a range of products or where the quality of the reagents varies significantly. In such cases, a fixed parameter controller may not be adequate.

6.8 *Conclusions*

This chapter has described how self-tuning controllers can be implemented, showing that practical considerations dictate the use of carefully conceived 'jacketing' software. The example of its application to the control of chemical batch reactors, together with the increasing number of other reported applications, shows that the self-tuner is rapidly maturing into a viable control method for industry.

Acknowledgments

The author would like to thank the Science Research Council for funds supporting the construction and use of SESAME, and Warren Springs Laboratory for their support of the development of the Control Basic interpreter.

The batch reactor trials form part of a continuing programme for investigating the application of self-tuners ably conducted by Dr P Gawthrop, P Brown and A Hodgson.

References

Åström, K J and Wittenmark, B (1973). 'On self-tuning regulators', Automatica, Vol 9, pp 185-199.

Bierman, G J (1977). 'Factorization Methods for Discrete System Estimation', Academic Press.

Borrison, U and Syding, R (1976). 'Self-tuning control of an ore-crusher', Automatica, Vol 12, pp 1-7.

Cegrell, T and Hedqvist, T (1975). 'Successful adaptive control of paper machines', Automatica, Vol 11, pp 53-59.

Clarke, D W (1980). 'Some implementation considerations of self-tuning controllers', in "Numerical Techniques for Stochastic Systems" edited by F Archetti and M Cugiani, North Holland.

Clarke, D W, Cope, S N and Gawthrop, P J (1975). 'Feasibility study of the application of microprocessors to self-tuning regulators', OUEL report, 1137/75.

Clarke, D W and Frost, P J (1979). 'Control BASIC for microcomputers', IEE Conference on "Trends in On-line Computer Control Systems", Sheffield.

Clarke, D W and Gawthrop, P J (1975). 'Self-tuning controller', Proceedings IEE, Vol 122, No 9, pp 929-934.

Clarke, D W and Gawthrop, P J (1979). 'Self-tuning control', Proceedings IEE, Vol 126, No 6, pp 633-640.

Dumont, G A and Bélanger, P R (1978). 'Self-tuning control of a titanium-dioxide kiln', IEEE Trans Aut Control, Vol AC-23, No 4, pp 588-592.

Edmunds, J M (1976). 'Digital adaptive pole-shifting regulators', PhD Thesis, UMIST.

Fortescue, T R, Kirshenbaum, L S and Ydstie, B E (1979). 'Implementation of self-tuning regulators with variable forgetting factors', Internal report, Dept of Chem Eng, Imperial College.

Jensel, L and Hansel, R (1974). 'Computer control of an enthalpy exchanger', Report 7417, Lund Institute of Technology.

Kallström, C G, Åström, K J, Thorel, N E, Eriksson, J and Sten, L (1977). 'Adaptive autopilots for steering of large tankers', Report LUTFD2/(TFRT-3145)/1-66/(1977), Lund Institute of Technology.

Kalman, R E (1958). 'Design of a self-optimising control system', Transactions ASME, pp 468-478.

Keviczky, L, Hetthessy, J, Hilger, M and Kolostori, J (1978). 'Self-tuning adaptive control of cement raw material blending', Automatica, Vol 14, pp 525-532.

Kučera, V (1979). 'Discrete Linear Control', Wiley.

Kurz, H (1979). 'Digital parameter-adaptive control of processes with unknown constant or varying dead time', IFAC Symposium on "Identification and System Parameter Estimation", Darmstadt.

Lam, K P (1980. 'Implicit and explicit self-tuning controllers', D Phil Thesis, Oxford University.

Ljung, L, Morf, F and Falconer, D (1978). 'Fast calculation of gain matrices for recursive estimation techniques', Int J of Control, Vol 27, pp 1-50.

Peterka, V (1975). 'A square-root filter for real-time multivariable regression', Kybernetika, Vol 11, pp 53-67.

Sastry, V A, Seborg, D E and Wood, R K (1977). 'A self-tuning regulator applied to a binary distillation column', Automatica, Vol 13, No 4, pp 417-424.

Thornton, C L and Bierman, G J (1978). 'Filtering and error analysis via the UDU covariance factorization', IEEE Trans Aut Control, Vol AC-23, No 5, pp 901-907.

Wittenmark, B (1973). 'A self-tuning regulator', Report 7311, Lund Institute of Technology.

Chapter Seven

Systematic design of discrete model reference adaptive systems

H UNBEHAUEN

7.1 *Introduction*

The most popular methods of designing adaptive control systems are based upon either the theory of self-tuning regulators or model reference adaptive systems. A schematic representation of these two approaches is shown in Fig 7.1.

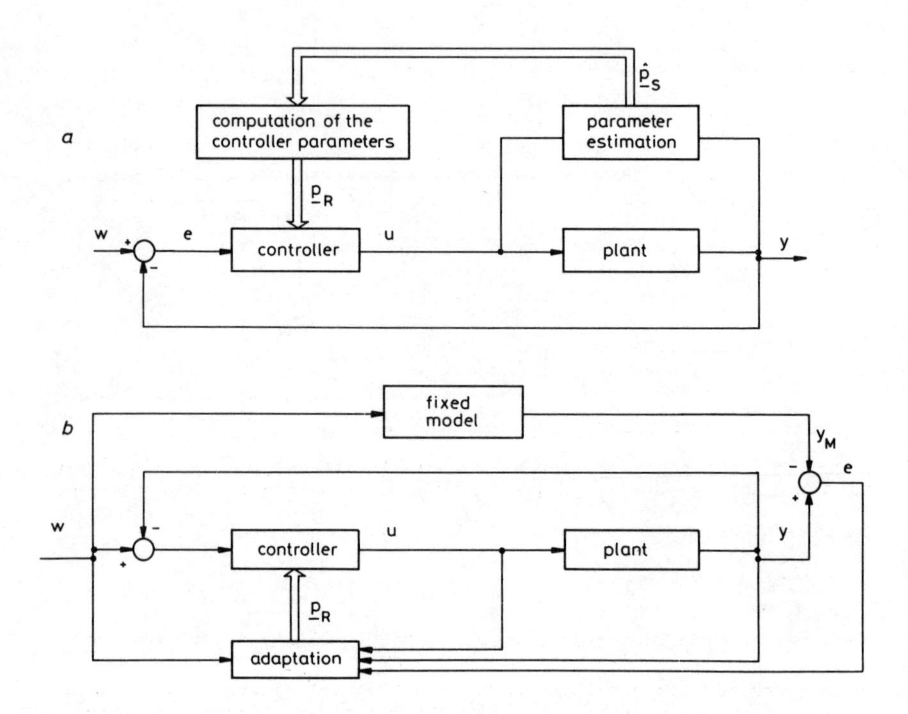

Fig 7.1 Control system using (a) a self-tuning regulator and (b) a model reference adaptive system

Professor Unbehauen is with the Department of Electrical Engineering, Ruhr University, Bochum.

The principle of self-tuning was originally introduced by Åström and Wittenmark (1973) and extended by Clarke (1975). It is based on a combination of a parameter estimation algorithm and an appropriate control scheme. The identification of the plant and the controller action can therefore be separated. According to Landau (1973a) model reference adaptive systems cannot be classified by such a general structure since the design may involve a comparison between the adaptive reference model and the plant only (explicit identification, implicit control) or between the reference model and a control loop with an adaptive controller (implicit identification, explicit control). In this chapter a fixed reference model of the dynamic behaviour of the control loop (explicit control) is assumed. Such adaptive systems are usually designed from a deterministic point of view.

Although self-tuning and model reference techniques are based on different design principles, it has recently been shown (Ljung (1978), Landau (1979a), Egardt (1979), Schmid (1979) and Narendra (1980)), that both control schemes are very similar and in some special cases even identical. Both design techniques result in nonlinear control schemes, which make it necessary to clarify the following problems:

- the stability of the whole system,

- the convergence of the controller parameters and

- the influence of disturbances.

The objective of this chapter is to outline a systematic design procedure for discrete model reference adaptive control systems, and to investigate the solution of the above problems. The systematic technique was originally developed by Schmid (1979), and can be applied to stable plants of arbitrarily high order and dead time for any reference model of the closed loop system. An important aspect of the proposed design allows the characteristic dynamics of the error signal driving the adaptation of the controller parameters, to be chosen arbitrarily fast to ensure that the adaptive control scheme allows a fast computation of the plant input.

7.2 *Some fundamentals of stability theory*

Consider the adaptive control scheme illustrated in Fig 7.1b where the plant output y is compared with the model output y_M, and the controller parameters are ideally adjusted such that the error

$$e(k) = y(k) - y_M(k) \tag{1}$$

converges to zero independently of the actual plant parameters. The adaptive control scheme should be designed so as to guarantee the globally asymptotic stability of the equilibrium point e = 0 of the error difference equation. To solve this problem, Lyapunov's second method or the hyperstability method of Popov (1973) can be applied. Whereas Lyapunov's second method requires the choice of an appropriate Lyapunov function, which may be difficult, the hyperstability method offers a systematic solution to the stability problem. Only the hyperstability approach will be considered in this chapter.

Adaptive systems can often be transformed to the standard non-linear feedback structure shown in Fig 7.2, with a linear feedforward path and non-linear feedback path. The linear subsystem can be described by the pulse transfer function

$$G(z) = \underline{c}^T(z\underline{I}-\underline{A})^{-1}\underline{b} + d^*, \tag{2}$$

or the discrete state space equations

$$\underline{x}(k+1) = \underline{A}\underline{x}(k) + \underline{b}\, u_1(k) \tag{3a}$$

$$e(k) = \underline{c}^T\underline{x}(k) + d^*u_1(k) \ . \tag{3b}$$

The nonlinear time-variant subsystem is represented by

$$y_2(k) = -f[e(\ell),k] \text{ with } \ell = 0,1,\ldots,k. \tag{4}$$

This system is asymptotically hyperstable for $\mu=0$, if G(z) is strictly positive real (s.p.r.) and if the nonlinearity fulfils the inequality (Landau, (1979b))

$$\eta(0,k_1 = \sum_{k=0}^{k_1} y_2(k)e(k) \geq -\gamma_o^2 \ , \ \forall\ k_1 \geq 0, \tag{5}$$

where γ_o represents a constant dependent on the initial conditions.

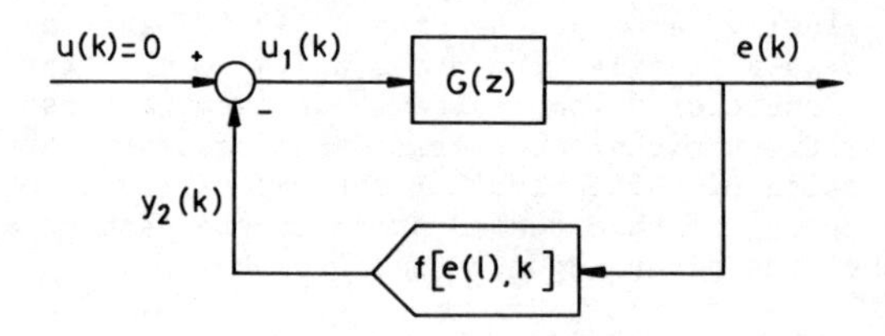

Fig 7.2 Standard structure of a nonlinear time-variant feedback system

Strictly positive real transfer functions are characterized by the following conditions:

(i) all poles of G(z) are situated inside the unit circle of the z-plane,

(ii) $Re\{G(z)\}>0$ for $\forall\ |z|= 1$.

Example

Fig 7.3 shows the parameter domains for the transfer function

$$G(z) = \frac{1}{1 + a_1 z^{-1} + a_2 z^{-2} + a_e z^{-3}} \qquad (6)$$

in the a_1, a_2-plane, in which for different values of a_3 the transfer function is stable and strictly positive real. The stability regions are limited by the dotted triangles, whereas the continuous lines inside these triangles indicate the limits of the hyperstability region in which the transfer function is strictly positive real, i.e.

$$\text{Re } G(z) > 0 \qquad \forall \ |z| = 1. \qquad (7)$$

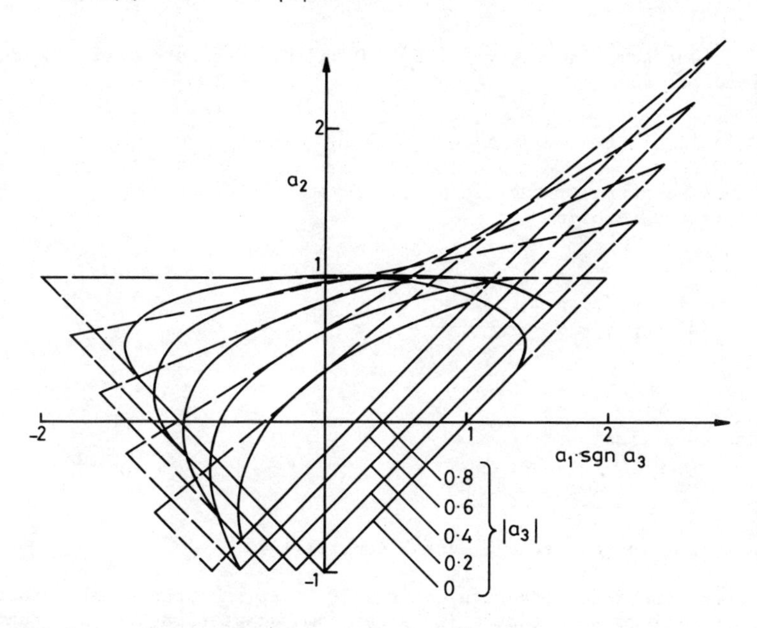

Fig 7.3 Region of stability (----) and hyperstability (——) for linear systems up to 3rd order

7.3 *The hyperstability design*

In any design of a model reference adaptive system three requirements should be considered:

(i) the state variables of the plant are (in nearly all cases) unavailable for measurement and therefore are not used for control,

(ii) the stability of the overall system should be guaranteed,

(iii) the adaptive controller will often be implemented on a process computer, and the design should therefore be considered in discrete time.

Starting from a continuous mathematical model of the plant of arbitrary (fixed) order n and dead time, T_t,

$$G_S(s) = \frac{B_S(s)}{A_S(s)} e^{-T_t s} \tag{8}$$

this can be transformed to discrete time

$$G_S(z) = \frac{Y(z)}{U(z)} = \frac{\sum_{i=0}^{m} b_i z^{-i}}{1 + \sum_{i=1}^{n} a_i z^{-i}} z^{-d} , \tag{9}$$

where T is the sampling interval, $b_o \neq 0$ and the discrete dead time d is defined by

$$(d-1)T \leq T_t < dT \qquad d > 0. \tag{10}$$

The desired response of the closed loop system is described by the parallel reference model

$$G_M(z) = \frac{Y_M(z)}{W(z)} = \frac{\sum_{i=0}^{q} b_{Mi} z^{-i}}{1 + \sum_{i=1}^{r} a_{Mi} z^{-i}} z^{-d_M} , \tag{11}$$

where the conditions

$$q \leq r \leq n \qquad d_M \geq d \tag{12}$$

must be fulfilled to ensure a causal control law.

The function of the adaptive controller is to make the plant output y(k) follow the model output $y_M(k)$ after some adaptation time, so that the error

$$e(k) = y(k) - y_M(k) \tag{13}$$

converges to zero:

$$\lim_{k\to\infty} e(k) = 0 \tag{14}$$

as rapidly as possible. To simplify the notation the polynomials in z^{-1} will be denoted as

$$P[n] = \sum_{i=0}^{n} p_i z^{-i} \quad \text{with } p_o \neq 0. \tag{15}$$

For $p_o=0$, throughout the corresponding polynomial of order n is defined by $P[n-1]z^{-1}$. With this notation equations (9) and (11) can be expressed

$$G_S(z) = \frac{Y}{U} = \frac{b_o + B[m-1]z^{-1}}{1 + A[n-1]z^{-1}} z^{-d} \tag{16}$$

and

$$G_M(z) = \frac{Y_M}{W} = \frac{B_M[q]}{1 + A_M[r-1]z^{-1}} z^{-d_M} . \tag{17}$$

Taking the z-transform of equation (13) and substituting $Y(z)$ and $Y_M(z)$ from equations (16) and (17) yields the *error equation*

$$(1+A_M[r-1]z^{-1})E = b_o z^{-d}U+B[m-1]z^{-d-1}U+\Delta A[n-1]z^{-1}Y-B_M[q]z^{-d_M}W = U_1 \tag{18}$$

with the abbreviation

$$\Delta A[n-1] = A_M[r-1] - A[n-1] . \tag{19}$$

In order to reach the equilibrium point $e(k) = 0$ for $k \geq 0$, the right hand side of equation (18) denoted with U_1 must vanish, i.e. $u_1(k) \equiv 0$. If b_o, B_M and ΔA are known, equation (18) can be solved for to U,

$$U = \frac{1}{b_o} \{B_M[q]z^{d-d_M}W - B[m-1]z^{-1}U - \Delta A[n-1]z^{d-1}Y\} , \tag{20}$$

to yield a possible control law. Unfortunately for $d>1$ this control law is not causal, because future values of y must be used. To avoid this, additional filters of order d-1 must be introduced, and the signals U and Y in equation (18) must be replaced by the filtered versions

$$Y_D = \frac{1}{D[d-1]} Y \tag{21}$$

$$U_D = \frac{1}{D[d-1]} z^{1-d} U \tag{22}$$

where

$$D[d-1] = \sum_{i-0}^{d-1} d_i z^{-i} . \tag{23}$$

The error equation can now be expressed

$$(1+A_M[r-1]z^{-1})E = U_1 \tag{24a}$$

where

$$U_1 = b_o d_o z^{-1} U_D + B_D[m+d-2]z^{-2}U_D + \Delta A_D[n-1]z^{-d}Y_D - B_M[q]z^{-d_M}W \equiv 0. \tag{24b}$$

The solution of this equation represents a possible *causal control law* in terms of the unknown plant parameters.

$$U_D = \frac{1}{b_o d_o} \{B_M[q]z^{1-d_M} W - B_D[m+d-2]z^{-1}U_D - \Delta A_D[n-1]z^{1-d} Y_D\} \quad (25)$$

The inverse z-transform of equation (24) yields

$$u_1(k) = b_o d_o u_D(k-1) + \underline{p}_D^T \underline{x}_D(k-d) = 0 \quad (26)$$

where the *controller parameter vector*

$$\underline{p}_D = [\underline{b}_D, \Delta\underline{a}_D, \underline{b}_M]^T , \quad (27)$$

contains the coefficients of the polynomials B_D, ΔA_D and B_M and the *signal vector* $\underline{x}_D$ is given by

$$\underline{x}_D(k-d) = [\underline{u}_D(k-2), \underline{y}_D(k-d), -\underline{w}(k-d_M)]^T \quad (28)$$

Since the elements of $\underline{b}_D$ and $\Delta\underline{a}_D$ in the parameter vector $\underline{p}_D$ are unknown and may be time-variant, a new vector $\hat{\underline{p}}_D$ of controller parameters is introduced, which in the adapted case (e=0) corresponds to the vector $\underline{p}_D/b_o d_o$. The *adaptive control law*

$$u_D(k) = - \hat{\underline{p}}_D^T(k)\ \underline{x}_D(k-d+1) \quad (29)$$

follows directly from equation (26). The parameter vector $\hat{\underline{p}}_D(k)$ is obviously dependent on the error d(k) and should therefore be expressed as

$$\hat{\underline{p}}_D(k) = \underline{f}_D[e(\ell), k] \quad \text{for } \ell \le k. \quad (30)$$

Inserting equations (29) and (30) into equation (26), and using equation (18) the *difference equation of the error*

$$\underbrace{\mathcal{Z}^{-1}\{(1+A_M[r-1]z^{-1})E\} = u_1(k)}_{\text{I}} = \underbrace{-\{b_o d_o \underline{f}_D[e(\ell),k-1]-\underline{p}_D\}^T \underline{x}_D(k-d) = -y_2(k)}_{\text{II}} \quad (31)$$

is obtained for $\ell<k$. Equation (31) describes a feedback system with both a linear and a nonlinear time-variant subsystem, as shown in Fig 7.4. This is equivalent to the canonical structure of the non-linear system introduced in section 7.2, and can be interpreted as a transformation of the original adaptive system into an "error system". The conditions for stability are invariant under this transformation. From the analysis given above the adaptive system of Fig 7.4 is asymptotically hyperstable, if the following conditions are valid:

$$\sum_{k=0}^{k_1} y_2(k)\ e(k) \ge -\gamma_0^2 , \ \forall\ k_1 \ge 0 \quad (32)$$

$$G_D(z) = \frac{E}{U_1} = \frac{1}{1 + A_M[r-1]z^{-1}} \quad \text{s.p.r.} \tag{33}$$

For a low model order r the transfer function $G_D(z)$ is normally strictly positive real (s.p.r.).

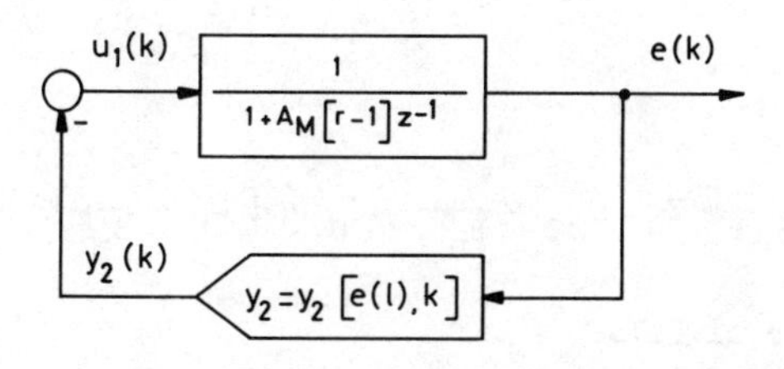

Figure 7.4 Block diagram of the transformed adaptive control system

As shown in Fig 7.3, the domain of hyperstability is almost as large as the domain of stability for low orders and the choice of the parameters of $G_D(z)$ is not restricted. However, additional steps must be considered to ensure that $G_D(z)$ is strictly positive real for high model orders and this is discussed in the following section.

7.4 *Design for an arbitrary model behaviour*

To ensure that the transfer function $G_D(z)$ in equation (33) is strictly positive real for any stable polynomial of the deonominator, an appropriate "stabilising polynomial" must be introduced into the numerator. Using this procedure equation (33) is modified to yield

$$G_H(z) = \frac{B_H[r]}{1 + A_M[r-1]z^{-1}} \quad \text{(s.p.r.)} . \tag{34}$$

The introduction of $B_H[r]$ requires a further transformation of the error difference equation, equation (31), and following Ionescu (1977) this can be achieved by defining a new augmented error signal

$$v(k) = e[v(\ell),k] + h[v(k)] , \quad \ell \le k-1. \tag{35}$$

The signal h is obtained by filtering the auxiliary signal $u_H(k)=f_H[v(k)]$ using

$$H = G_H(z)\ U_H .$$

Dividing equation (24) by $B_H[r]$, adding the z-transformed auxiliary signal U_H to both sides and using

$$\frac{V}{G_H} = \frac{E}{G_H} + U_H \tag{37}$$

from equations (35) and (36) yields the transformed error equation

$$\frac{1+A_M[r-1]z^{-1}}{B_H[r]} V = b_o d_o z^{-1} \frac{U_D}{B_H[r]} + B_D[m+d-2]z^{-2}U_{DH} + \Delta A_D[n-1]z^{-d}Y_{DH} -$$

$$- B_M[q]z^{-d_M} W_H + U_H , \tag{38}$$

where

$$Y_{DH} = \frac{1}{B_H[r]} Y_D, \ U_{DH} = \frac{1}{B_H[r]} U_D \text{ and } W_H = \frac{1}{B_H[r]} W \tag{39a-c}$$

are the filtered signals.

Substituting

$$\frac{U_D}{B_H[r]} = \frac{U_D}{b_{Ho}} - B^*[r-1]z^{-1}U_{DH} \tag{40}$$

in the first term on the right-hand side of equation (38) and summarizing the polynomials connected with U_{DH} yields

$$B_{DH}[m+d-2] = b_o d_o B^*[r-1] + B_D[m+d-2] \tag{41}$$

and hence from equation (38)

$$\frac{1+A_M[r-1]z^{-1}}{B_H[r]} V = \frac{b_o d_o}{b_{Ho}} z^{-1}U_D + B_{DH}[m+d-2]z^{-2}U_{DH} + \Delta A_D[n-1]z^{-d}Y_{DH} -$$

$$- B_M[q]z^{-d_M} W_H + U_H \tag{42}$$

Introducing a modified version of the vectors of equations (27) and (28)

$$\underline{p}_{DH} = [\underline{b}_{DH} \mid \Delta\underline{a}_D \mid \underline{b}_M]^T \tag{43}$$

and

$$\mathcal{Z}\{\underline{x}_{DH}(k)\} = \frac{1}{B_H[r]} \mathcal{Z}\{\underline{x}_D(k)\}, \tag{44}$$

it follows from equation (42) that

$$\frac{1+A_M[r-1]z^{-1}}{B_H[r]} V = \mathcal{Z}\{h_o u_D(k-1) + \underline{p}_{DH}^T \underline{x}_{DH}(k-d) + u_H(k)\} = U_{H1} , \tag{45}$$

where

$$h_o = \frac{b_o d_o}{b_{Ho}} \tag{46}$$

and

$$U_{H1} = \frac{1}{B_H[r]} U_1 + U_H \ . \tag{47}$$

The adaptive control signal in equation (45) is determined using the modified vector $\hat{\underline{p}}_{DH}^T$ of the controller parameters by defining

$$u_D(k) = \hat{\underline{p}}_{DH}^T(k)\ \underline{x}_{DH}(k-d-1) - \mu(k) \tag{48}$$

where $\mu_H(k)$ is an auxiliary signal to calculate $\mu_H(k)$. Introducing v(k) the requirement for asymptotical stability of the error in equation (14) gives

$$\lim_{k\to\infty} v(k) = \lim_{k\to\infty} h(k) = \lim_{k\to\infty} u_H(k) = \lim_{k\to\infty} \mu_H(k) = 0. \tag{49}$$

The adaptive law can now be formulated as a function of the augmented error signal v:

$$\hat{\underline{p}}_{DH}(k) = \underline{f}_{DH}[v(\ell),\ k] \text{ for } \ell \le k. \tag{50}$$

Inserting equations (48) and (50) into equation (45), the modified difference equation of the error is obtained after inverse z-transformation as

$$\mathcal{Z}^{-1}\{\frac{1+A_M[r-1]z^{-1}}{B_H[r]}\ V\} = u_{H1}(k) \tag{51}$$

with

$$u_{H1}(k) = -\{h_o\underline{f}_{DH}[v(\ell),\ k-1] - \underline{p}_{DH}\}^T\underline{x}_{DH}(k-d) - h_o\mu_H(k-1) + u_H(k) = -y_{H2}(k) \tag{52}$$

for $\ell \le k-1$.

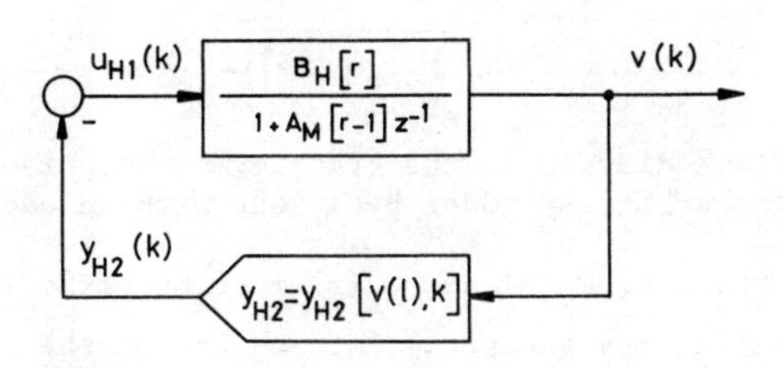

Fig 7.5 Block diagram of the transformed adaptive control system for arbitrary model behaviour.

It is easy to show that the equations (51) and (52) describe a time-variant nonlinear feedback system, which can be represented by Fig 7.5. The remaining step in this analysis is the derivation of an appropriate adaptation law according to equation (50). Before considering this in detail, methods of improving the rate of convergence of the adaptation will be discussed.

7.5 *Improving the convergence properties of the design*

The characteristic behaviour of the error signal v(k) in Fig 7.4 is dependent upon the poles of the model transfer function $G_M(z)$. In the presence of disturbances it is, however, desirable that the characteristic behaviour of v(k) becomes independent from the model behaviour. This can be achieved by an additional linear feedback. Because the characteristic behaviour of v(k) influences the rate of convergence of the adaptation, it should be made as fast as possible.

If an additional linear feedback is introduced into the system, as illustrated in Fig 7.6, the transfer function of the linear subsystem becomes

$$G_K(z) = \frac{V}{U_{H1}} = \frac{B_H[r]}{1 + A_H[r-1]z^{-1}} \tag{53}$$

where

$$A_H[r-1] = A_M[r-1] - A_K\ r-1 \ . \tag{54}$$

and an appropriate pole assignment can be made by selecting $A_H[r-1]$

$$A_H[r-1] = A_M[r-1] - A_K[r-1]$$

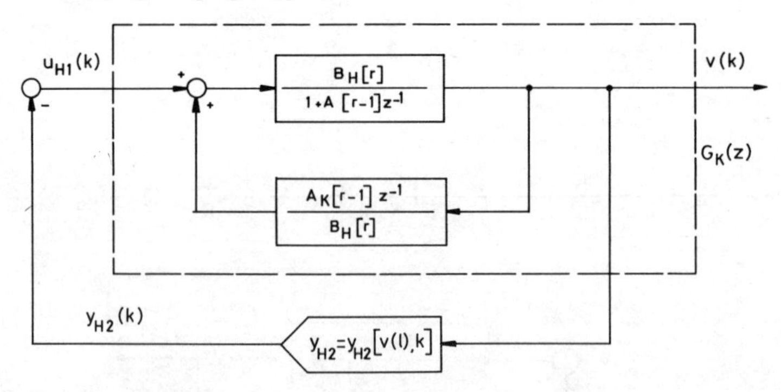

Fig 7.6 Block diagram of the transformed adaptive control system for arbitrary model behaviour with an additional feedback

Notice that instead of $G_H(z)$ now $G_K(z)$ must be strictly positive real. This can be realized by an appropriate choice of the numerator polynomial $B_H[r]$. The additionally introduced feedback in Fig 7.6 does not influence the hyperstability of the overall system, provided that the signal $u_H(k)$, is chosen appropriately. Using equations (45) and (47) the system without additional feedback is characterised by

$$(1 + A_M[r-1]z^{-1})\ V = U_1 + B_H[r]U_H \ . \tag{55}$$

The corresponding system with additional feedback is given by

$$(1 + A_H[r-1]z^{-1})\ V = U_1 + B_H[r]U_K \ . \tag{56}$$

If the signal U_H is determined by

$$U_H = U_K + \frac{A_K[r-1]z^{-1}}{B_H[r]} V \ , \tag{57}$$

then both systems, equations (55) and (56), are identical. The signal $u_K(k)$ represents a nonlinear function

$$u_K(k) = f_K[v(k)] \ , \tag{58}$$

and will be discussed later.

Summarising the steps of the design, the overall system is asymptotically hyperstable, if the following two conditions are valid:

$$\sum_{k=o}^{k_1} y_{H2(k)}\ v(k) \geq -\gamma_0^2 \ ; \ \forall\ k_1 \geq 0 \tag{59}$$

and

$$G_K(z) = \frac{V}{U_{H1}} = \frac{B_H[r]}{1 + A_H[r-1]z^{-1}} \quad \text{s.p.r.} \tag{60}$$

The only remaining problem involves the nonlinear functions $\underline{f}_{DH}$ and $\underline{f}_K$ of equations (50) and (58) which must be determined to ensure that the inequality in equation (59) is satisfied.

7.6 *The general adaptive control law*

Following Landau (1973b) the adaptive control law is selected to have a proprotional plus integral (PI) structure

$$\hat{\underline{p}}_{DH}(k) = \underline{f}_{DH}[v(\ell),k] = \underbrace{\underline{g}^P \otimes \underline{\psi}(k)}_{\text{P-term}} + \underbrace{\underline{g}^I \otimes \sum_{\ell=o}^{k} \underline{\psi}(\ell) + \hat{p}_{DH}^I(0)}_{\text{I-term}} \tag{61}$$

where

$$\underline{\psi}(k) = v(k)\ \underline{x}_{DH}(k-d) \ , \tag{62}$$

satisfies the inequality (59). The operator $\otimes$ defines the multiplication element by element of two vectors, and

$\underline{g}^P$ and $\underline{g}^I$ represent gain factors for the signals of the vector $\underline{\psi}(k)$.

The form of the nonlinear function $u_K = f_K[v(k)]$ in equation (58) is chosen by

$$u_K(k) = \hat{p}_o(k-1)\ \mu_H(k-1) - q(k)\ v(k) \tag{63}$$

where

$$q(k) = [(\underline{g}^I+\underline{q}) \otimes \underline{x}_{DH}(k-d)]^T \underline{x}_{DH}(k-d) + (g_o^I+q_o)\mu_H^2(k-1)\ . \tag{64}$$

A PI structure is also generated for the adaption of the parameter $\hat{p}_o$,

$$\hat{p}_o(k) = \underbrace{g_o^P\ \psi_o(k)}_{\text{P-term}} + \underbrace{g_o^I \sum_{\ell=o}^{k} \psi_o(\ell) + \hat{p}_o^I(0)}_{\text{I-term}}\ , \tag{65}$$

where

$$\psi_o(k) = -v(k)\ \mu_H(k-1)\ . \tag{66}$$

It has been shown by Schmid (1979), that the PI adaptation law described by equations (61) and (65) are asymptotically hyperstable, providing the following conditions for the gain vectors and gain factors are satisfied

$$\underline{g}^P \geq \underline{0},\ \underline{g}^I > \underline{0},\ g_o^P \geq 0,\ g_o^I > 0, \tag{67a-d}$$

$$\underline{q} \geq h_o\ \underline{g}^P + (\frac{h_o}{2} - 1)\ \underline{g}^I, \tag{67e}$$

$$q_o \geq g_o^P - \frac{1}{2}\ g_o^I. \tag{67f}$$

A maximum value of h_o in condition (67e) must be estimated using equation (46) and the known maximal value of b_o and d_o and b_{Ho}. The numerical calculation can be simplified by transforming equations (61) and (65) into an appropriate recursive form to yield the adaptive law

$$\hat{\underline{p}}_{DH}(k) = \underline{g}^P \otimes \underline{\psi}(k) + \hat{\underline{p}}_{DH}^I(k) \tag{68a}$$

where

$$\hat{\underline{p}}_{DH}^I(k) = \hat{\underline{p}}_{DH}^I(k-1) + \underline{g}^I \otimes \underline{\psi}(k) \tag{68b}$$

and

$$\hat{p}_o(k) = g_o^P\ \psi_o(k) + \hat{p}_o^I(k) \tag{69a}$$

where

$$p_o^I(k) = p_o^I(k-1) + g_o^I\ \psi_o(k)\ . \tag{69b}$$

Inspection of equation (66) shows that the augmented error signal v(k) must be available before the recursion in equation (69b) can be computed. From equation (37) follows

$$V = E + \frac{B_H[r]}{1 + A_M[r-1]z^{-1}} U_H , \tag{70}$$

and U_H can be inserted from equation (57). Utilising equation (54) gives

$$V = E + A_H[r-1]z^{-1} E - A_H[r-1]z^{-1}V + B_H[r]U_K \tag{71}$$

or by inverse z-transformation

$$v(k) = e(k) + \sum_{i=1}^{r} [a_{Mi} e(k-i) - a_{Hi-1} v(k-i) + b_{Hi} u_K(k-i)] + $$
$$+ b_{Ho} u_K(k) . \tag{72}$$

Inserting $u_K(k)$ from equation (63) and solving for v(k) yields

$$v(k) = \frac{e(k)+b_{Ho}\hat{p}_o(k-1)\mu_H(k-1)+\sum_{i=1}^{r}[a_{Mi} e(k-i)-a_{Hi} v(k-i)+b_{Hi} u_K(k-i)]}{1 + b_{Ho} q(k)} , \tag{73}$$

where apart from the auxiliary signal μ_H all constants and variables are known. An expression for μ_H will be developed in the following section.

7.7 *The general control law*

The relationship between the filtered control signal u_D and the plant input, i.e. equation (22), can be expressed as a difference equation

$$u(k-d+1) = \sum_{i=o}^{d-1} d_i u_D(k-i) . \tag{74}$$

Substituting for u_D from equation (48) yields

$$u(k-d+1) = - \sum_{i=o}^{d-1} d_i \hat{\underline{p}}_{DH}^T(k-i)\underline{x}_{DH}(k-d+1-i) - \sum_{i=o}^{d-1} d_i \mu_H(k-i) , \tag{75}$$

where all the signals except the auxiliary signal μ_H are known. Rearranging equation (75)

$$u(k-d+1)\sum_{i=o}^{d-1} d_i\mu_H(k-i) = -\sum_{i=o}^{d-1} d_i\,\hat{\underline{p}}_{DH}^T(k-i)\,\underline{x}_{DH}(k-d+1-i),$$

and expressing the right hand side in partial sums gives

$$\underline{u(k-d+1)} + \underline{\underline{\sum_{i=o}^{d-1} d_i\mu_H(k-i)}} = \underline{-\hat{\underline{p}}_{DH}^T(k-d+1-\ell)\sum_{i=o}^{d-1} d_i\underline{x}_{DH}(k-d+1-i)} - \underline{\underline{\sum_{i=o}^{d-1} d_i[\hat{\underline{p}}_{DH}(k-i)-\hat{\underline{p}}_{DH}(k-d+1-\ell)]^T\underline{x}_{DH}(k-d+1-i)}} \tag{76}$$

where $\ell=0,1$. Equating the underlined parts for $\ell = 1$ leads to the *control law*

$$u(k) = -\hat{\underline{p}}_{DH}^T(k-1)\sum_{i=o}^{d-1} d_i\underline{x}_{DH}(k-i) = -\hat{\underline{p}}_{DH}^T(k-1)\underline{x}_{DH}^*(k), \tag{77}$$

where

$$\underline{x}_{DH}^*(k) = \mathcal{Z}^{-1}\{D[d-1]\mathcal{Z}\{\underline{x}_{DH}(k)\}\} = \mathcal{Z}^{-1}\{\frac{D[d-1]}{B_H[r]}\{\underline{x}_D(k)\}\} \tag{78}$$

and the *auxiliary signal*

$$\mu_H(k) = -\hat{\underline{p}}_{DH}^T(k)\underline{x}_{DH}(k-d+1) + \hat{\underline{p}}_{DH}^T(k-d)\underline{x}_{DH}(k-d+1)\ . \tag{79}$$

Comparison with equation (48) shows that for the filtered control signal u_D is given by

$$u_D(k) = -\underline{p}_{DH}^T(k-d)\underline{x}_{DH}(k-d+1), \tag{80}$$

although in practice $u_D(k)$ is determined by solving equation (74)

$$u_D(k) = \frac{1}{d_o}[u(k-d+1) - \sum_{i=o}^{d-1} d_i\ u_D(k-i)]\ . \tag{81}$$

Combining the general adaptation law of equations (61) and (65) with the general control law equation (77) defines the structure of the model reference adaptive system illustrated in Fig 7.7. Inspection of Fig 7.7 shows that the *basic control loop* includes the control signal u, the plant output y and in addition the model input (reference signal) w each feedback via a filter. These filtered signals which are represented by the signal vector $\underline{x}_{DH}^*$ replace the non-measurable state variables in the control law. The multiplication of these signals with the vector of controller parameters $\hat{\underline{p}}_{DH}$ yields the control signal u.

The subsystem which realizes the adaption law in Fig 7.7 can be divided into two components. The first is used to adapt the parameters

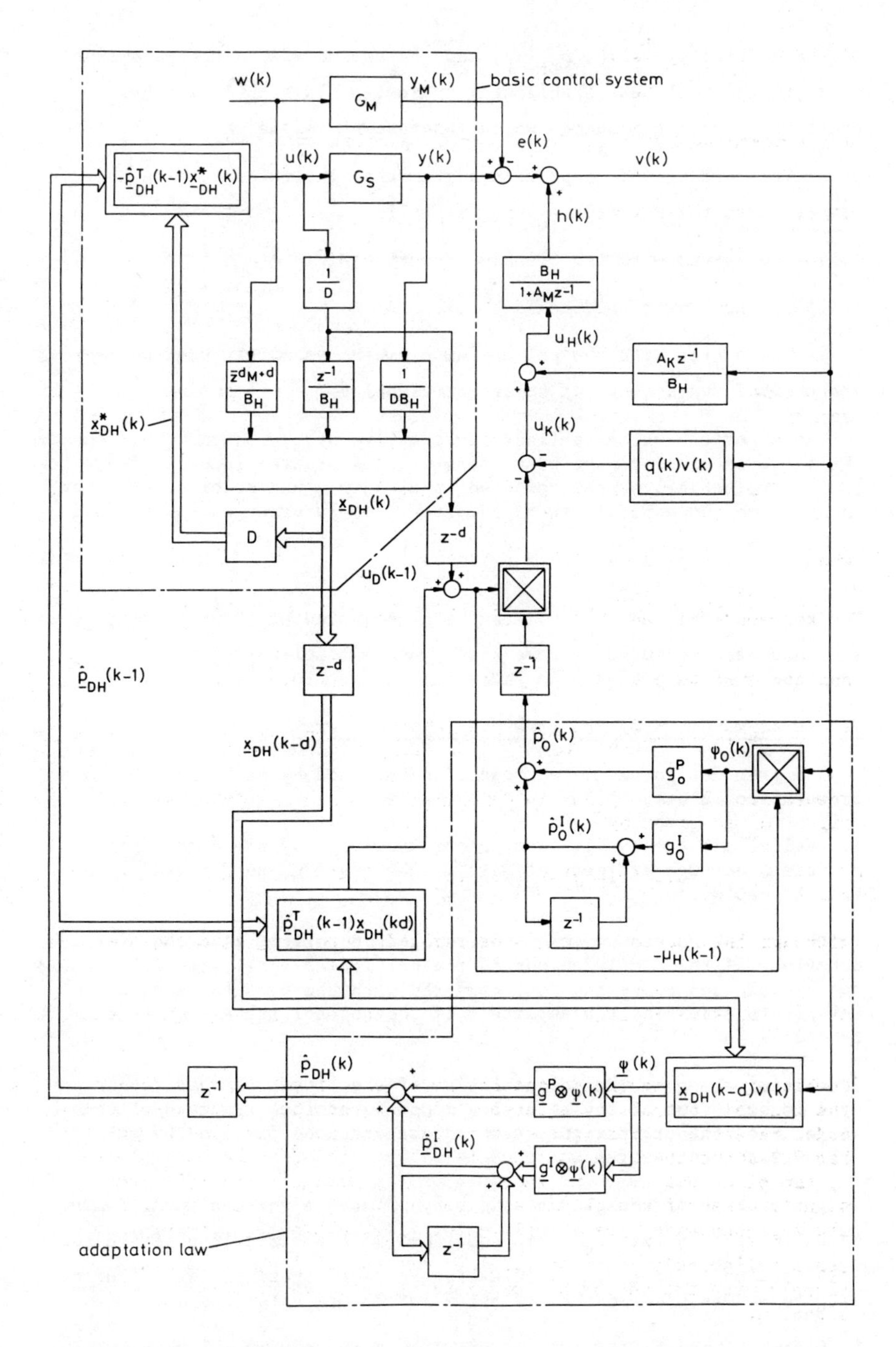

Fig 7.7 Block diagram of the complete hyperstable model reference adaptive system (non-linear blocks have double lines)

of the controller vector $\hat{\underline{p}}_{DH}(k)$, and the second adapts the parameter $p_o(k)$ in the nonlinear stabilising system. This stabilising system consists of four components which generate the signal h:

(i) The part of the adaptation system that generates $\hat{p}_o$. In the adapted case the equation $\hat{p}_o = h_o$ is valid.

(ii) The system with the transfer function $B_H/(1 + A_M z^{-1})$.

(iii) The nonlinear feedback $q(k)v(k)$.

(iv) The system with the polynomial A_K which is mainly used to improve the rate of convergence of the error signal v.

The algorithm which realizes the adaptive system can now be deduced. The sequence of single program steps will be ordered in such a way, that the control signal can be computed in a minimum number of steps. For this reason the control law of equation (77) is arranged in the form

$$u(k) = p_a(k-1)\, y(k) + p_b(k-1)\,. \tag{82}$$

The terms $p_a(k-1)$ and $p_b(k-1)$ can be precomputed at the time k-1. Once y(k) has been measured at time k only one multiplication and one addition must be executed to determine the control signal u(k).

7.8 *Determination of the free design parameters*

The free design parameters can be determined by following the guidelines given below:

1. Utilize all the available *a priori knowledge of the plant dynamics*, to obtain a rough estimate of $G_S(z)$. The order n and the dead time d must be estimated.

2. Design the *reference model* according to equation (11): the dynamic behaviour of the model depends on the particular application and should not be selected to be too fast compared with the dynamic behaviour of the plant, otherwise the amplitude of the control signal may become too large.

3. Determination of the polynomial A_K of the *linear feedback*: the dynamic behaviour of the reference model determines the rate of convergence of the error signal v which can often be improved by an appropriate choice of A_K.

4. Computation of the *stabilizing polynomial* B_H: this polynomial must be calculated so that $G_K(z)$ in equation (53) is strictly positive real. This is usually achieved by selecting the characteristic behaviour of $1/B_H$ to be faster than that of the plant or the reference model.

5. Design of the *filter polynomial* D: only polynomials with real eigenvalues

$$D[d-1] = d_o(1 + z_D z^{-1})^{d-1} \tag{83}$$

have been found to be appropriate, where d_o is selected to ensure that the transfer function $1/D$ always has a gain factor $K_D = 1$. The parameter z_D should be negative and much smaller than the magnitude of the smallest eigenvalue of the plant and the reference model.

6. Determination of the *initial values* $\hat{\underline{p}}^I_{DH}(0)$ and $\hat{p}^I_o(0)$: it is of course possible to choose any bounded values. However, all a priori information of the plant should be used to determine $\hat{\underline{p}}^I_{DH}(0)$. Set $\hat{p}^I_o(0) = h_o$.

7. Determination of the *adaptive gains* g^P_o, g^I_o, $\underline{g}^I$ and $\underline{q}$: these values have a strong influence on the dynamics of the overall system, especially the convergence of the error signal and the controller parameters. An optimal convergence behaviour, independent of the mean of the signals, is obtained if the adaptive gains are normalized by the variances of the corresponding signals, according to Schmid (1979):

$$\lambda_{u_D} = \overline{u^2_D(k)}\ , \quad \lambda_{u_{DH}} = \overline{u^2_{DH}(k)} \tag{84a,b}$$

$$\lambda_{y_{DH}} = \overline{y^2_{DH}(k)}\ , \quad \lambda_{w_H} = w^2_H(k) \tag{84c,d}$$

$$\left.\begin{array}{lll} g^x_i = g^x/\lambda_{u_D} & \text{for } i = 0 & \\ g^x_i = g^x/\lambda_{u_{DH}} & \text{for } i = 1,2,\ldots,m+d-1 & \\ g^x_i = g^x/\lambda_{y_{DH}} & \text{for } i = m+d,\ldots,m+n+d-1 & \\ g^x_i = g^x/\lambda_{w_H} & \text{for } i = m+n+d,\ldots,m+n+d+q & \end{array}\right\} \text{for } x = P,\ I$$

$$\left.\begin{array}{ll} q_i = g^q/\lambda_{u_D} & \text{for } i = 0 \\ q_i = g^q/\lambda_{u_{DH}} & \text{for } i = 1,2,\ldots,m+d-1 \\ q_i = g^q/\lambda_{y_{DH}} & \text{for } i = m+d,\ldots,m+n+d-1 \\ q_i = g^q/\lambda_{w_H} & \text{for } i = m+n+d,\ldots,m+n+d+q \end{array}\right\} \tag{85}$$

Appropriate values of g^P, g^I and g^q can be obtained by using the cost functional

$$I_e = \frac{1}{N}\sqrt{\sum_{k=0}^{N} e^2(k)} \tag{86}$$

where e(k) = 0 for k>N. Fig 7.8 shows that suitable values are

$$g^I = 1 \div 10, \tag{87a}$$

$$g^P = (0,5 \div 0,8)\ g^I, \tag{87b}$$

$$g^q = (1 \div 1,5)\ g^P. \tag{87c}$$

Notice that even small disturbances can influence the value of the cost functional, equation (86), significantly although the overall system remains stable in practice.

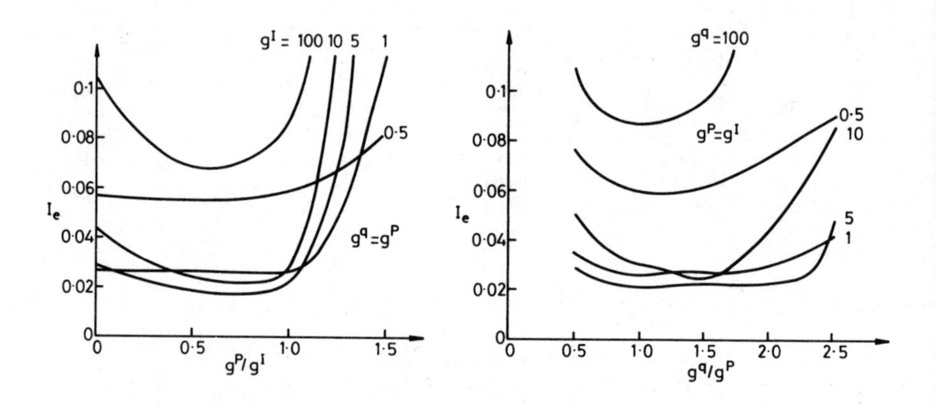

Fig 7.8 Diagrams for determining appropriate values of g^I, g^P and g^q

Although the disturbances are filtered by the integral component of the adaptive law, they bypass the proportional part. This yields a better transient response, and as a compromise the proportional gains are usually reduced in the adapted case.

7.9 *An example of the general approach*

Consider the control of a plant with time varying parameters to illustrate the algorithm described above. These plant parameter variations are described by the following three continuous transfer functions.

$$G_{S1}(s) = \frac{1+3s}{(1+s)(1+4s)}\ e^{-2s} \tag{88}$$

$$G_{S2}(s) = \frac{1+3s}{(1+s)(1+9s)} e^{-2s} \tag{89}$$

$$G_{S3}(s) = \frac{2(1+3s)}{(1+s)(1+4s)} e^{-2s} . \tag{90}$$

The design criteria specifies that the behaviour of the closed-loop system should follow the parallel reference model,

$$G_M(s) = \frac{1}{(1+s)(1+0.5s)} e^{-2s} , \tag{91}$$

independently of the plant parameter variations. The corresponding step responses are shown in Fig 7.9.

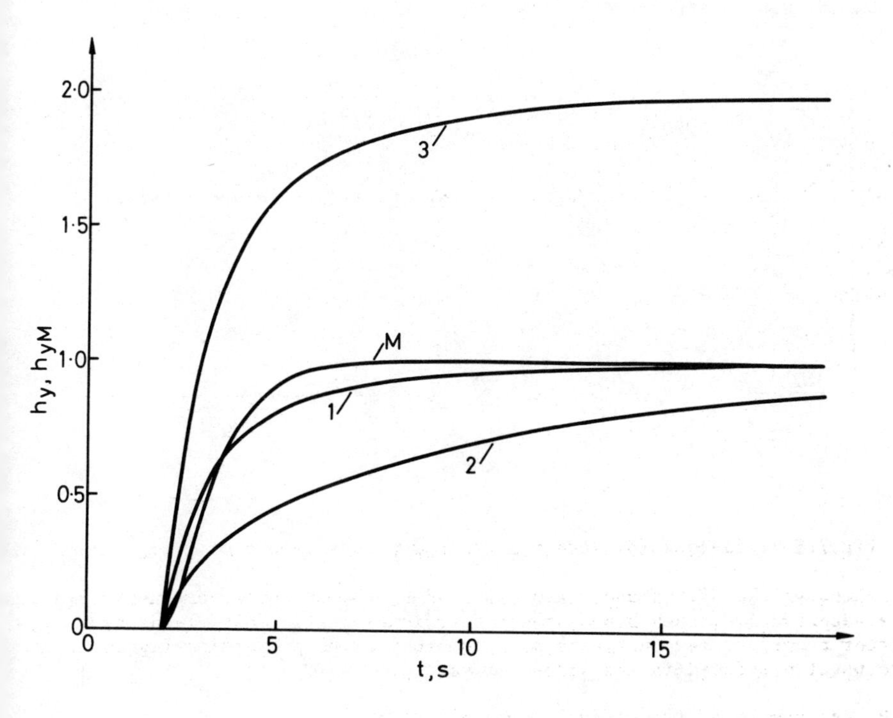

Fig 7.9 Step responses of the plant: (1) according to equation (88), (2) according to equation (89), (3) according to equation (90), and of the parallel reference model: (M) according to equation (91). (Notice: all delays and time constants are measured in seconds).

For the discrete operation of the adaptive system the sampling time T=1 sec was chosen. Thus d=3 and the filter polynomial D[d-1] equation (83) is second order, with a double pole of $z_D = -0.1$. The discrete system representation yields the transfer function

$$G_K = \frac{B_H[r]}{1 + A_H[r-1]z^{-1}} = \frac{1.83 - 0.92\, z^{-1} + 0.09\, z^{-2}}{1 - 0.50\, z^{-1} + 0.05\, z^{-2}} , \qquad (92)$$

The orders of the polynomials in equations (16) and (17) are given by

$$m = 1,\ n = 2 \text{ and } 9 = 2, \qquad (93)$$

and the gains of equations (68) and (69) are chosen from Fig 7.8 as

$$g^I = 5,\ g^P = g^q = g^I/2 = 2\ 5. \qquad (94)$$

The calculated signal variances are

$$\lambda_{u_D} = 0.0270 \qquad \lambda_{u_{DH}} = 0.0052$$

$$\lambda_{y_{DH}} = 0.0031 \qquad \lambda_{w_H} = 0.0050 .$$

Finally from equation (85) the following gain factors are obtained (with d=3, m=1, n=2 and q=1):

$$\begin{aligned} g_o^I &\approx 185 , \\ g_1^I = g_2^I = g_3^I &\approx 1000, \\ g_4^I = g_5^I &\approx 1612 , \\ g_6^I = g_7^I &\approx 963 , \end{aligned} \qquad (95)$$

and from equation (94)

$$g_i^P = q_i = g_i^I/2 \quad \text{for } i=0,1,\ldots,7. \qquad (96)$$

Hence $d_o/b_{Ho} = 0.67$, $b_{o\,max} = 1$ and from equation (46)

$$h_{o\,max} = 0.67 .$$

With this value and the above gain factors the hyperstability of the overall adaptive system can be checked, using equations (67). This yields the following results:

(a) $q_o \geq g_o^P - g_o^I/2 = g_o^I/2 - g_o^I/2 = 0,$

thus $q_o \geq 0$ is satisfied

(b) for all other coefficients q_i (i-1,2,...7)

$$q_i \geq h_{o\ max}\ g_i^P + (h_{o\ max}/2 - 1)g_i^I = - (1-h_{o\ max})g_i^I = -0.33\ g_i^I;$$

and from equations (95) and (96) $q_i \geq 0$ always such that conditions are also satisfied.

Fig 7.10 shows the time-behaviour of the adaptive control system during the time interval $0 \leq t \leq 1000$ sec with the plant output y, the control signal u, the reference signal w and the model error e. During this time interval the parameters of the plant are changed stepwise at $t \approx 60$ sec from $G_{S3}(s)$ to $G_{S2}(s)$, at $t \approx 350$ sec from $G_{S2}(s)$ to $G_{S1}(s)$, at $t \approx 540$ sec from $G_{S1}(s)$ to $G_{S2}(s)$ and at $t \approx 740$ sec from $G_{S2}(s)$ to $G_{S3}(s)$. During the same time the reference signal is also changed stepwise from w=0 to w=0.2 and visa versa. It can be clearly seen that in all the cases of parameter variation stable behaviour of the overall system is guaranteed. Furthermore the desired behaviour of the reference model is always achieved after a short adaptation time.

7.10 *Simplifications of the general approach*

The general approach, described in the previous sections, yields a relatively complicated adaptive control law, which can be simplified by considering the special features of actual plants, (e.g. plants without dead time). Other simplifications are obtained by modifying the general design procedure. In some instances this leads to well-known approaches of adaptive control which can be considered as special cases of the general approach.

7.10.1 Simplification for plants without dead time

In this case d=1 holds. Thus the filter polynomial D[d-1] is reduced to

$$D[0] = d_o = 1\ , \tag{97}$$

and therefore, no filtering of the actuating and controlled signals is necessary. Thus

$$U_D = U,\ Y_D = Y,\ B_D[m-1] = B[m-1],\ \Delta A_D[n-1] = \Delta A[n-1]\ , \tag{98}$$

and equation (77) reduces to

$$u(k) = -\underline{p}_{DH}^T(k-1)\ \underline{x}_{DH}(k)\ . \tag{99}$$

Similarly the auxiliary signal of equation (79) yields

$$\mu_H(k) = -\underline{p}_{DH}^T(k)\ \underline{x}_{DH}(k) - u(k)\ . \tag{100}$$

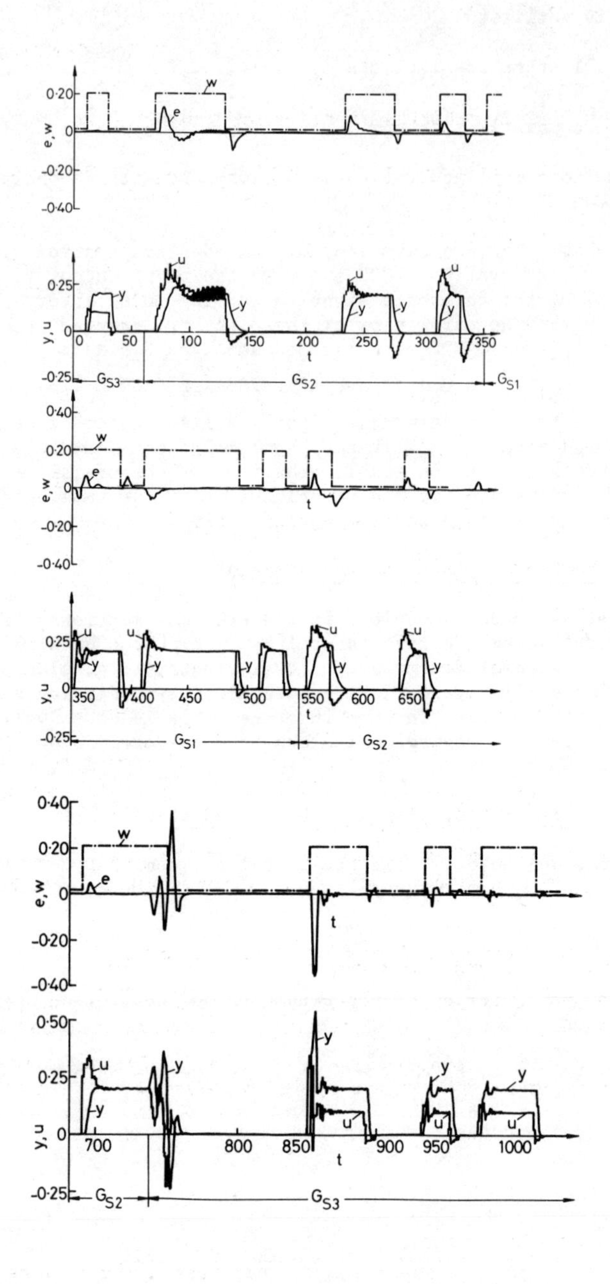

Fig 7.10 Dynamic behaviour of the adaptive control system for step changes of the plant parameters and the reference signal

It can be shown that even for $\mu_H(k) = 0$ the overall adaptive systems still remains hyperstable, and further simplifications can therefore be obtained.

7.10.2 Simplified calculation of the error signal v(k)

The error signal to equation (73) contains the nonlinear function q(k), which is calculated from equation (64) using the gain factors $\underline{g}^I$, g_o^I, $\underline{q}$ and q_o. In order to obtain hyperstable behaviour of the overall system, the inequalities (67) must be satisfied. For simplification of q(k) all elements of the gain factors in equation (64) are chosen to be equal. Thus, with a scalar K_q

$$q(k) = K_q[\underline{x}_{DH}^T(k-d)\ \underline{x}_{DH}(k-d) + \mu_H^2(k-1)] \ . \tag{101}$$

In order to guarantee hyperstability, the factor K_q must be greater than or equal maximum of the largest element of the vector $(\underline{g}^I+\underline{q})$ and the factor $(g_o^I+q_o)$. If the maximal value $h_{o\ max}$ of the parameter h_o is used in the inequalities (67e,f) then follows

$$K_q \geq \max[h_{o\ max} \max_{1\leq i\leq N}(g_i^P + \frac{1}{2}g_i^I),\ (g_o^P + \frac{1}{2}g_o^I)]. \tag{102}$$

For plants without dead time (d=1) this relation simplifies to

$$K_q \geq h_{o\ max} \max_{1\leq i\leq N}(g_i^P + \frac{1}{2}g_i^I) \ . \tag{103}$$

Although this simplified error signal reduces the realization effort considerably, the influence on the rate of convergence, will be limited.

7.10.3 Simplification by splitting the reference model

In the general approach all (q+1) coefficients of the numerator polynomial $B_M[q]$ of the parallel reference model are included in the adaptation law. This is actually not necessary. The reference model according to equation (11) can be split as follows:

$$G_M(z) = B_V[q_V]\frac{b_{Mo}\, z^{-d_M}}{1 + A_M[r-1]z^{-1}} = B_V[q_V]G_M'(z) \ , \tag{104}$$

where $B_V[q_V] = B_M[q]/b_{Mo}$, and the block diagram in Fig 7.11 results.

This system represents a special case of the general approach, when q=0, $G_M(z)$ is replaced by $G_M'(z)$, and w(k) by w'(k), respectively.

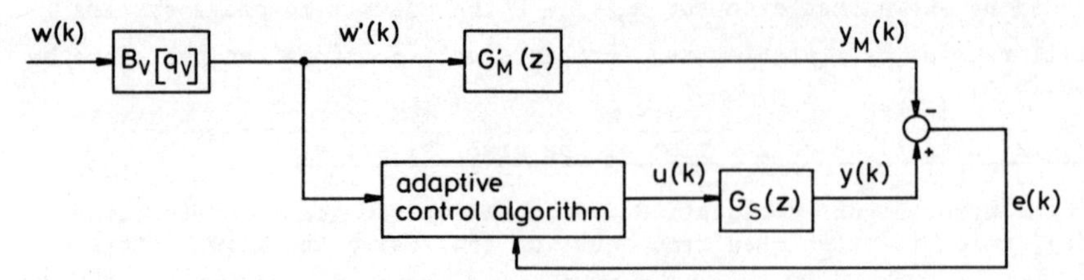

Fig 7.11 Block diagram of the basic structure of the adaptive system with split reference model

7.10.4 Simplification by introducing a series model

In the previous section it was shown that the zeros of the model transfer function $G_M(z)$ can be ignored in the design of the actual adaptive system. A further reduction of the realisation effort is possible, if the polynomials of the numerator and denominator of the parallel model are taken from the parallel path and inserted in the series path as shown in Fig 7.12,

where

$$G_V(z) = \frac{B_M[q]}{1 + A_M[r-1]z^{-1}} \tag{105}$$

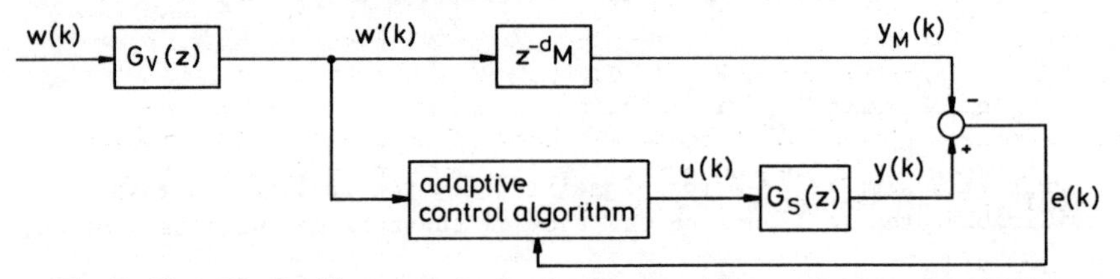

Fig 7.12 Block diagram of the basic structure of the adaptive system with model parts in the series and parallel path

represents the series model. The characteristic behaviour of the error equation remains independent of this model.

This modified system represents a special case of the general approach, when q=0 and r=0, and if $G_H(z)$ is replaced by z^{-d_M} and w(k) by w'(k) respectively.

Since

$$G_D(z) = G_H(z) = G_K(z) = 1 \tag{106}$$

is valid, asympototical hyperstability is always guaranteed. Furthermore, $B_H[r] = 1$ and $A_K[r-1] = 0$ and these simplifications reduce the realisation effort considerably. Further simplifications are obtained for systems without dead time (d=1).

7.10.5 Monopoli's (1974) "augmented error"-method as a special case

The "augmented error"-method, originally developed by Ionescu and Monopoli (1977) is based on Liapunov's second method. The error difference equations are, however, very similar to the hyperstability design. In this approach it must be assumed that the order and dead time of the plant and the reference model are the same, (i.e. r=n and $d=d_M$ and q=0) and that the compensation polynomial B_H has order n-1. With these assumptions an error difference equation can be developed from equation (53), which must be transformed into an appropriate state space representation, in order to prove the stability by the second method of Liapunov. This stability proof can be conducted only for adaptation laws with integral action, (i.e. without proportional action). Thus an adaptation law is obtained which can be considered as a special case of the general approach of equations (68) and (69 with $\underline{g}^P=0$ and $g_o^P=0$.

7.11 *Conclusions*

Based on the hyperstability theory a general design procedure has been developed for stable discrete adaptive control systems using a parallel reference model. This general approach can be applied for any model behaviour to plants of arbitrary order and dead time. The adaptive control law can be readily implemented on a process-computer, and its effectiveness has been illustrated via a simulation example. The general, relatively complicated design procedure can be simplified by considering various modifications. For plants without dead time the design effort reduces considerably. Further simplifications can be obtained by splitting the reference model into a parallel and series model. It can also be shown that well-known adaptive methods, (e.g. Monopoli's "augmented error'-method) can be considered as a special case of the general approach.

References

Åström, K J and Wittenmark, B (1973). 'On self-tuning regulators', Automatica 9, pp 185-198.

Clarke, D W and Gawthrop, P J (1975). 'Self-tuning controller', Proc IEE 122, pp 929-934.

Egardt, B (1979). 'Unification of some continuous-time adaptive control schemes', IEEE Trans on Automatic Control, AC-24, pp 588-592.

Ionescu, T and Monopoli, R V (1977). 'Discrete model reference adaptive control with an augmented error signal', Automatica 13, pp 507-517.

Kalman, R E and Bertram J E (1960). 'Control system analysis and design via the second method of Liapunov: II. Discrete-time systems', ASME J Basic Eng 82, series D, pp 394-400.

Landau, I D (1973a). 'A survey of model reference adaptive techniques (theory and applications)', Proc IFAC-Symposium on Sensitivity, Adaptivity and Optimality, Ischia, pp 15-42.

Landau, I D (1973b). 'Design of discrete model reference adaptive systems using the positivity concept', Proc IFAC-Symposium on Sensitivity, Adaptivity and Optimality, Ischia, pp 307-314.

Landau, I D (1979a). 'Model reference adaptive control and stochastic self-tuning regulators - towards cross-fertilization. Presented at AFOSR Workshop on Adaptive Control, Univ of Illinois Urbana-Champaign.

Landau, I D (1979b). 'Adaptive control - the model reference approach', Marcel Dekker Inc, New York.

Ljung, L and Landau, I D (1978). 'Model reference adaptive systems and self-tuning regulators - some connections', Proc 7th IFAC Congress, Vol 3, pp 1973-1980.

Monopoli, R V (1973). 'The Kalman-Yacubowich-Lemma in adaptive control system design', IEEE Trans on Automatic Control, AC-18, pp 527-529.

Monopoli, R V (1974). 'Model reference adaptive control with an augmented error signal', IEEE Trans on Automatic Control, AC-19, pp 474-483.

Narendra, K S and Lin, Y H (1980). 'Stable discrete adaptive control', IEEE Trans Automatic Control, AC-24, pp 456-460.

Popov, V M (1973). 'Hyperstability of control systems', Springer-Verlag, Berlin.

Schmid, Chr (1979). 'Ein Beitrag zur Realisierung adaptiver Regelungssysteme mit dem Prozeßrechner', Dissertation Ruhr-Universität Bochum.

Chapter Eight

Design of set-point tracking and disturbance rejection controllers for unknown multivariable plants

B PORTER

8.1 *Introduction*

In the case of many industrial plants, very little is known about the detailed dynamical properties of the processes involved. The design of controllers for such plants which ensure that satisfactory set-point tracking occurs simultaneously with disturbance rejection is accordingly of great practical importance. It is shown that, provided only that plants satisfy the fundamental conditions of Porter and Power (1970) and Power and Porter (1970) for the preservation of stabilisability in the presence of integral action, effective error-actuated controllers can be readily designed for open-loop asymptotically stable unknown plants. These controllers are fundamentally different from those proposed by Davison (1976) and Davison, Taylor, and Wright (1980) in that both the error between command input and plant output, together with the integral of error, are processed. Such error-actuated controllers for unknown plants frequently result in superior performance to that achievable by controllers in which only the integral of error is processed (Davison 1976; Davison, Taylor and Wright 1980) and are more generally applicable than those proposed by Pentinnen and Koivo (1980) which exist only for the restricted class of plants with first Markov parameters of maximal rank. Finally, these error-actuated controllers can be equally readily designed in either analogue or digital form.

However, in spite of such differences of structure or of ranges of applicability, all these recently developed design methodologies for controllers for unknown plants (Davison 1976; Pentinnen and Koivo 1980; Davison, Taylor, and Wright 1980; Porter 1980; and Owens and Chotai, chapter 10) constitute closely related attempts to provide effective controllers for strongly interactive multivariable plants which are based upon the common recognition (Foss 1973; Peterka and Åström 1973) that endeavours to develop accurate models for many industrial plants are often either impossible or impractical. These design methodologies are accordingly motivated by the same considerations as inspired many of the recently proposed techniques for the design of self-tuning controllers (see, for example, Åström and Wittenmark (1973); Åström, Borisson, Ljung, and Wittenmark (1977); Wellstead and Zanker (1978); Clarke and Gawthrop (1979)), but nevertheless yield fixed-structure controllers which are often simpler to implement than self-tuning

Professor B Porter is with the Department of Aeronautical and Mechanical Engineering, Salford University.

controllers. The performance achievable by self-tuning controllers may, in the event, be significantly superior to that achievable by fixed-structure controllers; but the possibility of implementing simple fixed-structure controllers for unknown plants can be appraised so very easily that it behoves control engineers to at least contemplate the use of such fixed-structure controllers when faced with complex industrial plants before resorting to self-tuning controllers or, indeed, to any class of variable-structure self-optimalising adaptive controllers.

8.2 *Analogue Controllers*

8.2.1 Analysis

The plants under consideration are assumed to be governed on the continuous-time set $T = [0,+\infty)$ by state and output equations of the respective forms

$$\dot{x} = Ax + Bu + Dd \tag{1}$$

and

$$y = Cx \tag{2}$$

where the state vector $x \in R^n$, the input vector $u \in R^m$, the output vector $y \in R^\ell$, the unmeasurable constant disturbance vector $d \in R^p$, and all the eigenvalues of the plant matrix $A \in R^{n\times n}$ lie in the open left half-plane C^-. Furthermore, it is assumed that A, B, C, D and n are *unknown* but that the steady-state transfer function matrix $G = G(0) = -CA^{-1}B$ is *known* from 'off-line' tests performed on the plant (see Appendix) where the transfer function matrix $G(\lambda) = C(\lambda I_n - A)^{-1}B$. Finally, it is assumed that the introduction of integral action preserves stabilisability and therefore that (Porter and Power 1970; Power and Porter 1970)

$$\operatorname{rank} G = \ell \tag{3}$$

This condition clearly requires that $\ell \leq m$ and that $G(\lambda)$ has no zero-valued transmission zeros.

The state equations of such plants under the action of error-actuated analogue controllers governed on $T = [0,+\infty)$ by control-law equations of the form

$$u = \alpha\varepsilon Ke + \varepsilon Kz \tag{4}$$

clearly assume the form

$$\begin{bmatrix} \dot{x} \\ \dot{z} \end{bmatrix} = \begin{bmatrix} A-\alpha\varepsilon BKC\ , & \varepsilon BK \\ -C & 0 \end{bmatrix} \begin{bmatrix} x \\ z \end{bmatrix} + \begin{bmatrix} \alpha\varepsilon BK \\ I_\ell \end{bmatrix} v + \begin{bmatrix} D \\ 0 \end{bmatrix} d \tag{5}$$

where the error vector $e = v - y \in R^\ell$, the unmeasurable constant set-point input vector $v \in R^{\ell}$, $z = \int_0^t e dt \in R^{\ell}$, the scalars α, $\varepsilon \in R^{+}$, and the controller matrix $K \in R^{m \times \ell}$. Therefore, provided only that α, ε, and K are such that all the eigenvalues of the closed-loop plant matrix in equation (5) lie in the open left half-plane C^{-}, $\lim_{t\to\infty} \dot{z} = 0$ and consequently $\lim_{t\to\infty} e = 0$ so that set-point tracking occurs simultaneously with disturbance rejection.

The closed-loop characteristic equation can readily be expressed in the form (Porter and Bradshaw 1979)

$$\phi_c(\lambda) = |S(\lambda)|\,|\lambda I_{\ell} + \varepsilon(\alpha\lambda+1)S^{-1}(\lambda)U(\lambda)K| \tag{6}$$

where $S(\lambda)$ and $U(\lambda)$ are polynomial matrices such that $G(\lambda) = S^{-1}(\lambda)U(\lambda)$ and $|S(\lambda)| = |\lambda I_n - A|$. Hence, in case

$$K = G^T(GG^T)^{-1}\Sigma = [S^{-1}(0)U(0)]^T[[S^{-1}(0)U(0)][S^{-1}(0)U(0)]^T]^{-1}\Sigma \tag{7}$$

where $\Sigma = \mathrm{diag}\,\{\sigma_1, \sigma_2, \ldots, \sigma_{\ell}\}$, $\sigma_i > 0$ $(i=1,2,\ldots,\ell)$, it follows from equations (6) and (7) that $Z_c = Z_1 \cup Z_2$ is the set of closed-loop characteristic roots where

$$Z_1 = \{\lambda \in C : |\lambda I_n - A + 0(\varepsilon)| = 0\} \tag{8}$$

and

$$Z_2 = \{\lambda \in C : |\lambda I_{\ell} + \varepsilon\Sigma + 0(\varepsilon^2)| = 0\} \tag{9}$$

These expressions indicate that, provided ε is sufficiently small, both $Z_1 \subset C^{-}$ and $Z_2 \subset C^{-}$ since $|\lambda I_n - A| = 0$ is the characteristic equation of the open-loop asymptotically stable plant and since $\varepsilon\Sigma$ is a positive diagonal matrix. The introduction of error-actuated analogue controllers governed by equations (4) and (7) accordingly ensures that set-point tracking occurs simultaneously with disturbance rejection when the gain parameter $\varepsilon \in (0, \varepsilon^*]$, where $\varepsilon^* = \varepsilon^*(\alpha, \Sigma)$ can be readily determined by simple 'on-line' tuning (Davison 1976; Davison , Taylor and Wright 1980). Furthermore, because of the presence of the additional parameter α and the additional matrix Σ, such error-actuated controllers can obviously be subjected to further tuning and therefore frequently result in superior performance to that achievable by controllers in which $\alpha = 0$ and $\Sigma = I_{\ell}$.

8.2.2 Illustrative Example

In the case of the open-loop asymptotically stable plant governed on $T = [0, +\infty)$ by the state and output equations (Davison 1976)

$$\dot{x} = \begin{bmatrix} -1 , & 0 , & 0 \\ 0 , & -2 , & 0 \\ 0 , & 0 , & -3 \end{bmatrix} x + \begin{bmatrix} 1 , & -1 \\ 1 , & 1 \\ 1 , & 0 \end{bmatrix} u + \begin{bmatrix} 3 , & -1 \\ 1 , & 5 \\ 1 , & 1 \end{bmatrix} d \qquad (10)$$

and

$$y = \begin{bmatrix} 1 , & 0 , & 0 \\ 0 , & 1 , & 0 \end{bmatrix} x \quad , \qquad (11)$$

it can be readily found by 'off-line' tests that

$$G = \begin{bmatrix} 1 & , & -1 \\ 1/2 & , & 1/2 \end{bmatrix} \qquad (12)$$

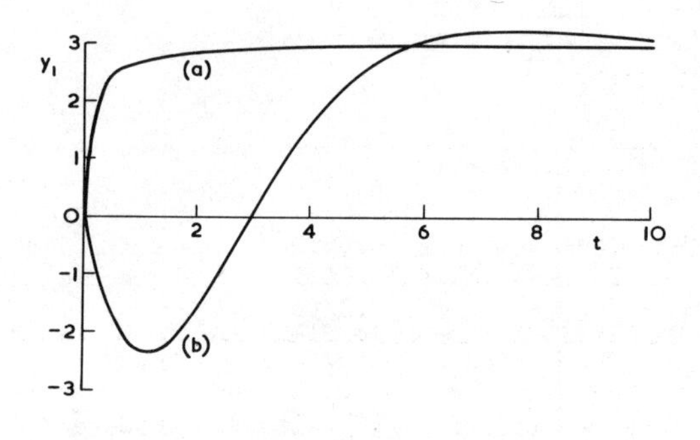

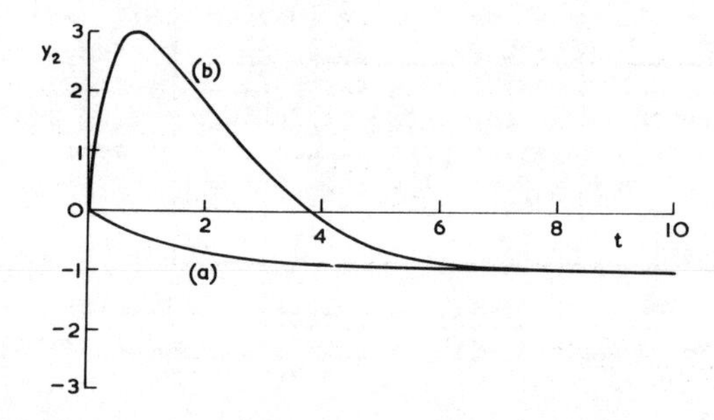

Fig 8.1 Responses of controlled plant: (a) error-actuated analogue controller with $\alpha = 1.5$, $\varepsilon = 3$, $\Sigma = \text{diag}\{2,1\}$; (b) analogue controller with $\alpha = 0$, $\varepsilon = 0.5$, $\Sigma = \text{diag}\{1,1\}$ (Davison 1976).

It is thus immediately evident that equation (3) is satisfied and therefore that, in view of equation (7), the required analogue controller matrix is given by

$$K = G^{-1}\Sigma = \begin{bmatrix} 1/2 \ , & 1 \\ -1/2 \ , & 1 \end{bmatrix}\Sigma = \begin{bmatrix} \sigma_1/2 \ , & \sigma_2 \\ -\sigma_1/2 \ , & \sigma_2 \end{bmatrix} \ . \qquad (13)$$

The behaviour of this initially quiescent plant under the action of the error-actuated analogue controller governed by equation (4) when $\alpha = 1.5$, $\varepsilon = 3$, $\Sigma = \text{diag}\ \{2,1\}$, $d = [-1\ ,\ 2]^T$, and $v = [3\ ,\ -1]^T$ is shown in Fig 8.1, and it is evident that the response consists of a fast non-oscillatory approach to the desired steady output $y = [3\ ,\ -1]^T$. This behaviour is obviously superior to the much slower oscillatory response of the same plant under the action of a controller with the fundamentally different structure corresponding to $\alpha = 0$ (Davison 1976) which is shown also in Fig 8.1 for purposes of comparison. The sets Z_c of closed-loop characteristic roots when $\alpha = 1.5$ and $\Sigma = \text{diag}\ \{2,1\}$ are presented in Table 8.1 when $\varepsilon = 0.01$, 0.1, and 3.0, and it is clear that $Z_c \subset \mathcal{C}^-$ in each case. Furthermore, in consonance with equations (8) and (9), it is evident from Table 8.1 that $Z_1 = \{-1+0(\varepsilon), -2+0(\varepsilon), -3+0(\varepsilon)\}$ and $Z_2 = \{-\varepsilon+0(\varepsilon^2), -2\varepsilon+0(\varepsilon^2)\}$.

Table 8.1

Gain Parameter ε	Characteristic Roots Z_c
0.01	-1.0102 -2.0201 -3.0000 -0.0099 -0.0198
0.1	-1.1217 -2.2095 -3.0000 -0.0905 -0.1783
3.0	-9.3589 -10.4244 -3.0000 -0.5756 -0.6411

8.3 *Digital Controllers*

8.3.1 Analysis

In order to design digital error-actuated controllers for unknown plants governed on the continuous-time set $T = [0,+\infty)$ by state and output equations of the respective forms (1) and (2), it is convenient to consider the behaviour of such plants on the discrete-time set $T_T = \{0,T,2T,...\}$. This behaviour is governed by state and output equations of the form (Kwakernaak and Sivan 1972)

$$x_{k+1} = \Phi x_k + \Psi u_k + \Delta d_k \tag{14}$$

and

$$y_k = \Gamma x_k \tag{15}$$

where $x_k = x(kT) \in R^n$, $u_k = u(kT) \in R^m$, $y_k = y(kT) \in R^\ell$,

$$\Phi = \exp(AT) \quad , \tag{16}$$

$$\Psi = \int_0^T \exp(At)Bdt \quad , \tag{17}$$

$$\Delta = \int_0^T \exp(At)Ddt \quad , \tag{18}$$

$$\Gamma = C \quad , \tag{19}$$

and T is the sampling period.

The state equations of such plants under the action of error-actuated digital controllers governed on $T_T = \{0,T,2T,...\}$ by control-law equations of the form

$$u_k = \tilde{\alpha}T\tilde{K}e_k + T\tilde{K}z_k \tag{20}$$

clearly assume the form

$$\begin{bmatrix} x_{k+1} \\ z_{k+1} \end{bmatrix} = \begin{bmatrix} \Phi-\tilde{\alpha}T\Psi\tilde{K}\Gamma , & T\Psi\tilde{K} \\ -T\Gamma , & I_\ell \end{bmatrix} \begin{bmatrix} x_k \\ z_k \end{bmatrix} + \begin{bmatrix} \tilde{\alpha}T\Psi\tilde{K} \\ I_\ell \end{bmatrix} v_k + \begin{bmatrix} \Delta \\ 0 \end{bmatrix} d_k \tag{21}$$

where $e_k = e(kT) = v_k - y_k \in R^\ell$, $v_k = v(kT) \in R^\ell$, $z_k = z(kT) \in R^\ell$, and $z_{k+1} = z_k + Te_k$. Therefore, provided only that $\tilde{\alpha}$, T, and $\tilde{K}$ are such that all the eigenvalues of the closed-loop plant matrix in equation

(21) lie in the open unit disc D^-, $\lim_{k\to\infty} \Delta z_k = 0$ and consequently $\lim_{k\to\infty} e_k = 0$ so that set-point tracking occurs simultaneously with disturbance rejection.

The closed-loop characteristic equation can readily be expressed in the form (Bradshaw and Porter 1980)

$$\psi_c(\mu) = |\tilde{S}(\mu,T)|\,|\mu I_\ell + T(\tilde{\alpha}\mu+1)\tilde{S}^{-1}(\mu,T)\tilde{U}(\mu,T)\tilde{K}| \qquad (22)$$

where $\mu = (\lambda-1)/T$, $\lim_{T\to 0} \tilde{U}(\mu,T) = U(\mu)$, $\lim_{T\to 0} \tilde{S}(\mu,T) = S(\mu)$, and $G(\mu) = C(\mu I_n - A)^{-1}B = S^{-1}(\mu)U(\mu)$. Hence, in case

$$K = G^T(GG^T)^{-1}\tilde{\Sigma} = [S^{-1}(0)U(0)]^T[[S^{-1}(0)U(0)][S^{-1}(0)U(0)]^T]^{-1}\tilde{\Sigma} \qquad (23)$$

where $\tilde{\Sigma} = \text{diag}\,\{\tilde{\sigma}_1, \tilde{\sigma}_2, \ldots, \tilde{\sigma}_\ell\}$, $\tilde{\sigma}_i > 0$ $(i=1, 2, \ldots, \ell)$, it follows from equations (22) and (23) that $\tilde{Z}_c = \tilde{Z}_1 \cup \tilde{Z}_2$ is the set of closed-loop characteristic roots where

$$\tilde{Z}_1 = \{\lambda \in C : |\lambda I_n - I_n - TA + 0(T^2)| = 0\} \qquad (24)$$

and

$$\tilde{Z}_2 = \{\lambda \in C : |\lambda I_\ell - I_\ell + T^2\tilde{\Sigma} + 0(T^3)| = 0\} \qquad (25)$$

These expressions indicate that, provided T is sufficiently small, both $\tilde{Z}_1 \subset D^-$ and $\tilde{Z}_2 \subset D^-$ since $\{\lambda \in C : |\lambda I_n - A| = 0\} \subset C^-$ because the open-loop plant is asymptotically stable on the continuous-time set T and since $T^2\tilde{\Sigma}$ is a positive diagonal matrix. The introduction of error-actuated digital controllers governed by equations (20) and (23) accordingly ensures that set-point tracking occurs simultaneously with disturbance rejection when the sampling period $T \in (0,T^*]$, where $T^* = T^*(\tilde{\alpha},\tilde{\Sigma})$ can be readily determined by simple 'on-line' turning. Furthermore, because of the presence of the additional parameter $\tilde{\alpha}$ and the additional matrix $\tilde{\Sigma}$, such error-actuated controllers can obviously be subjected to further tuning and therefore frequently result in superior performance to that achievable by controllers in which $\tilde{\alpha} = 0$ and $\tilde{\Sigma} = I_\ell$.

8.3.2 Illustrative Example

In the case of the plant governed on $T = [0,+\infty)$ by the state and output equations (10) and (11), it follows from equations (12) and (23) that the required digital controller matrix is given by

$$\tilde{K} = G^{-1}\tilde{\Sigma} = \begin{bmatrix} 1/2 \ , & 1 \\ -1/2 \ , & 1 \end{bmatrix} \tilde{\Sigma} = \begin{bmatrix} \tilde{\sigma}_1/2 \ , & \tilde{\sigma}_2 \\ -\tilde{\sigma}_1/2 \ , & \tilde{\sigma}_2 \end{bmatrix} \tag{26}$$

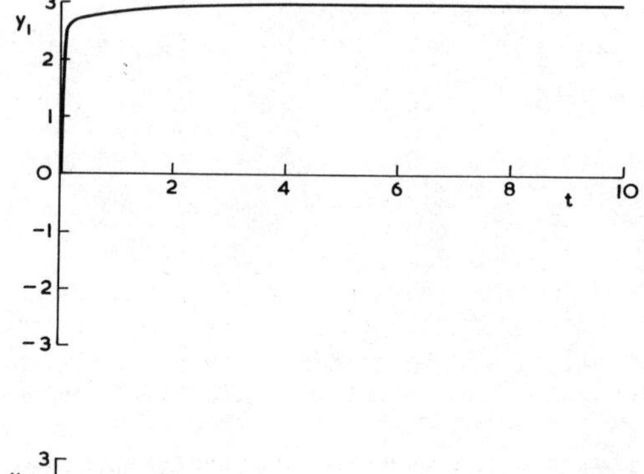

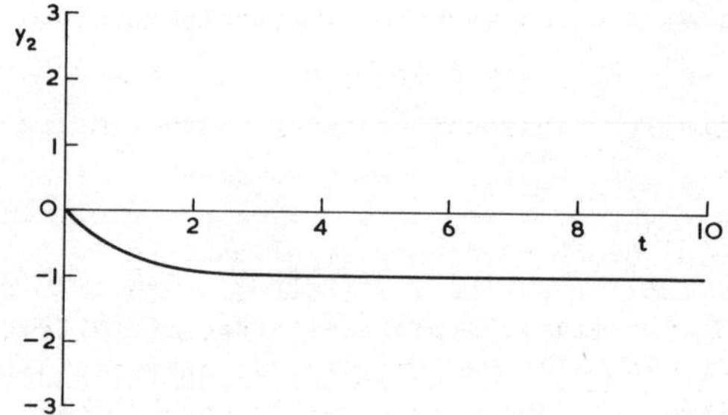

Fig 8.2 Responses of controlled plant: error-actuated digital controller with $\tilde{\alpha} = 0.9$, $T = 0.1$, $\tilde{\Sigma} = \text{diag}\{120,60\}$.

The behaviour of this initially quiescent plant under the action of the error-actuated digital controller governed on $T_T = \{0,T,2T,\ldots\}$ by equation (10) when $\tilde{\alpha} = 0.9$, $T = 0.1$, $\tilde{\Sigma} = \text{diag}\ \{120,60\}$, $d = [-1\ ,\ 2]^T$, and $v = [3\ ,\ -1]^T$ is shown in Fig 8.2, and it is evident that the response again consists of a fast non-oscillatory approach to the desired steady output $y = [3,\ -1]^T$. The sets $\tilde{Z}_c$ of closed-loop characteristic roots when $\tilde{\alpha} = 0.9$, and $\tilde{\Sigma} = \text{diag}\ \{120,60\}$ are presented in Table 8.2 when the sampling period T = 0.001, 0.02, and 0.1,

Table 8.2

Sampling Period T	Characteristic Roots $\tilde{Z}_c$
0.001	0.9990 0.9980 0.9970 0.9999 0.9999
0.02	0.9861 0.9634 0.9324 0.9418 0.9740
0.1	0.8971 0.8869 0.7408 -0.0573 -0.0098

and it is clear that $\tilde{Z}_c \subset D^-$ in each case. Furthermore, in consonance with equations (24) and (25), it is evident from Table 8.2 that $\tilde{Z}_1 = \{1-T+O(T^2), 1-2T+O(T^2),\ 1-3T+O(T^2)\}$ and $\tilde{Z}_2 = \{1-120T^2+O(T^3),\ 1-60T^2+O(T^3)\}$.

8.4 *Conclusions*

These results indicate that, for a large class of multivariable plants, the design of both analogue and digital error-actuated controllers which ensure that excellent set-point tracking occurs simultaneously with disturbance rejection can be readily effected even though the detailed dynamical properties of the processes involved are unknown. Such error-actuated controllers for unknown plants frequently result in superior performance to that achievable by controllers in which only the integral of error is processed (Davison 1976; Davison, Taylor and Wright 1980) and are more generally applicable than the controllers proposed by Pentinnen and Koivo (1980).

The theoretical basis for the design of controllers for unknown plants is accordingly now in an advanced state of development, following the seminal contribution of Davison (1976), and readily yields effective fixed-structure controllers for strongly interactive multivariable industrial plants without the necessity for endeavouring to determine accurate models of such plants. These developments in the design of controllers for unknown plants, when taken in conjunction with recent developments in the design of self-tuning controllers, constitute a significant response to the withering critique by Foss (1973) of the impotence of much of 'modern' control theory in the face of the complexity of industrial plants.

Acknowledgment

This research was sponsored by the Air Force Flight Dynamics Laboratory (AFSC), United States Air Force, under contract F49620-80-C-0035.

8.5 *Appendix*

The matrix $G = G(0) = -CA^{-1}B \in R^{\ell \times m}$ can be conveniently determined (Davison 1976) by performing a sequence of m separate 'off-line' tests on the open-loop asymptotically stable plant in the absence of the disturbance vector, i.e., with $d \equiv 0$. Thus, in the jth test, an input vector u(t) is applied to the open-loop plant with the property that

$$\lim_{t\to\infty} u(t) = u_j$$

where the constant vector $u_j \in R^m$ is chosen to be linearly independent of $\{u_1, u_2, \ldots u_{j-1}\}$ and u_1 is chosen to be non-null. The resulting output vector $y_j \in R^\ell$ is measured, where

$$\lim_{t\to\infty} y(t) = y_j$$

and

$$y_j = Gu_j \ .$$

It is therefore evident that, after m such separate tests have been performed, the required matrix is given by the simple formula (Davison 1976)

$$G = [y_1, y_2, \ldots, y_m][u_1, u_2, \ldots, u_m]^{-1} \ .$$

In the special case when, in the jth test, a unit step function is applied to the jth input channel of the open-loop plant and the remaining (j-1) input channels are not excited, it is obvious that

$$[u_1, u_2, \ldots, u_m] = I_m = [u_1, u_2, \ldots, u_m]^{-1}$$

and therefore that G is obtained directly by merely assembling the ordered output vectors.

References

Åström, K J and Wittenmark, B (1973). 'On self-tuning regulators', Automatica, 9, pp 185-199.

Åström, K J, Borisson, U, Ljung, L and Wittenmark, B (1977). 'Theory and application of self-tuning regulators', Automatica, 13, pp 457-476.

Bradshaw, A and Porter, B (1980). 'Design of linear multivariable discrete-time tracking systems incorporating fast-sampling error-actuated controllers', Int J Systems Sci, 11, pp 817-826.

Clarke, D W and Gawthrop, P J (1979). 'Self-tuning control', Proc IEE, 126, pp 633-640.

Davison, E J (1976). 'Multivariable tuning regulators: the feed-forward and robust control of a general servomechanism problem', IEEE Trans, AC-21, pp 35-47.

Davison, E J, Taylor, P A and Wright, J D (1980). 'On the application of tuning regulators to control a commercial heat exchanger', IEEE Trans, AC-25, pp 361-375.

Foss, A S (1973). 'Critique of chemical process control theory', IEEE Trans, AC-18, pp 646-652.

Kwakernaak, H and Sivan, R (1972). 'Linear Optimal Control Systems', Wiley-Interscience, New York.

Pentinnen, J and Koivo, H M (1980). 'Multivariable tuning regulators for unknown systems', Automatica, 16, pp 393-398.

Peterka, V and Åström, K J (1973). 'Control of multivariable systems with unknown but constant parameters', Proc IFAC Symposium on Identification and System Parameter Estimation, The Hague, pp 535-544.

Porter, B (1980). 'Design of error-actuated controllers for unknown multivariable plants', Report USAME/DC/107/80, University of Salford.

Porter, B and Bradshaw, A (1979). 'Design of linear multivariable continuous-time tracking systems incorporating high-gain error-actuated controllers', Int J Systems Sci, 10, pp 461-469.

Porter, B and Power, H M (1970). 'Controllability of multivariable systems incorporating integral feedback', Electron Lett, 6, pp 689-690.

Power, H M and Porter, B (1970). 'Necessary and sufficient conditions for controllability of multivariable systems incorporating integral feedback', Electron Lett, 6, pp 815-816.

Wellstead, P E and Zanker, P (1978). 'The technique of self-tuning', Control Systems Centre Report 432, UMIST.

Chapter Nine

Controller design using the theory of variable structure systems

A S I ZINOBER

9.1 *Introduction*

Variable Structure Systems (VSS) have been described in the literature by numerous Soviet authors including Emel'yanov (1964, 1967), Utkin (1971, 1974, 1977) and Itkis (1976). The basic philosophy is that the structure of the feedback control is altered as the state crosses discontinuity surfaces in the state space with the result that certain desirable properties are achieved. The switching of the control function yields total (or selective) invariance to system parameter variations and disturbances, and closed-loop eigenvalue placement in time-varying and uncertain systems. VSS design entails the choice of the switching control and the associated switching logic to yield sliding motion on the switching surfaces. This will be described in later sections.

The theory of scalar VSS is well documented and deals with systems having time-delay (Bakakin and Utkin 1968), imprecise switching logic (Utkin 1971), nonlinearities and state variable constraints (Bakakin et al 1964). The theory has been extended to the multivariable case with vector control.

The switching surfaces are generally fixed hyperplanes in the state space. The design yields asympotically stable sliding motion on these hyperplanes. The system in the sliding mode can be described by a linear system of lower order than the original system. The state asymptotically approaches the state origin on the intersection of the hyperplanes.

Some VSS systems have been designed to have only non-sliding motion yielding a bilinear system, but usually VSS systems exploit the properties of sliding motion (Flügge-Lotz 1953, André and Seibert 1956, Weissenberger 1966, Fuller 1967).

The scalar problem will first be described and the concept of sliding motion introduced and illustrated using a second-order system. The main features of multivariable VSS theory with constant switching hyperplanes will next be discussed. The application of VSS design theory to model-following control systems and adaptive near-time-optimal control with time-varying hyperplanes will be considered.

Dr Zinober is with the Department of Applied and Computational Mathematics, University of Sheffield.

9.2 *A Second-order Scalar Problem*

To illustrate the basic concepts of VSS consider the second-order system in phase canonic form with scalar control u

$$\begin{cases} \dot{x}_1 = x_2 \\ \dot{x}_2 = -a_1x_1 - a_2x_2 + bu \qquad (b > 0) \end{cases} \tag{1}$$

where x_i are the state variables. Let the state vector be $\underset{\sim}{x} = (x_1 \; x_2)^T$. The a_i and b are constants or time-varying parameters and their precise values may be unknown.

Consider the discontinuous control

$$u = \begin{cases} u^+ & cx_1 + x_2 > 0 \\ u^- & cx_1 + x_2 < 0 \end{cases} \tag{2}$$

with $c > 0$ and $u^+ \neq u^-$.

The switching function is

$$v = cx_1 + x_2$$

and the line

$$v = 0$$

is the surface on which the control u has a discontinuity. Suppose at time $t = 0$, $v > 0$. It can be readily shown that the state $\underset{\sim}{x}$ reaches the switching line $v = 0$ in a finite time τ as depicted in Figure 1.

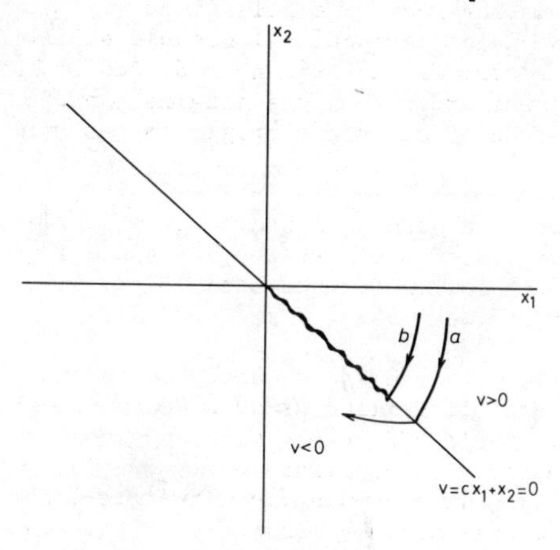

Fig 9.1 State paths a. 'bang-bang'
b. sliding

The state $\underset{\sim}{x}$ crosses the switching line and enters the region $v < 0$. resulting in the value of u being altered from u^+ to u^-. Depending upon the values of the system parameters and c the state trajectory may continue in the region $v < 0$, yielding bang-bang control. Alternatively the state trajectory may immediately re-cross the switching line and enter the region $v > 0$. This yields *sliding* (or chatter) motion (Flügge-Lotz 1953, André and Seibert 1956, Weissenberger 1966, Fuller 1967). Assuming that the switching logic works infinitely fast, the state $\underset{\sim}{x}$ is constrained to remain on the switching line $v = 0$ by the control which oscillates between the values u^+ and u^-.

For sliding motion to occur we need on opposite sides of the switching line

$$\lim_{v \to 0^+} \dot{v} < 0 \quad \text{and} \quad \lim_{v \to 0^-} \dot{v} > 0. \tag{3}$$

This ensures that the motion of the state $\underset{\sim}{x}$ on either side of the switching line $v = 0$ in the neighbourhood of the switching line is towards the switching line. The two conditions may be combined to give

$$v\dot{v} \leq 0 \tag{4}$$

in the neighbourhood of the switching line.

During sliding motion v remains zero so we have the differential equation

$$v = cx_1 + \dot{x}_1 = 0$$

governing the system dynamics, with solution

$$x_1(t) = x_1(\tau)e^{-c(t - \tau)}.$$

Thus the second-order system behaves like an asymptotically stable first-order system with time constant c^{-1} during the sliding mode. The dynamics are *independent* of the system (1) and (2). $x_2(t)$ may be obtained by differentiating $x_1(t)$.

9.3 *Phase Canonic System with Scalar Control*

We next consider the nth order system in phase canonic form

$$\begin{cases} \dot{x}_i = x_{i+1} & i = 1, 2, \ldots, n-1 \\ \dot{x}_n = -\sum_{j=1}^{n} a_j x_j + bu & (b > 0) \end{cases} \tag{5}$$

where the a_j and b may be time-varying and imprecisely known.

Consider the control given by

$$u = - \sum_{j=1}^{k} \psi_i x_i \qquad (1 \le k \le n) \tag{6}$$

where

$$\psi_i = \begin{cases} \alpha_i & x_i v > 0 \\ \beta_i & x_i v < 0 \end{cases} \tag{7}$$

and $$v = \underset{\sim}{c}^T \underset{\sim}{x} \; (\underset{\sim}{c} = (c_1 \; c_2 \; \ldots \; c_n)^T, \; \underset{\sim}{x} = (x_1 \; x_2 \; \ldots \; x_n)^T). \tag{8}$$

The parameters c_i are constants.

For sliding on the hyperplane $v = 0$ we require from (3) (or (4))

$$\begin{aligned} &\alpha_j \ge \frac{1}{b}(c_{j-1} - a_j - c_j c_{n-1} + c_j a_n) && j = 1, 2, \ldots k \\ &\beta_j \le \frac{1}{b}(c_{j-1} - a_j - c_j c_{n-1} + c_j a_n) && c_0 = 0 \\ &c_{j-1} - a_j = c_j(c_{n-1} - a_n) && j = k+1, \ldots n-1 \end{aligned} \tag{9}$$

Sliding motion can be attained for suitably chosen values of the α_i and β_i providing the range of the expected parameter values, a_j and b are known. The parameters c_i should be chosen to yield asymptotically stable motion in the sliding mode (Utkin 1977, Itkis 1976).

It is essential to ensure that the state trajectory $\underset{\sim}{x}(t)$ reaches the sliding hyperplane $v = 0$ from all initial conditions in the state space. Sufficient and necessary reachability conditions are that all the real eigenvalues of the system (5) with $a_i = \psi_i$ $(i = 1, 2, \ldots k)$ be non-negative if the switching hyperplane $v = 0$ is a sliding hyperplane, and that the sliding mode is asymptotically stable.

In the sliding mode $v = 0$ giving

$$\dot{v} = c_1 x_1 + c_2 x_2 + \ldots + c_{n-1} x_{n-1} + c_n \dot{x}_{n-1} = 0 .$$

So the (n-1)th order differential equation

$$\dot{x}_{n-1} = - (c_1 x_1 + c_2 x_2 + \ldots + c_{n-1} x_{n-1})$$

describes the dynamics of the system. Thus the system dynamics are independent of the system parameters a_i in the sliding mode. The parameters c_j specify the (n-1) eigenvalues of the system in the sliding mode.

9.4 *Multivariable Variable Structure Systems*

The multivariable system to be considered satisfies in its most general form the differential equations

$$\dot{x}(t) = f(x, t) + Bu(t) + Dh(x, t) \tag{10}$$

where $x \in R^n$, h represents disturbances and the effects of time-varying parameters, and $u \in R^m$ is the vector control. The discontinuous control has the form

$$u_i = \begin{cases} u_i^+(x) & v_i(x) > 0 \\ \\ u_i^-(x) & v_i(x) < 0 \end{cases} \tag{11}$$

where u_i is the ith component of u and $u_i^+ \neq u_i^-$. $v_i(x) = 0$ is the ith of the m switching hyperplanes

$$v(x) = Gx = 0. \tag{12}$$

9.4.1 Sliding Motion

For sliding on the ith hyperplane we need

$$v_i \dot{v}_i \leq 0 \tag{13}$$

in the neighbourhood of $v_i(x) = 0$. In the sliding mode the system satisfies the equations

$$v_i(x) = 0 \text{ and } \dot{v}_i(x) = 0$$

and the system has invariance properties (Draženovic, 1969) yielding motion which is independent of certain system parameters and disturbances. Thus variable structure systems are usefully employed in systems with uncertain and time-varying parameters.

Before studying the nature of the control u enforcing sliding for the case of constant G, i.e. fixed switching hyperplanes, the detailed behaviour of the sliding system dynamics needs to be considered. Suppose the sliding mode exists on all the hyperplanes. Then, during sliding

$$v = Gx = 0 \text{ and } \dot{v} = G\dot{x} = 0$$

The system equations (10) and (11) have discontinuous right hand sides and do not satisfy the classical theorems on the existence and uniqueness of the solutions. In the idealized equations nonideal properties such as switching time-delays and hysteresis are not considered. These properties determine the system dynamics in the neighbourhood of the discontinuity surfaces. As the non-ideal properties tend to zero, the motion tends to the ideal sliding mode on the intersection of the switching hyperplanes (Utkin 1977).

Utkin has formulated the equivalent control technique to study the sliding mode. One sets

$$\dot{v} = 0$$

and solves the resulting equations for the control vector termed the equivalent control, u_{eq}. u_{eq} is the function u which yields the sliding mode equations for the system. The value of u_{eq} is effectively the average value of u which maintains the state on the discontinuity surfaces $v = 0$. The actual control u consists of a low frequency (average) and a high frequency component. The equivalent control is the control without the high frequency component. Further discussion of this topic may be found in Utkin (1971). Substituting the equivalent control u_{eq} for the original control u gives the sliding mode equations

$$\dot{x} = f + Bu_{eq} + Dh.$$

Solving for u_{eq} from $\dot{v} = G\dot{x} = 0$, yields

$$u_{eq} = -(GB)^{-1} G(f + Dh)$$

assuming $(GB)^{-1} \neq 0$, and substitution into the original equation yields the nth order equation

$$\dot{x} = [I - B(GB)^{-1} G] (f + Dh) \qquad (14)$$

Utkin (1974) has considered the nonsingular case $(GB)^{-1} = 0$.

If $f = Ax$ and $h = 0$ (the disturbance-free case) then

$$u_{eq} = -(GB)^{-1} GAx$$

and

$$\dot{x} = [I - B(GB)^{-1} G]Ax. \qquad (15)$$

The state vector x remains on the (n-m)-dimensional manifold of the intersection of the m discontinuity surfaces. Therefore we can obtain an (n-m)th order set of system equations describing the sliding mode by eliminating m state variables using the equations $v = 0$. Young et al (1977) have studied high-gain feedback systems and obtained results linking sliding motion and high-gain systems. The motion of the equivalent system (15) is identical to that of the 'slow' and fast subsystems of a high-gain system. Using the transformation

$$\underset{\sim}{x}^* = \begin{pmatrix} x_1^* \\ x_2^* \end{pmatrix} = M\underset{\sim}{x} \qquad (x_1^* \in R^{n-m}, \; x_2^* \in R^m) \tag{16}$$

where

$$M = \begin{pmatrix} M_1 \\ M_2 \end{pmatrix}$$

$$M_1 B = 0$$

$$MAM^{-1} = \begin{pmatrix} A_{11} & A_{12} \\ A_{21} & A_{22} \end{pmatrix} \tag{17}$$

$$B = \begin{pmatrix} B_1 \\ B_2 \end{pmatrix}$$

and $\quad GM = (G_1 \quad G_2)$

an expression for the equivalent control system is obtained. In terms of $\underset{\sim}{x}_1^*$ and $\underset{\sim}{x}_2^*$

$$\underset{\sim}{u}_{eq} = -(G_2 B_2)^{-1} \{(G_1 A_{11} + G_2 A_{21})\underset{\sim}{x}_1^* + (G_1 A_{12} + G_2 A_{22})\underset{\sim}{x}_2^*\}$$

and $\quad \underset{\sim}{v} = G_1 \underset{\sim}{x}_1^* + G_2 \underset{\sim}{x}_2^* = \underset{\sim}{0}.$

Thus

$$\dot{\underset{\sim}{x}}_1^* = A_{11}\underset{\sim}{x}_1^* + A_{12}\underset{\sim}{x}_2^*, \; \underset{\sim}{v} = \underset{\sim}{0} \text{ and } \dot{\underset{\sim}{v}} = \underset{\sim}{0},$$

yield the (n-m)th order system

$$\dot{\underset{\sim}{x}}_1^* = (A_{11} - A_{12} G_2^{-1} G_1)\underset{\sim}{x}_1^* \tag{18}$$

corresponding to the 'slow' subsystem. The eigenvalues of this system are the transmission zeros of the triple (G, A, B) considering G as the output matrix. The eigenvalues are determined by the choice of switching hyperplanes. Assuming (A, B) to be a controllable pair, then so is (A_{11}, A_{12}) (Young et al 1977) and the eigenvalues of (18) can be placed arbitrarily in the complex plane by suitable choice of G. The 'fast' subsystem is given by

$$\dot{\underset{\sim}{x}}_2^* = A_{21}\underset{\sim}{x}_1^* + A_{22}\underset{\sim}{x}_2^* + B_2\underset{\sim}{u} .$$

The matrix G specifies the location of the hyperplanes and the associated eigenvalues of the sliding system. Alternatively Utkin and Young (1978) have shown that the hyperplanes may be chosen to minimise the (infinite-time) optimal control functionals

$$I_1 = \int_{t_s}^{\infty} \underset{\sim}{x}^T Q \underset{\sim}{x} \, dt \tag{19}$$

or

$$I_2 = \int_{t_s}^{\infty} (\underset{\sim}{x}^T Q \underset{\sim}{x} + \underset{\sim}{u}_{eq}{}^T R \underset{\sim}{u}_{eq}) \, dt \tag{20}$$

where Q and R are positive semi-definite symmetric matrices and t_s is the starting time of the sliding mode.

Some of the invariance properties will now be described. Suppose the disturbance vector $\underset{\sim}{h}(\underset{\sim}{x}, t)$ to be non-zero. Then, from (14)

$$\underset{\sim}{u}_{eq} = -(GB)^{-1}(A\underset{\sim}{x} + D\underset{\sim}{h})$$

and

$$\dot{\underset{\sim}{x}} = [I - B(GB)^{-1}G](A\underset{\sim}{x} + D\underset{\sim}{h}).$$

For total disturbance rejection during the sliding mode (Draženović 1969) we require that

$$[I - B(GB)^{-1}G]D\underset{\sim}{h} = \underset{\sim}{0} \tag{21}$$

This equation is satisfied for any value of $\underset{\sim}{h}$ if we can solve $D\underset{\sim}{h} = B\underset{\sim}{w}$ giving the rank condition

$$\text{rank } (B, D) = \text{rank } B \tag{22}$$

which is independent of A and $\underset{\sim}{x}$.

If $A = A_c + \Delta A$ and ΔA includes all the parameter variations in A, the equivalent control system is not influenced by the parameter variations in A if

$$\text{rank } (B, \Delta A\, T) = \text{rank } B \tag{23}$$

where the columns of matrix T are constructed from basis vectors of the space R^{n-m}. Mita (1976) has termed this property zero sensitivity. For a system in phase canonic form both the disturbance and parametric invariance conditions are satisfied.

For $B = B_c + \Delta B$, with ΔB representing parameter variations in B, if

$$\text{rank } (B_c, \Delta B) = \text{rank } B_c = m, \tag{24}$$

M_1 may be chosen such that $M_1B_c = 0$ and the system is not influenced by ΔB. If (24) is not satisfied, a stable system (18) may still be obtained if the bounds on ΔB are known (Young 1978). For M_1 such that $M_1B_c = 0$, we have

$$\dot{\underset{\sim}{x}}_1^* = [Z + M_1\Delta B(GB)^{-1} \{G_1A_{11} + G_2A_{21} -$$

$$(G_1A_{12} + G_2A_{22})G_2^{-1}G_1\}]\underset{\sim}{x}_1^*$$

where the eigenvalues of Z are the transmission zeros of the triple (G, A, B_c) which can be placed arbitrarily in the complex plane by suitable choice of the switching hyperplanes, $\underset{\sim}{v} = G\underset{\sim}{x} = \underset{\sim}{0}$. Providing the bounds on ΔB are known we can ensure stability by placing the eigenvalues of Z sufficiently far into the left-hand half of the complex plane.

9.4.2 Choice of control function

The variable structure system requires a suitable control function u to ensure that sliding motion on the m switching hyperplanes is achieved. The components u_i of $\underset{\sim}{u}$ are given by

$$u_i = - \sum_{j=1}^{n} \psi_{ij}|x_j| - \delta_i \tag{25}$$

where

$$\psi_{ij} = \begin{cases} \alpha_{ij} & v_i(\underset{\sim}{x}) > 0 \\ \beta_{ij} & v_i(\underset{\sim}{x}) < 0 \end{cases}$$

and

$$\delta_i = \begin{cases} \delta_i^+ & v_i(\underset{\sim}{x}) > 0 \\ \delta_i^- & v_i(\underset{\sim}{x}) < 0 \end{cases}$$

The reachability conditions, ensuring that the state $\underset{\sim}{x}(t)$ reaches all the m hyperplanes within a finite time, are

$$v_i\dot{v}_i < 0 \quad (i = 1, 2, \ldots m) \tag{26}$$

and this influences the choice of the ψ_{ij} and δ_i.

To ensure that sliding motion exists simultaneously on the intersection of the m switching hyperplanes, an (n−m) subspace, Utkin (1974) has proposed three methods, namely the diagonalization method, the quadratic form design technique and the method of control hierarchy. The first two techniques require a precise knowledge of the system parameters, so we shall concentrate on the final technique since parameter uncertainties are directly taken into account. The design assumes that sliding motion occurs successively on the switching hyperplanes $v_i(\underset{\sim}{x}) = 0$, $i = 1, 2, \ldots m$.

9.4.3 Control hierarchy algorithm

Take successively $i = m, m-1, \ldots 2, 1$. Suppose sliding occurs on the first $i - 1$ switching planes. Solve for the equivalent control $\underset{\sim}{u}_{eq}^{i-1} = (u_1 \; u_2 \; u_{i-1})^T$ as a function of u_i and $\underset{\sim}{u}^{i+1} = (u_{i+1} \ldots u_m)^T$ from the equations $\dot{v}_j = 0$, $j = 1, 2, \ldots i-1$, giving

$$\underset{\sim}{u}_{eq}^{i-1} = P_{i-1}\,\underset{\sim}{x} + Q_{i-1}\,\underset{\sim}{u}^{i+1} + d_i u_i$$

where matrices P_{i-1}, Q_{i-1} and scalar d_i depend on the first $i - 1$ rows of GA and GB. Choose u_i^+ and u_i^- to satisfy the reachability condition $v_i \dot{v}_i < 0$ which is also the condition for sliding motion on $v_i(\underset{\sim}{x}) = 0$ (13). Thus u_i must satisfy

$$\begin{aligned} \gamma_i u_i^+ &\le - \min_{\underset{\sim}{u}^{i+1}} \left[\underset{\sim}{p}_i^{\,T}\,\underset{\sim}{x} + \underset{\sim}{g}_i^{\,T}\,\underset{\sim}{u}^{i+1} \right] \\ \gamma_i u_i^- &\ge - \max_{\underset{\sim}{u}^{i+1}} \left[\underset{\sim}{p}_i^{\,T}\,\underset{\sim}{x} + \underset{\sim}{g}_i^{\,T}\,\underset{\sim}{u}^{i+1} \right] \end{aligned} \qquad (27)$$

where $\gamma_i \neq 0$ and vectors $\underset{\sim}{p}_i$ and $\underset{\sim}{g}_i$ depend upon P_{i-1}, Q_{i-1} and d_i. Simplification results if the δ_i (25) are zero, since this allows the inequalities to be considered componentwise. The above inequalities allow uncertain system parameters to be considered.

In a recent paper Calise (1980) has used a technique employing hyperstability theory yielding more relaxed sufficient conditions for the reachability and the sliding conditions in linear time-invariant systems.

9.5 *Model-following control systems*

Linear model-following control (LMFC) is an efficient control method that avoids the difficulty of specifying a performance index which is usually encountered in the application of optimal control to multivariable control systems. The model that specifies the design objective is part of the system. However, LMFC systems are inadequate when there are large parameter variations or disturbances. This has led to the development of so-called adaptive model-following control systems (AMFC) (Landau 1979). There are two well-known approaches to the design of AMFC systems using stability conditions. The first is based upon Lyapunov functions (Grayson 1965 and Hang and Parks 1973), while the second is based upon the hyperstability concept (Landau 1974 Courtiol and Landau 1975. Both approaches guarantee that the error goes to zero as $t \to \infty$ but neither offer any direct quantitative design of the error transient.

In model-following systems the plant is controlled in such a way that its dynamic behaviour approximates that of a specified model. The model plant is a part of the system and it specifies the design objectives. The adaptive controller should force the error between the model and the plant to zero as time tends to infinite, i.e. $\lim_{t\to\infty} \underset{\sim}{e}(t) = \underset{\sim}{0}$. The plant and model are described by

$$\dot{\underset{\sim}{x}}_p = A_p(t)\underset{\sim}{x}_p + B_p(t)\underset{\sim}{u}_p \tag{28}$$

$$\dot{\underset{\sim}{x}}_m = A_m\underset{\sim}{x}_m + B_m\underset{\sim}{u}_m \tag{29}$$

where $\underset{\sim}{x}_p, \underset{\sim}{x}_m \in R^n$, $\underset{\sim}{u}_p \in R^g$ and $\underset{\sim}{u}_m \in R^r$.

$\underset{\sim}{u}_m$ is the input and $\underset{\sim}{u}_p$ is the control. The error vector is $\underset{\sim}{e} = \underset{\sim}{x}_m - \underset{\sim}{x}_p$. We shall assume that the pairs (A_p, B_p) and (A_m, B_m) are stabilizable and that A_m is a stable matrix. It can easily be shown that

$$\dot{\underset{\sim}{e}} = A_m\underset{\sim}{e} + (A_m - A_p)\underset{\sim}{x}_p + B_m\underset{\sim}{u}_m - B_p\underset{\sim}{u}_p \tag{30}$$

Certain model reference control systems designed by using Lyapunov theory have a discontinuous control law. The last section of the trajectories have been shown to consist of sliding motion (Devaud and Caron 1975, Young 1977) in the case of single-input single-output systems. Young (1977) discusses the VSS derivation of the control law giving an asymptotically stable sliding mode.

Lyapunov design requires the choice of a matrix Q giving a unique solution of the matrix P in the equation

$$C^TP + PC = -Q \tag{31}$$

where the positive definite C is defined by the model dynamics. The last column of P, the vector $\underset{\sim}{p}$, corresponds to the vector $\underset{\sim}{p}$ of the switching hyperplane

$$v = \underset{\sim}{p}^T\underset{\sim}{e} = 0$$

of the VSS design (assuming the model system (29) is in phase canonical form). The designer using Lyapunov theory alters Q, which changes $\underset{\sim}{p}$, until a satisfactory time response is obtained. However, using VSS theory the designer may specify in advance the desired error transient response by choice of the vector $\underset{\sim}{p}$ which yields the pole placement corresponding to the sliding mode on the switching hyperplane.

We shall next consider in detail the VSS design of the multivariable model-following control problem (Young 1978). For perfect model-following the error $\underset{\sim}{e}$ and its derivative $\dot{\underset{\sim}{e}}$ should be zero for any input $\underset{\sim}{u}_m \in R^r$ and $\underset{\sim}{x}_p \in R^n$ if $\underset{\sim}{e}(0) = \underset{\sim}{0}$. Consider the linear control law

$$\underset{\sim}{u}_p = K_p\underset{\sim}{x}_p + K_e\underset{\sim}{e} + K_m\underset{\sim}{u}_m \tag{32}$$

(We shall see later (equation (40)) that in the sliding mode $\underset{\sim}{u}_{eq}$ may be written in this form). Then (30) becomes

$$\dot{\underset{\sim}{e}} = (A_m - B_pK_e)\underset{\sim}{e} + (A_m - A_p - B_pK_p)\underset{\sim}{x}_p + (B_m - B_pK_m)\underset{\sim}{u}_m. \tag{33}$$

For perfect model following we require (Erzberger 1968, Chen 1973)

$$(A_m - A_p - B_pK_p)\underset{\sim}{x}_p + (B_m - B_pK_m)\underset{\sim}{u}_m = \underset{\sim}{0}. \tag{34}$$

The conditions for the existence of K_e, K_p and K_m are that

$$\text{rank } (B_p) = \text{rank } (B_p, B_m) = \text{rank } (B_p, A_m - A_p). \tag{35}$$

In addition $(A_m - B_pK_p)$ should be a Hurwitz matrix to yield an asymptotically stable error system

$$\dot{\underset{\sim}{e}} = (A_m - B_pK_e)\underset{\sim}{e} . \tag{36}$$

It will be assumed below that conditions (35) and (36) hold.

The VSS design is characterized by the discontinuous control

$$u_i = \begin{cases} u_i^+ (\underset{\sim}{x}_p, \underset{\sim}{e}, \underset{\sim}{x}_m) & v_i(\underset{\sim}{e}) > 0 \\ \\ u_i^- (\underset{\sim}{x}_p, \underset{\sim}{e}, \underset{\sim}{x}_m) & v_i(\underset{\sim}{e}) < 0 \end{cases} \tag{37}$$

where u_i is the ith component of $\underset{\sim}{u}_p$ and $v_i(e) = 0$ is the ith of the g switching hyperplanes

$$\underset{\sim}{v}(\underset{\sim}{e}) = G\underset{\sim}{e} = \underset{\sim}{0}, \tag{38}$$

in the error state space $\underset{\sim}{e} \in R^n$.

Suppose the sliding mode exists on all the hyperplanes. Then, during sliding

$$\underset{\sim}{v} = G\underset{\sim}{e} = \underset{\sim}{0} \text{ and } \dot{\underset{\sim}{v}} = G\dot{\underset{\sim}{e}} = \underset{\sim}{0}. \tag{39}$$

The equations governing the system dynamics may be obtained by substituting the equivalent control $\underset{\sim}{u}_{eq}$ for the original control $\underset{\sim}{u}_p$. From $\dot{\underset{\sim}{v}} = \underset{\sim}{0}$

$$\underset{\sim}{u}_{eq} = - (GB_p)^{-1}G \left\{A_m\underset{\sim}{e} + (A_m - A_p)\underset{\sim}{x}_p + B_m\underset{\sim}{u}_m\right\} \tag{40}$$

assuming $(GB_p)^{-1} \neq 0$, and substitution into the original equation yields the nth order equation

$$\dot{\underset{\sim}{e}} = [I - B_p(GB_p)^{-1}G][A_m\underset{\sim}{e} + (A_m - A_p)\underset{\sim}{x}_p + B_m\underset{\sim}{u}_m] . \tag{41}$$

For the perfect model-following case we obtain

$$\dot{\underset{\sim}{e}} = (I - B_p(GB_p)^{-1}G)A_m\underset{\sim}{e} . \tag{42}$$

The insensitivity of the error dynamics governed by (42) with respect to $\underset{\sim}{x}_p$ and $\underset{\sim}{u}_m$ is due to the fact that the perfect model-following conditions (35) coincide with the invariance conditions (22) of the sliding mode. If $\underset{\sim}{x}_p$ and $\underset{\sim}{u}_m$ are considered as 'disturbances' to the error dynamics, then the perfect model-following conditions guarantee that the behaviour of the VSS when sliding motion occurs on $\underset{\sim}{v} = G\underset{\sim}{e} = \underset{\sim}{0}$, is insensitive to these 'disturbances'. The error vector $\underset{\sim}{e}$ remains on the (n-g)-dimensional manifold of the intersection of the g discontinuity surfaces. Therefore we can obtain an (n-g)th order set of system equations from (39) describing the sliding mode.

The results of section 9.4 may be systematically applied to the $\underset{\sim}{e}$ system. The motion of the equivalent reduced order sliding system is identical to that of the 'slow' subsystem of a high-gain system. Using the transformation

$$\underset{\sim}{e}^* = \begin{bmatrix} \underset{\sim}{e}_1^* \\ \underset{\sim}{e}_2^* \end{bmatrix} = M\underset{\sim}{e} \tag{43}$$

and the defined terms (17),

an expression for the equivalent control system is obtained. In terms of $\underset{\sim}{e}_1^*$ and $\underset{\sim}{e}_2^*$

$$\underset{\sim}{u}_{eq} = -(G_2B_2)^{-1}\left\{(G_1A_{11} + G_2A_{21})\underset{\sim}{e}_1^* + (G_1A_{12} + G_2A_{22})\underset{\sim}{e}_2^*\right\} \tag{44}$$

and

$$\underset{\sim}{v} = G_1\underset{\sim}{e}_1^* + G_2\underset{\sim}{e}_2^* = \underset{\sim}{0}. \tag{45}$$

Thus $\dot{\underset{\sim}{e}}_1^* = A_{11}\underset{\sim}{e}_1^* + A_{12}\underset{\sim}{e}_2^*$, $\underset{\sim}{v} = \underset{\sim}{0}$ and $\dot{\underset{\sim}{v}} = \underset{\sim}{0}$, yielding the (n-g)th order equivalent system

$$\dot{\underset{\sim}{e}}_1^* = (A_{11} - A_{12}G_2^{-1}G_1)\underset{\sim}{e}_1^* \tag{46}$$

corresponding to the 'slow' subsystem. The eigenvalues of this system are the transmission zeros of the triple (G, A_m, B_p) considering G as the output matrix. Assuming (A_m, B_p) to be a controllable pair, then the eigenvalues of GB_p and those of (46) can be placed arbitrarily in the complex plane by suitable choice of G (Young et al 1977). If $B_p = B_c + \Delta B$, and (24) holds, with m = g, the system is not influenced by ΔB.

Linear model-following control has the form

$$\underset{\sim}{u}_p = K_p \underset{\sim}{x}_p + K_e \underset{\sim}{e} + K_m \underset{\sim}{u}_m. \tag{47}$$

In the VSS design we replace the K's by ψ's which are defined to be

$$\begin{aligned} \underset{\sim}{\psi}_p &= \tilde{\underset{\sim}{\psi}}_p \operatorname{diag}(\operatorname{sgn} x_p^1, \ldots, \operatorname{sgn} x_p^n),\ \underset{\sim}{x}_p = (x_p^1, \ldots, x_p^n)^T \\ \underset{\sim}{\psi}_e &= \tilde{\underset{\sim}{\psi}}_e \operatorname{diag}(\operatorname{sgn} e^1, \ldots, \operatorname{sgn} e^n),\ \underset{\sim}{e} = (e^1, \ldots, e^n)^T \\ \underset{\sim}{\psi}_m &= \tilde{\underset{\sim}{\psi}}_m \operatorname{diag}(\operatorname{sgn} u_m^1, \ldots, \operatorname{sgn} u_m^r),\ \underset{\sim}{u}_m = (u_m^1, \ldots, u_m^r)^T \end{aligned} \tag{48}$$

where

$$\tilde{\underset{\sim}{\psi}}_i = (\psi_{i1}, \ldots, \psi_{in},)^T \text{ and } \psi_{ij} = \begin{cases} \underset{\sim}{\alpha}_{ij} & v_j(\underset{\sim}{e}) > 0 \\ & \qquad j=1, \ldots, g \\ \underset{\sim}{\beta}_{ij} & v_j(\underset{\sim}{e}) < 0 \end{cases}$$

and i takes the symbols p, e and m respectively.

The design objective is to choose G and the ψ to ensure sliding so that the error tends to zero with suitable transient motion on the switching hyperplane. The matrices A_p and B_p are assumed uncertain but the lower and upper bounds of their elements are known. The reduced order system (46) should have eigenvalues in the left hand half plane of the complex plane to ensure stable motion giving $\lim_{t\to\infty} \underset{\sim}{e}(t) = \underset{\sim}{0}$.

For the state to reach the ith sliding hyperplane and for sliding to occur on the intersection of the m hyperplanes the method of control hierarchy (for g > 1), which takes into account the expected range of parameter variation, may be used (see section 9.4).

The design of model-following variable structure systems is the same as that for multivariable VSS once the problem is presented in terms of the error vector. The designer first specifies the stable eigenvalues of the closed-loop error system by the choice of sliding hyperplanes and then chooses a suitable control law to ensure sliding motion. The design technique ensures that the desired error transients are obtained. The resulting system is insensitive to a class of disturbances and parameter variations, and the design can allow for a range of uncertain parameter values in the control hierarchy algorithm.

Since the VSS controller does not require any parameter identification, the requirement for sufficient "richness" of the input is avoided. The VSS control design incorporates an assumed range of values of the time-varying $A_p(t)$ and $B_p(t)$. The stability proofs for other AMFC structures assume time-invariant A_p and B_p. The zeros of the plant must lie in the open left half of the complex plane (as required in other design techniques).

9.6 *Model-following control examples*

A second-order system with scalar control is designed in Zinober et al (1981). In this section we shall consider the design of a fourth-order system with two controls $\underset{\sim}{u}_p = (u_1\ u_2)^T$. The plant is represented by the differential equations

$$\dot{\underset{\sim}{x}}_p(t) = \begin{bmatrix} 0 & 1 & 0 & 0 \\ 0 & 0 & 1 & 0 \\ 0 & 0 & 0 & 1 \\ -6 & -6 & -11 & -10 \end{bmatrix} \underset{\sim}{x}_p(t) + \begin{bmatrix} 0 & 0 \\ 0 & 0 \\ 0 & 1 \\ 1 & 1 \end{bmatrix} \underset{\sim}{u}_p(t) \tag{49}$$

and the model plant to be followed by the actual plant is represented by

$$\dot{\underset{\sim}{x}}_m(t) = \begin{bmatrix} 0 & 1 & 0 & 0 \\ 0 & 0 & 1 & 0 \\ 0 & 0 & 0 & 1 \\ -3 & -12 & -19 & -8 \end{bmatrix} \underset{\sim}{x}_m(t) + \begin{bmatrix} 0 \\ 0 \\ 0 \\ 3 \end{bmatrix} u_m(t) \tag{50}$$

where $u_m(t)$ the input is

$$u_m(t) = \begin{cases} 4 & 0 < t < 15 \\ 0 & t \geq 15 \end{cases}$$

The initial conditions are $\underset{\sim}{x}_p(0) = (-1\ \ 0\ \ 0\ \ 0)^T$ and $\underset{\sim}{x}_m(0) = (0\ 0\ 0\ 0)^T$.

The error transients can be specified by the choice of switching hyperplanes. For the given initial conditions the choice $v_1 = e_4 = 0$ ensures that sliding motion occurs immediately at $t = 0$ on $v_1 = 0$. The plane $v_2 = c_1e_1 + 2c_2e_2 + e_3 = 0$ yields the error differential equation

$$\ddot{e}_1 + 2c_2\dot{e}_1 + c_1e = 0 \tag{51}$$

giving an error transient in the sliding mode corresponding to the second-order system (51). The model-following conditions (35) are satisfied.

Having selected the desired error responses the control $\underset{\sim}{u}_p$ needs to be chosen so that the switching surfaces are reached and that sliding motion occurs for the expected range of parameter variations. We may choose

$$u_i = -\sum_{j=1}^{4} \psi^i_{x_{p_j}} |x_{pj}| - \sum_{j=1}^{4} \psi^i_{e_j} |e_j| - \psi^i_m|u_m| \quad (i=1,2)$$

where the ψ coefficients have the form

$$\psi_{ij} = \begin{cases} \alpha_{ij} & v_i > 0 \\ \beta_{ij} & v_i < 0 \end{cases}$$

for $i = 1, 2$.

Sliding on $v_1 = e_4 = 0$ yields from (49) and (50)

$$\dot{\underset{\sim}{e}} = \begin{bmatrix} 0 & 1 & 0 \\ 0 & 0 & 1 \\ 0 & 0 & 0 \end{bmatrix} \underset{\sim}{e}$$

with

$$u_{1_{eq}} = (3 \quad -6 \quad -8 \quad 2)\underset{\sim}{x}_p - (3 \quad 12 \quad 19 \quad 8)\underset{\sim}{e} + 3u_m - u_2 \; .$$

We must ensure sliding successively on the hyperplanes in the order v_1, v_2 by satisfying the inequality conditions of the control hierarchy method (27). The inequality constraints yield for the second hyperplane for instance the requirements

$$\alpha_{e_2} < - c_1, \; \alpha_{e_3} < - 2c_2$$

with the remaining α's negative. Symmetric results arise for the β's.

The plant (49) has a highly oscillatory transient response without any adaptive control. The choice of desired model transients (50) *and* desired error transients yield satisfactory motion in the model-following system.

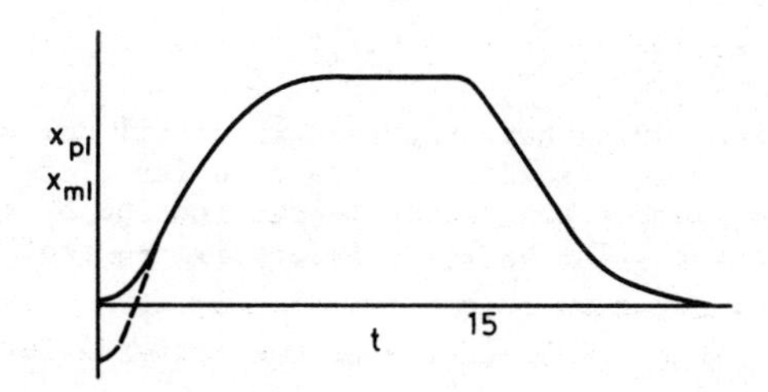

Fig 9.2 Fourth-order system -----x_{p1} ——x_{m1}

Equation (51) indicates a critically damped error transient $e_1(t)$ based on the dynamics of the unforced second-order system (51) for the parameter choice $c_1 = c_2 = 1$. The trajectories $x_{p_1}(t)$ and $x_{m_1}(t)$ are given in Figure 9.2. The error state $e_1(t) = x_{m_1}(t) - x_{p_1}(t)$ follows the expected transient. The eigenvalues of the error system are specified in the design by the selection of the two switching hyperplanes $v_1 = 0$ and $v_2 = 0$. For $t \geq 15$ in Figure 9.2 $\underset{\sim}{e}(t) = 0$ and the sliding mode continues.

Landau and Courtiol (1972) have used the hyperstability theory to design the model-following control system of a fourth-order control system aircraft with three control variables. The VSS design (Young 1978) has three switching hyperplanes and yields error state trajectories which have better transient response than those of Landau and Courtiol (1972). The error trajectories for given initial conditions, input function and a randomly varying plant parameter are compared in Figure 9.3. In these examples the straightforward VSS design technique may be carried out without any computer calculations.

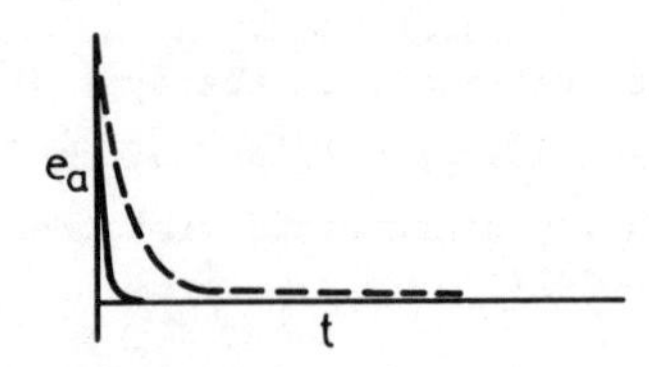

Fig 9.3 Aircraft control system
e_a pitch angle error
——— VSS design
----- Hyperstability design

9.7 *The self-adaptive controller*

For linear systems of the form

$$\dot{\underset{\sim}{z}} = A\underset{\sim}{z} + \underset{\sim}{b}u \tag{52}$$

the author (Zinober 1974) has described in detail self-adaptive control strategies which yield state trajectories close to the time-optimal responses without the direct identification of the plant parameters in A and $\underset{\sim}{b}$. The basic self-adaptive controller has the control law

$$u = -\operatorname{sgn} v \tag{53}$$

where v is the linear switching function

$$v = \underset{\sim}{k}^T\underset{\sim}{z} = q^{n-1}a_{n-1}z_1 + q^{n-2}a_{n-2}z_2 + \ldots + a_o z_n \quad (q > 0,\ a_i > 0) \tag{54}$$

the z_i are the chosen state variables, q is the variable control parameter and the a_i are constants. The necessary conditions for sliding (or chatter) motion on v = 0 with q constant have already been described. The self-adaptive strategies to be described below have time-varying switching hyperplanes and this necessitates the analysis of systems with non-linear switching hypersurfaces.

9.7.1 The basic strategy

At t = 0 the control parameter q is set to the value δ, a small positive number. The position of the switching hyperplane v = 0 (54) in the state space depends on the parameter q. The value of q is continually adjusted by the following rules:

If sliding motion occurs on the current switching hyperplane, the next position of the switching hyperplane is chosen to be just ahead of the state point on its trajectory in the state space by setting

$$q(t+) = q(t)\,\{1 + \varepsilon\}, \tag{55}$$

where ε is a small positive number. Otherwise q is kept constant.

The state point can be shown to move from the initial conditions to a certain sliding boundary surface $S_o(\underset{\sim}{z}, \underset{\sim}{A}, \underset{\sim}{b}) = 0$ on which it remains until the state origin is reached. The reachability of the initial and subsequent positions of the switching hyperplane can be checked by verifying the necessary condition $v\dot{v} < 0$. The self-adaptive controller identifies at a given time instant during the sliding phase the local position of the nonlinear surface $S_o(\underset{\sim}{z}, A, \underset{\sim}{b}) = 0$, which is a function of the chosen state variables and the plant parameters. The sliding boundary surface S_o is the boundary between the sliding and non-sliding regions of the state space. In the sliding region we have $v\dot{v} < 0$ in the neighbourhood of the switching hyperplane $v = 0$ for a certain positive value of q. In the non-sliding region $v\dot{v} > 0$. Thus on the boundary of these two regions $v\dot{v} = 0$. Therefore points on the nonlinear surface S_o are generated by the equations

$$v = \underset{\sim}{k}^T(q)\underset{\sim}{z} = 0 \quad \text{and} \quad \left(\frac{\partial v}{\partial t}\right)_q = \underset{\sim}{k}^T(q)\dot{\underset{\sim}{z}} = 0 \tag{56}$$

for values of positive q. The system's dynamic behaviour on S_o is given by the nonlinear equations

$$\frac{d}{dt}(\underset{\sim}{k}^T(q)\underset{\sim}{z}) = 0 \quad \text{and} \quad \frac{d}{dt}(\underset{\sim}{k}^T(q)\dot{\underset{\sim}{z}}) = 0 \tag{57}$$

which can be solved for $\dot{q}$ and $\dot{\underset{\sim}{z}}$. The technique may also be applied to certain nonlinear and slowly time-varying systems. S_o may be regarded as the 'equivalent switching surface'. The system parameters need not be directly identified to determine the local position of S_o.

The author has analysed the properties of second-order systems (Zinober 1975) and a third-order system (Zinober 1977) using nonlinear switching surface theory. The analysis of higher-order systems is rather complex. To illustrate the technique the dynamic behaviour of the double integrator system

$$\dot{x}_1 = x_2, \ \dot{x}_2 = u/b, \ (b > 0, \ |u| \le 1) \tag{58}$$

is analysed and the properties associated with the sliding boundary surface are derived. Sliding motion occurs on the switching line

$$v = qx_1 + x_2 = 0 \tag{59}$$

when q is held constant, if $v\dot{v} \le 0$, yielding the condition

$$|x_2| \le (qb)^{-1}. \tag{60}$$

Thus the sliding region of the x_1x_2 state space consists of points on the switching line satisfying (60) which lie between points A and A' in Fig 9.4. Suppose the state coordinates satisfy (59) and

$$q = (b|x_2|)^{-1}. \tag{61}$$

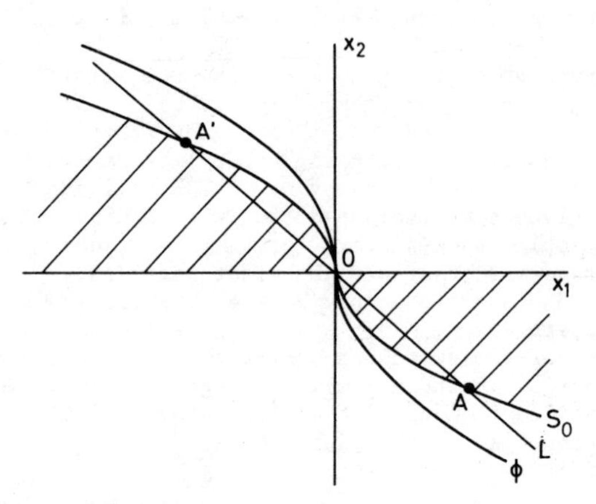

Fig 9.4 Second order system
L switching line $qx_1 + x_2 = 0$
S_o sliding boundary curve $x_1 + bx_2|x_2| = 0$
ϕ time-optimal switching curve $x_1 + \frac{1}{2}bx_2|x_2| = 0$
sliding region shaded

Then the point (x_1, x_2), A or A', lies on the boundary of the sliding region. If q is allowed to vary, the point (x_1, x_2) describes the sliding boundary surface, from (59) and (61)

$$S_o(x_1, x_2, b) = x_1 + bx_2|x_2| = 0, \tag{62}$$

which is the boundary between the sliding and non-sliding regions.

The self-adaptive controller initially drives the state with $u = \text{sgn}(-\delta x_1 - x_2)$ to the point A on the switching line L_1

$$\delta x_1 + x_2 = 0$$

in Fig 9.5, since $v\dot{v} < 0$ (satisfying the reachability condition). For δ sufficiently small sliding begins and the switching line is rotated to position L_2 by altering the value of q to $q = \delta(1 + \varepsilon)$, $\varepsilon > 0$. The state slides on the switching line L_1 from the point A to A'. Assuming that sliding motion can be detected within an infinitesimal time interval, A' is coincident with A. The new control law is

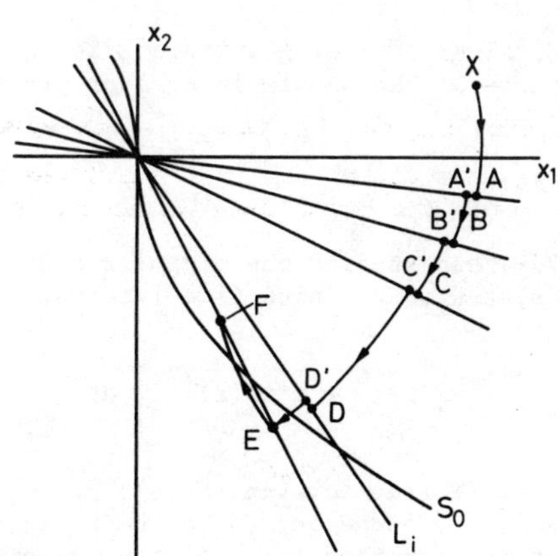

Fig 9.5 Second order system
Adaptive control
⟶ state path

$$u = \operatorname{sgn}\{-\delta(1 + \varepsilon)x_1 - x_2\}$$

and the control value at point A is now

$$u_v = \operatorname{sgn}(-x_1) = \operatorname{sgn}(x_2),$$

which drives the state to point B on the current switching line L_2

$$\delta(1 + \varepsilon)x_1 + x_2 = 0.$$

Sliding recommences and the switching line is rotated further to position L_3. After repeated rotation of the switching line whenever sliding occurs, the state reached the point D in the neighbourhood of the sliding boundary surface, S_o. The current switching line L_i is rotated to the position L_{i+1},

$$q_i(1 + \varepsilon)x_1 + x_2 = 0.$$

The control u_v drives the state to the point E on the line L_{i+1} in the non-sliding region. Sliding does not occur at point E and the state enters the region

$$x_1\{q_i(1 + \varepsilon)x_1 + x_2\} < 0.$$

The state path returns to the line L_{i+1} at point F in the sliding region and sliding recommences. As $\varepsilon \to 0$ the overshoot of the surface S_o becomes negligibly small and the state moves effectively on the sliding boundary surface to the state origin within a finite time. The effective value of the control maintaining the state on the surface S_o is $|u_{eq}| = \frac{1}{2}$ and $\dot{q} > 0$ from (57). The surface S_o, which may be regarded as the 'equivalent switching surface', is in fact the time-optimal switching curve of the double-integrator plant with the parameter 2b, i.e.

$$x_1 + bx_2|x_2| = 0.$$

Ryan (1979) has studied the singular optimal control of the double-integrator system on a finite time interval minimising the functional

$$J = \int_0^T \left\{ |x_1(t)|^\nu + \mu|x_2(t)|^{2\nu} \right\} dt$$

where $\underset{\sim}{x}(T) = \underset{\sim}{0}$ (T free terminal time), $\mu \geq 0$, $\nu \geq 1$. The special case with $\mu(2\nu - 1) = 1$ turns out to be the functional minimised by the above adaptive control strategy for a wide range of initial conditions.

The analysis of an nth order system requires the study of motion n on a nonlinear sliding boundary surface $S_o(\underset{\sim}{z}, A, \underset{\sim}{b})$ in an n-dimensional state space, $\underset{\sim}{z} \in R^n$.

9.7.2 The modified strategy

A modified strategy allows the controller to be reset whenever a new set of initial conditions or an error condition occurs. For initial conditions in certain regions the settling time is decreased. The following steps are carried out consecutively in the regions of the state space where the switching hyperplane assumes positive q values.

Step (i) : The hyperplane is rotated until it passes through the current state point.

Step (ii) : The controller tests for sliding at this point.

Step (iii) : If sliding occurs the hyperplane is rotated further, to be just ahead of the state point (q is reset in accordance with (55)).

Step (iv) : If sliding does not occur, the hyperplane is kept fixed until sliding recurs (or until Step I below causes Step (i) to rotate the hyperplane.

Step I : An auxiliary switching hyperplane

$$v(Q, \underset{\sim}{z}) = 0$$

is continuously rotated so as to pass through the state point $\underset{\sim}{z}(t)$. If the inequality

$$q \leq Q < q(1 + \varepsilon)$$

does not hold, then we reset the value of the control parameter q to that of the auxiliary parameter Q

$$q = Q$$

i.e. rotate the switching hyperplane until it passes through the current state point (Step (i) above). The purpose of the auxiliary hyperplane is to reset the control parameter q whenever a sudden disturbance gives rise to a new set of initial conditions. In those regions where v = 0 yields non-positive q values the control is identical to the basic strategy.

9.7.3 The switching hyperplane

The design choice of the vector $\underset{\sim}{k}$ with components $q^i a_i$ in (54) differs from that of variable structure system design with fixed hyperplane. For n = 2 we have $\underset{\sim}{a} = (1 \quad 1)^T$. Any other choice simply rescales the value of q. Stability results for second-order systems are given in Zinober (1975). The stability of a triple-integrator system has been studied (Zinober 1977) and the fastest transient motion is attained with $\underset{\sim}{a} = (1 \quad 4\sqrt{2}/3 \quad 1)^T$. For higher-order systems the vector $\underset{\sim}{a}$ with components a_i being the binomial coefficients has been found to yield stable motion for a wide variety of systems up to tenth order. The (n-1) repeated eigenvalues associated with the sliding hyperplane have the value -q.

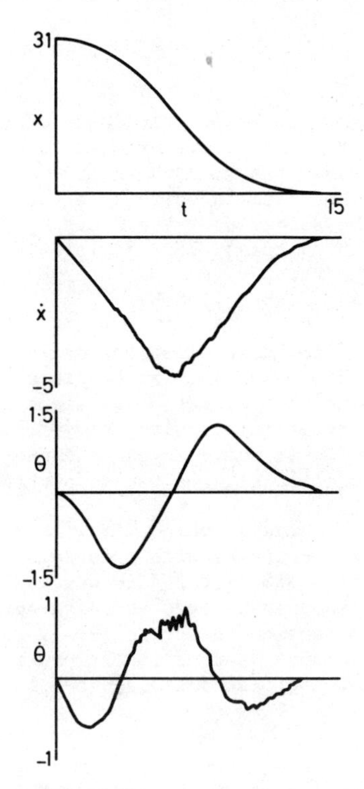

Fig 6 Digital simulation of crane system with varying hoist length
x = position
θ = angle of hoist to vertical
Hoist length $L(x) = 0.5 + |x-15.5|/15.5$

The self-adaptive control of a variety of systems has been studied by digital simulation. The adaptive control of overhead crane operations (which yields a fourth-order system) has yielded trajectories close to the open-loop time-optimal state paths (Zinober 1979). The self-adaptive control system can handle external disturbances, parameter variations, plant nonlinearities, time-varying hoist length and a velocity magnitude constraint. None of these aspects can be satisfactorily incorporated into the open-loop time-optimal bang-bang controller (see Figure 9.6).

The control of linear and nonlinear diffusion equations has also been studied and the adaptive control is exceedingly close to the time-optimal control. A fourth-order approximation yielding the control law (54) with $n = 4$ and suitably chosen state variables is sufficiently accurate (Zinober 1980a).

Since the self-adaptive controller identifies the sliding boundary surface of the system by studying the sliding mode behaviour, an accurate knowledge of the system parameters is not needed and in many instances the values of certain parameters are not needed. This obviates the need for the accurate identification of the system model and its parameters.

9.8 *Conclusions*

In this tutorial paper the basic principles of the VSS design method have been detailed and the invariance properties of multivariable VSS have been described. The theory has been applied to the adaptive model-following control problem in which both the model and the error system transients can be selected by the designer. The self-adaptive control of a system with a scalar control yields near-time-optimal state trajectories. The strategy may be incorporated into a multivariable control system (Zinober, 1980b).

The implementation of the control switching logic can be achieved using microprocessors (Young 1978). The problems associated with the high-frequency chattering control may be overcome by incorporating suitable low-pass filters or relays with delayed action as shown in Sabanović and Izosimov (1979) where an experimental squirrel-cage induction motor is controlled using a variable structure controller.

There are many other fields in which VSS with sliding motion plays a part. Inaccessible plant states with noise disturbances may be computed using an adaptive VSS filter (Zinober 1979, Golembo et al 1976). The technique can also be used to solve optimization problems and to determine system parameters (Utkin 1974). Gutman (1979) has studied generalized dynamical systems in the context of uncertain systems using a Lyapunov approach and has shown links with VSS.

References

Emel'yanov, S V (1964). 'Design of variable structure control systems with discontinuous switching functions'. Eng Cybern, 1, 156-160.

Emel'yanov, S V (1967). 'Automatic Control Systems of Variable Structure'. Nauka, Moscow.

Utkin, V I (1971). 'Equations of sliding mode in discontinuous systems I'. Automat Remote Contr, 21, 1897-1907.

Utkin, V I (1974). 'Sliding modes and their application in variable structure systems'. MIR, Moscow. (English translation, 1978).

Utkin, V I (1977). 'Variable structure systems with sliding modes'. IEEE Trans, AC-22, 212-222.

Itkis, U (1976). 'Control systems of variable structure'. Keter Publishing House, Jerusalem.

Bakakin, A A and Utkin, V I (1968). 'Variable structure systems with pure delay in switching devices', in Variable Structure Systems and their Application to Flight Automation, Nauka. Moscow, 64-71.

Bakakin, A A, Bermant, M A and Ezerov, V B (1964). 'Application of variable structure systems to stabilization of time varying plants in the presence of restrictions on the movement of the controlling device'. Automat Remote Contr, 7, 1016-1021.

Flügge-Lotz, I (1953). 'Discontinuous automatic control'. Princeton University Press.

André, J and Seibert, P (1956). 'Über stückweise linear Differentialgleichungen, die bei Regelungsproblemen auftreten, I and II'. Arch Math, 7, 148-165.

Weissenberger, S (1966). 'Stability-boundary approximation for relay-control systems with a steepest ascent construction of Lyapunov functions'. J Bas Engrg, 88, 419-428.

Fuller A T (1967). 'Linear control of non-linear systems'. Int J Contr, 5, 197-243.

Draženović, B (1969). 'The invariance conditions in variable structure systems'. Automatica, 5, 287-295.

Young, K, Kokotović, P V and Utkin, V I (1977). 'A singular perturbation analysis of high-gain feedback systems'. Trans IEEE, AC-22, 931-938.

Utkin, V I and Young, K (1978). 'Methods for constructing discontinuity planes in multidimensional variable structure systems'. Automat Remote Contr, 26, 1466-1470.

Mita, T (1976). 'Design of a zero-sensitive system'. Int J Contr, 24, 75-81.

Young, K (1978). 'Design of variable structure model-following control systems'. IEEE Trans, AC-23, 1079-1085.

Calise, A J (1980). 'Hyperstability in variable structure systems'. Proc 19th IEEE Conf on Decision and Control, Albuquerque, USA.

Landau, I D (1979). 'Adaptive Control: The Model Reference Approach'. Marcel Dekker, New York.

Grayson, L (1965). 'The status of synthesis using Lyapunov's Method'. Automatica, 3, 91-125.

Hang, C C and Parks, P C (1973). 'Comparative studies of model reference adaptive control systems'. Trans IEEE, AC-18, 419-428.

Landau, I D (1974). 'A survey of model reference adaptive techniques - theory and applications'. Automatica, 10, 353-379.

Courtiol, B and Landau, I D (1975). 'High speed adaptation system for controlled electrical drives'. Automatica, 11, 119-127.

Devaud, F M and Caron, J Y (1975). 'Asymptotic stability of model reference systems with bang-bang control'. Trans IEEE, Ac-20, 694-696.

Young, K-K D (1977). 'Asymptotic stability of model reference systems with variable structure control'. Trans IEEE, AC-21, 279-281.

Erzberger, H (1968). 'Analysis and design of model following control systems by state space techniques'. Proc JACC, Ann Arbor, 572-581.

Chen, Y T (1973). 'Perfect model following with real model'. Proc JACC, 287-293.

Zinober, A S I, El-Ghezawi, O M E and Billings, S A (1981). 'Variable Structure Control of Adaptive Model-Following Systems'. Proc IEE Conference on Control and its Applications, Warwick, 123-127.

Landau, I D and Courtiol, B (1972). 'Adaptive model-following for flight control and simulation'. J Aircraft, 9, 668-674.

Zinober, A S I (1974). 'Relay control of plants subject to parameter uncertainty'. PhD thesis, University of Cambridge.

Zinober, A S I (1975). 'Adaptive relay control of second-order systems'. Int J Contr, 21, 81-98.

Zinober, A S I (1977). 'Analysis of an adaptive third-order relay control system using non-linear switching surface theory'. Proc R Soc Edinburgh, 76A, 239-254.

Ryan, E P (1979). 'Singular optimal controls for second-order saturating systems'. Int J Contr, 30, 549-564.

Zinober, A S I (1979). 'Self-adaptive control of overhead crane operations'. Proc 5th IFAC Symp on Identification and system parameter estimation, Darmstadt, Pergamon Press, Oxford, 1161-1167.

Zinober, A S I (1980). 'Self-adaptive near-optimal control of diffusion equations'. Proc IEE, 127D, 290-295.

Zinober, A S I (1980). 'Adaptive variable structure systems'. Proc 3rd IMA Conf on Control Theory, Sheffield.

Young, K-K D (1978). 'Controller design for a manipulator using theory of variable structure systems'. IEEE Trans, SMC-8, 101-109.

Sabanović, A and Izosimov, D (1979). 'Application of sliding modes to induction motor control'. Proc IEEE Ind Appl Soc Meeting, Cleveland, USA, 793-801.

Golembo, B Z, Emel'yanov, S V, Utkin, V I and Shubladze, A M (1976). 'Application of piecewise - continuous dynamic systems to filtering problems'. Autom Remote Contr, 37, 369-377.

Gutman, S (1979). 'Uncertain dynamical systems - a Lyapunov min-max approach'. Trans IEEE, AC-24, 437-443.

Chapter Ten

Simple models for robust control of unknown or badly defined multivariable systems

D H OWENS and A CHOTAI

10.1 *Introduction*

Almost without exception frequency domain methods for the design of feedback control schemes for both scalar (Raven (1978)) and multivariable systems (Rosenbrock (1974), Owens (1978), Harris and Owens (1979) and MacFarlane (1980)) rely upon the existence of a model of the process to be controlled (the plant) in a form suitable as a basis for design calculations such as simulation, transfer function matrix or frequency response evaluation, calculation of poles and zeros etc. There are many instances however when a plant model is not known or the available plant model (obtained perhaps from a detailed analytical modelling exercise) is so complex that design calculations other than simulation are not feasible with available computing facilities. In either situation the plant model is (at least partially) unknown for the purposes of controller design yet the problem of constructing the control system still remains.

At the present time, there appear to be three general philosophies providing possible solutions to the problem:

(a) Identification (Eykhoff (1974)) of a low-order approximate system model from off-line analysis of input/output data obtained from plant records or simulation of a more complicated plant model. The resulting model is then used as the basis of controller design and the success of the approach assessed by on-line tuning at the commissioning stage or by extensive simulations of the controller using the real plant model.

(b) Self-tuning control of the unknown system (as described in chapters 2, 3 and 5) using a control strategy based on an assumed low-order parametric system model and on-line identification of the required controller parameters.

(c) Robust design of the control system in a manner ensuring that closed-loop stability and performance are insensitive to the unknown components of system dynamics.

All three philosophies have their own problems and areas of applicability and it is not the purpose of this chapter to make abstract judgements. We will however restrict attention to the notion of robust

Dr Owens and Dr Chotai are with the Control Engineering Department, Sheffield University.

controller design for unknown systems and highlight its place in the scheme of things using the following observations:

(i) If a controller designed on the basis of an identified plant model produces satisfactory closed-loop performance from this model, it is not necessarily true that the real plant is even stable and, if stable, the design is not necessarily insensitive to modelling errors, time-variation of parameters or nonlinearities. This is particularly true if high performance specifications are demanded for the closed-loop system.

(ii) Self-tuning controllers are known to be capable of providing useful solutions to practical design problems and a number of stability conditions are known (see chapter 5) when the plant model and the identified model have the same order. Little is known, however, of the general effect of order mismatch on the performance of the algorithms or of the effect of nonlinearities.

(iii) Both identification and self-tuning concepts require either access to sophisticated identification software or the use of high-level control hardware and software.

In this chapter it will be shown that it is possible to identify a class of unknown multivariable process plant for which robust proportional plus integral control systems can be designed without encountering the difficulties (i)-(iii) above. Clearly if a given piece of plant belongs to this class, robust design is a powerful alternative to the other strategies. This power is, however, obtained at a price - it is (in theory) necessary to have certain *a priori* information on the system structure. In many cases this structural information may well be self-evident from the physical laws governing dynamic behaviour, but, in other cases, it may be necessary to assume that the structure is correct and assess the validity of the assumption by the success (or failure) of the final design.

The conceptual basis of the ideas, as illustrated by the well-known technique from classical control of approximating plant dynamics by a first-order lag, is described in section 10.2 together with its theoretical justification from root-locus ideas. Section 10.3 extends the notion of a first-order lag to the case of multivariable process plant (Owens 1978, 1979) and outlines how simple two term controllers ensuring fast system responses with the required overshoot and damping characteristics and small loop interaction effects are easily designed without the need for other than 'back of envelope' computing facilities. In section 10.4 the use of the 'first-order controller' for the control of higher order plant is described (Owens 1979, Owens and Chotai 1980a) with emphasis on prediction of stability and transient performance and an evaluation of its sensitivity/robustness (Owens and Chotai 1981a) to data errors. Section 10.5 described some illustrative applications of the theory and, in section 10.6, it is noted that the design technique is easily extended to cope with assessment of the effect of measurement nonlinearities (Boland and Owens 1980, Owens 1981a) such as quantization and deadzone. Finally, in section 10.7, a brief review of current work in the area is given.

Throughout the chapter attention is focussed primarily on the intuitive source of the ideas, the form of the rigorous mathematical results and their interpretation in practice. The proofs (which rely on extensive use of linear systems theory and functional analysis) can be found in the references.

10.2 *The Single-input/Single-output Case: A Motivation*

In order to illustrate the basic notion underlying the technical content of this chapter, consider the single-input/single-output system described by an n-dimensional linear, time-invariant model of the state-space form

$$\dot{x}(t) = Ax(t) + Bu(t)$$
$$y(t) = Cx(t) \tag{1}$$

or the equivalent differential equation

$$P(D)y(t) = Q(D)u(t) \tag{2}$$

over the range of signal amplitudes of interest. Here A,B,C are real matrices of the appropriate dimension and P,Q are polynomials in the 'D-operator' D = d/dt. Denote the system transfer function by

$$G(s) \triangleq C(sI_n-A)^{-1}B \equiv \frac{Q(s)}{P(s)} \tag{3}$$

and consider the problem of designing a two-term controller K(s) for G(s) to ensure the required stability, transient performance and tracking characteristics from the closed-loop system shown in Fig 1(a).

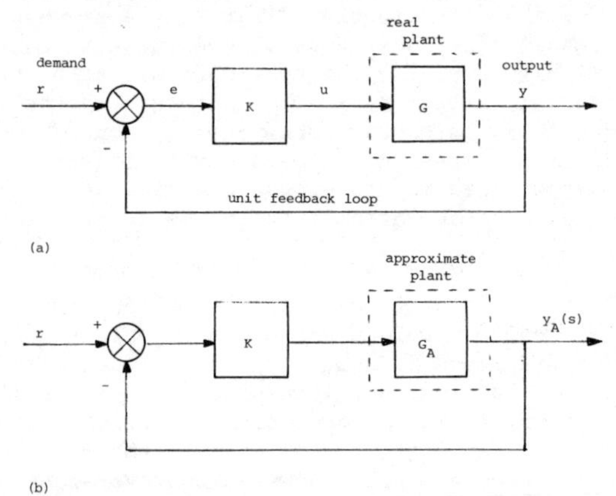

Fig 10.1 Real (a) and Approximating (b) Feedback Control Schemes

It is clear that controller design for the plant is possible in the normal manner if G(s) is known, or at least, if the frequency response $G(i\omega)$, $\omega \geq 0$, and the number of closed right-half-plane poles of G is known. Suppose that, for reasons such as those outlined in the introduction, G(s) is not known but that the system response y(t) to a unit step input has been obtained from plant tests or a simulation of an available complex model. A 'rough and ready' approximate model of the process is easily obtained from the well-known graphical construction shown in Fig 10.2 and can be represented by the first-order transfer function

$$G_A(s) \triangleq \frac{a}{1 + sJ} \tag{4}$$

which is defined and non-trivial if $a \neq 0$ and $y(t)$ has non-zero derivative at $t = 0+$.

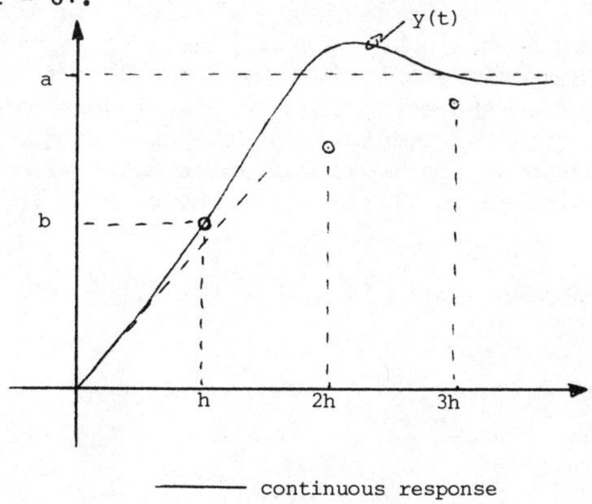

Fig 10.2 Construction of Approximate First Order Model from Time Constant and Steady State Data

An alternative form is

$$G_A^{-1}(s) = sA_o + A_1 \quad , \quad A_o \neq 0 \tag{5}$$

where, from the initial and final value theorems, we have the useful identities

$$A_o^{-1} = \frac{a}{J} = \left.\frac{dy_A(t)}{dt}\right|_{t=0+} = \left.\frac{dy(t)}{dt}\right|_{t=0+} = \lim_{s\to\infty} sG(s) = CB \tag{6}$$

(from which we note immediately that $CB \neq 0$ and that the system transfer function must have rank unity) and

$$A_1^{-1} = a = y_A(+\infty) = y(+\infty) = \lim_{s\to 0} G(s) \tag{7}$$

The unit step response $y_A(t)$ of the approximate first-order model may be very similar to $y(t)$, particularly if there are a large number of dipoles present in $G(s)$, but, in general, it should not be anticipated that the approximation is good in *open-loop conditions*.

Suppose now that the two-term controller

$$K(s) = \frac{pJ}{a}\left(1 + \frac{1}{sT}\right) \quad , \quad T>0 \tag{8}$$

is designed for the *approximate* plant G_A to ensure the required stability and transient response characteristics from the approximating feedback system shown in Fig 10.1b. This is a straightforward exercise but it is of no great value unless the results enable us to make useful predictions about the stability and performance of the *real* closed-loop system Fig 10.1a incorporating the final design. It is at this stage of the theoretical work that it is necessary to make some assumption concerning the structure of the unknown real plant. More precisely, we will suppose that, by means unknown, it is known that *the plant G is minimum-phase*. Remembering from the above that we also require that G has rank unity, a simple classical argument yields the observation that the root-locus of the configuration Fig 10.1a for gains $p \geq 0$ has one first order asymptote and n dipoles in the left-half plane at high gains and hence that

(a) the real closed-loop system is stable for all *high enough gains* p, and

(b) the closed-loop response has a first order character at *high gains*.

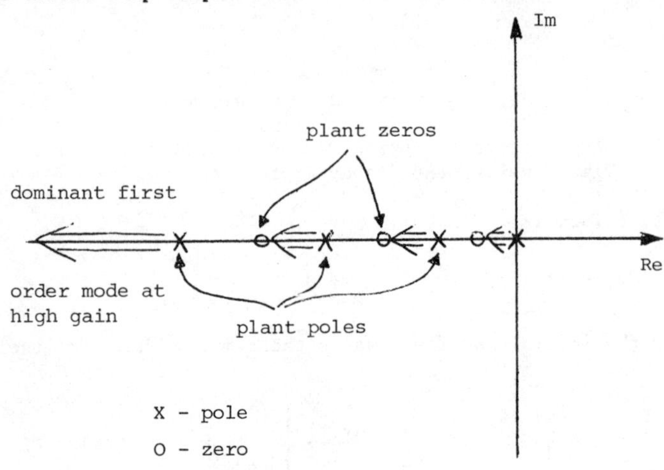

Fig 10.3 Typical Root-locus Plot

These ideas are illustrated in Fig 10.3. The first observation is reassuring as it indicates that controller design on the basis of a rough and ready first order model can be successful. The second observation provides some intuitive justification of the use of the first order model for prediction of *closed-loop* transient performance.

The justification of the above ideas can be obtained from the rigorous multivariable arguments outlined in Owens (1978), Edwards and Owens (1977) or, in more detail, in Owens and Chotai (1980a). Our main concern is to consider the generalization of this procedure to the case of multivariable sampled-data systems (Owens 1979, Owens and Chotai 1980a). This is described in the following sections but it is useful at this point to highlight the essential features in the context of the scalar case.

Suppose that the system input and output are actuated and sampled synchronously with sample interval h. Assuming piecewise constant inputs, the input and output $u_k \triangleq u(kh)$, $y_k = y(kh)$ are related by the discrete state-space model

$$x_{k+1} = \Phi x_k + \Delta u_k$$

$$y_k = Cx_k \tag{9}$$

or by an equivalent difference equation

$$\tilde{P}(z^{-1})y_k = \tilde{Q}(z^{-1})u_k \tag{10}$$

where $\tilde{P},\tilde{Q}$ are polynomials in z^{-1} and

$$\Phi = e^{Ah} \quad , \quad \Delta = \Phi\int_o^h e^{-At}Bdt \tag{11}$$

The system z-transfer function is denoted

$$G(z) \triangleq C(zI_n-\Phi)^{-1}\,\Delta \equiv \frac{Q(z^{-1})}{P(z^{-1})} \tag{12}$$

A 'rough and ready' approximate model of the process can now be obtained from a known response from zero initial conditions to a unit step input by the construction illustrated in Fig 10.4 to yield a first-order z-transfer function

$$G_A(z) = \frac{a}{1 + (z-1)a/b} \tag{13}$$

which is defined and non-trivial if $a \neq 0$ and $y(h) \neq 0$. An alternative form is

$$G_A^{-1}(z) = (z-1)B_o + B_1 \quad , \quad B_o \neq 0 \tag{14}$$

where

$$B_o^{-1} = y_A(h) = y(h) = \lim_{|z|\to\infty} zG(z) = C\Delta \tag{15}$$

and

$$B_1^{-1} = a = y_A(\infty) = y(\infty) = \lim_{z\to 1} G(z) \tag{16}$$

Clearly the real and approximate plants have identical initial and steady state response characteristics.

Consider now the question of identifying situations when a two-term controller designed using the simple first-order model will produce a stable closed-loop system with the required performance characteristics when applied to the real plant G. This problem is non-trivial (Owens 1979, Owens and Chotai 1980a) but the intuitive basis can be motivated using the results for the continuous case. More precisely, we have seen that the use of 'high' control gains is sufficient to ensure success in the continuous case. Interpreting the use of 'high' gains

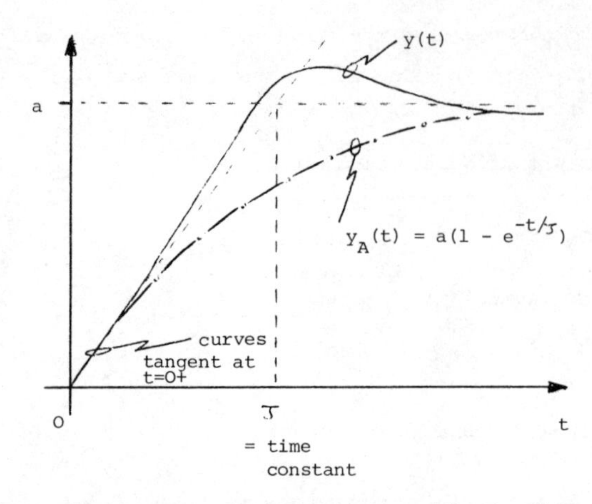

Fig 10.4 Graphical Construction for Approximate First-order Discrete System

as the generation of 'fast' closed-loop responses and noting that fast closed-loop responses will require fast sampling conditions in the discrete case, it is intuitively obvious that the use of 'fast' sample rates will form part of a set of sufficient conditions in the discrete case. Noting also that, under fast sampling conditions, a discrete system will tend to behave (roughly speaking) like a continuous system we must also expect that, as in the continuous case, the underlying continuous system will need to be minimum-phase with rank one transfer function. In summary, on intuitive grounds, the following conditions can be expected to be sufficient to guarantee the success of controller design based on the first-order model:

(a) The sample rate h^{-1} must be 'reasonably high', and

(b) The underlying continuous system should be minimum-phase with rank one transfer function.

These statements are in fact correct and *carry over to the multivariable case* provided that a little care is taken over the choice of control system! The generalization is described in the following sections.

10.3 *Discrete First-order Lags: The Multivariable Case*

The generalization of the material of section 10.2 to the multi-input/multi-output case relies upon the generalization of the idea of a first-order lag (Owens 1979, 1981b and Owens and Chotai 1980a). By analogy with the scalar case (in the form of equation (14) an m-input/m-output system with *synchronous* output sampling and control actuation of frequency h^{-1} and with mxm z-transfer function matrix $G_A(z)$ is an mxm discrete first-order lag if, and only if,

$$G_A^{-1}(z) = (z-1)B_o + B_1 \tag{17}$$

for some mxm matrices B_o, B_1 with $|B_o| \neq 0$. In the case of m = 1, this definition clearly reduces to the familiar scalar first-order lag.

The suggested controller for G_A has the multivariable proportional plus summation form

$$K(z) = B_o \operatorname{diag}\{1-k_jc_j + \frac{(1-k_j)(1-c_j)z}{(z-1)}\}_{1\le j\le m} - B_1 \tag{18}$$

parameterized in terms of the parameter matrices B_o, B_1 and scalars $k_1, \ldots, k_m$ and $c_1, \ldots, c_m$. The controller can be realized via the state-variable model

$$\begin{aligned} q_{k+1} &= q_k + e_k \\ u_k &= B_o\{\operatorname{diag}\,(1-k_j)(1-c_j)\}_{1\le j\le m}q_k \\ &\quad + (B_o \operatorname{diag}\{2-k_j-c_j\}_{1\le j\le m} - B_1)e_k \end{aligned} \tag{19}$$

or, in cases where $(1-k_j)(1-c_j) = 0$ for a number of indices j, a minimal realization of this model obtained by deleting the states in q that correspond to these indices.

After a little manipulation, the closed-loop z-transfer function matrix

$$\begin{aligned} H_c(z) &\triangleq (I_m+G(z)K(z))^{-1}G(z)K(z) \\ &\equiv \operatorname{diag}\{\frac{1}{(z-k_j)(z-c_j)}\}_{1\le j\le m} \quad (z\,(\operatorname{diag}\{2-k_j-c_j\}_{1\le j\le m} \\ &\quad - B_o^{-1}B_1) + (B_o^{-1}B_1 - \operatorname{diag}\{1-k_jc_j\}_{1\le j\le m})) \end{aligned} \tag{2}$$

which takes the form

$$H_c(z) = \operatorname{daig}\{\frac{z(2-k_j-c_j) - (1-k_jc_j)}{(z-k_j)(z-c_j)}\}_{1\le j\le m} \tag{21}$$

in the special (and very important) case of $B_o^{-1}B_1 = 0$. Clearly the closed-loop system is stable if

$$-1 < k_j < 1 \quad , \quad -1 < c_j \le 1 \qquad (1\le j\le m) \tag{22}$$

with rise-times and reset times in loop j obtained by suitable choice of 'tuning' parameters k_j, c_j respectively and zero steady state errors in response to steps if $c_j \neq 1$. In particular, if $B_o^{-1}B_1 \simeq 0$, equation (21) indicates that the closed-loop system also possesses small interaction effects between its loops.

In summary it is seen from the above that the given controller (18) is capable of producing the required rise and steady state responses from G_A by suitable choice of parameters, and, in a given special case, the closed-loop system is also approximately non-interacting. A more detailed analysis (Owens 1979, Owens and Chotai 1981a) also indicates that the special condition $B_o^{-1}B_1 \simeq 0$ can *always* be achieved under fast sampling conditions and hence will frequently be encountered in practice.

10.4 *Control Design for Unknown Discrete Multivariable System using a First-order Approximate Model*

Consider an m-input/m-output sampled-data system with synchronous input actuation and output sampling of frequency h^{-1} and described by a state-variable model of the form of (9) which is supposed to be generated by the underlying continuous model (1)

10.4.1 Construction of First Order Approximate Model

The approximate model (17) is characterized by the matrices B_o and B_1. The method of estimating them depends upon whether or not the large, complex model (9) is known. If it is then B_o can be defined by the natural generalization of (15) to the multivariable case,

$$B_o^{-1} = \lim_{|z|\to+\infty} zG(z) = C\Delta \tag{23}$$

provided that $C\Delta$ is nonsingular. It turns out (Owens and Chotai 1980a, 1981) that the choice of B_1 is not too critical. If we require the approximate model to reproduce the same steady state characteristics as the plant, the natural choice is the generalization of (16) namely

$$B_1^{-1} = \lim_{z\to1} G(z) = C(I_n-\Phi)^{-1}\Delta \tag{24}$$

if $I_n-\Phi$ is nonsingular. Alternatively the choice of $B_1 = 0$ can be acceptable (Owens 1979, Owens and Chotai 1980a) and has the advantage of considerably simplifying the form of the approximate model. In more general situations it may be possible (Owens and Chotai 1980a) to choose B_1 to achieve other desirable properties but such considerations are outside the scope of this chapter.

Suppose now that the model (9) is not known but that plant tests are undertaken to estimate the output vector sequence $\{y_1^{(i)}, y_2^{(i)}, \ldots\}$ generated by a unit step input in the ith plant input from zero conditions and that these experiments are repeated for all indices, $1\le i\le m$. Defining

$$Y_k = [y_k^{(1)}, y_k^{(2)}, \ldots, y_k^{(m)}] \quad , \quad k\ge0$$

then it is clear that

$$C\Delta = Y_1 \tag{25}$$

and, if the plant is stable, that

$$C(I_n-\Phi)^{-1}\Delta = \lim_{k\to\infty} Y_k \qquad (26)$$

Relations (25) and (26) can then be used in (23) and (24) respectively to obtain computed estimates of B_o and B_1.

10.4.2 Stability and Performance of the Real Feedback System

Consider now the problem of predicting the stability and performance characteristics of the feedback system of Fig 10.1a (with the controller (18) deduced from a first order approximate plant model) in terms of the characteristics of the approximating feedback system of Fig 10.1b. Several approaches are possible (Owens and Chotai 1980a) based upon refinements of the following result:

Theorem 10.1: An unknown m-input/m-output discrete multivariable plant (9), known to be generated from an unknown continuous multivariable plant (1) that is both minimum-phase and of uniform rank one (i.e. CB is nonsingular (Owens 1978)), will be stable in the presence of unity negative feedback with forward path controller (18) deduced from a first order approximate model if

(a) the tuning parameters k_j, c_j, $1\le j\le m$, satisfy (22) (i.e. the approximating feedback system Fig 10.1b is stable),

(b) the sampling rate h^{-1} is sufficiently fast.

Moreover, under these simple conditions, we have, for each reference demand sequence $\{r_k\}_{k\ge 0}$, the relation

$$\lim_{h\to 0+} (y_k - (y_A)_k) = 0 \qquad (27)$$

holds uniformly on the integers $k\ge 0$.

Remarks:

(i) In more general situations (Owens and Chotai 1980a) it is necessary to add in the condition that the procedure for choosing B_1 is such that $B_o^{-1}B_1\to 0$ as $h\to 0+$. This will always be the case if the procedures of section 10.4.1 are followed!

(ii) It is not necessary that the real plant is stable, nor that the real and approximate plants have similar stability characteristics.

Despite its simple form, the theorem requires careful interpretation for the purpose of application. More precisely, the result states that, if direct computation (using a model when available) or a combination of physical insight and intelligent guess work suggests that the discrete plant under consideration has underlying continuous dynamics possessing the *structural* properties of being minimum-phase (Owens 1978) with CB nonsingular, then the procedure of approximating open-loop plant dynamics by a multivariable discrete first order lag using the techniques of section 10.4.1, designing the controller using this model using the results of section 10.3 and finally implementing the designed controller on the real plant will be successful if (a) the controller stabilizes the approximate plant and (b) the sampling rate used in the implementation of the control scheme is fast enough. Success is, of course, measured by the consequent stability of the final design implemented on the real plant and, in particular, the fact (deduced from (27)) that, under fast sampling conditions, the closed-loop responses of real and approximating feedback systems will be almost *identical*.

The application of the result is illustrated in later sections. It is important to point out that, except in certain special cases (Owens and Chotai 1980a), it is not possible to provide computable estimates of the 'slowest' sampling rate $(h^*)^{-1}$ that will guarantee stability. This is a direct consequence of our assumed ignorance of those parts of the dynamics of the plant that cannot be modelled by first order dynamics and cannot be removed unless more detailed information is used. For a given choice of sampling rate it is clear therefore that the success or failure of the approach will only be discovered when the controller is finally tried out on plant or plant model. If the response characteristics obtained are not satisfactory, the theorem at least points out that the use of an increased sample rate will improve matters.

10.4.3 Sensitivity and Robustness

An important interpretation of the above results (Owens and Chotai 1981) is that careful design of the proportional plus summation control system and the use of reasonably fast sampling will yield a closed-loop system that is *insensitive* to a large class of perturbations to plant dynamics that leave the chosen values of B_o and B_1 unchanged and also do not violate the minimum-phase requirement. The control system is said to be *robust* with respect to this class of perturbations. The available proof of the results however (Owens and Chotai 1980a) does rely on *exact* evaluation of the B_o and B_1. Clearly this is not going to be possible in practice either because of computational errors, the effect of noise or errors in plant measurement equipment. Fortunately it can be shown (Owens and Chotai 1981a) that stability of the final closed-loop design will be achieved even if large errors are introduced and, in particular, that stability is most sensitive to errors in estimation of B_o.

10.5 *Illustrative Examples*

In each of the following examples the 'unknown' system will be represented by a known linear model for the purposes of obtaining comparative responses. This fact will not be used in the controller design however.

10.5.1 Level Control: A Single-input/single-output Example

Consider the problem of the construction of a proportional plus summation digital controller for the system illustrated in Fig 10.5 in order to regulate the liquid level in vessel one to a specified equilibrium level. It is assumed that input actuation and output sampling are synchronous.

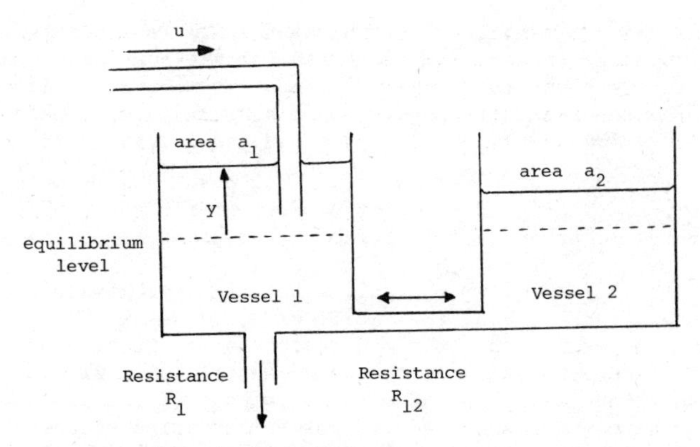

Fig 10.5 **Liquid Level System**

Assuming linear dynamics the plant has a (assumed unknown) model of the second order form

$$\dot{x}(t) = \begin{bmatrix} -\frac{1}{a_1}\left(\frac{1}{R_1} + \frac{1}{R_{12}}\right) & \frac{1}{a_1 R_{12}} \\ \frac{1}{a_2 R_{12}} & -\frac{1}{a_2}\frac{1}{R_{12}} \end{bmatrix} x(t) + \begin{bmatrix} \frac{1}{a_1} \\ 0 \end{bmatrix} u(t)$$

$$y(t) = [1 \quad 0]x(t) \tag{28}$$

which has $CB = a_1^{-1} \neq 0$ and one zero at the point $-a_2^{-1}R_{12}^{-1}$ in the left-half complex plane i.e. this system is always of uniform rank one and minimum phase. Taking, for numerical simplicity, the case of $a_1 = a_2 = 1$ and $R_1 = R_{12} = 1$, the system has poles at $s = -0.3$ and

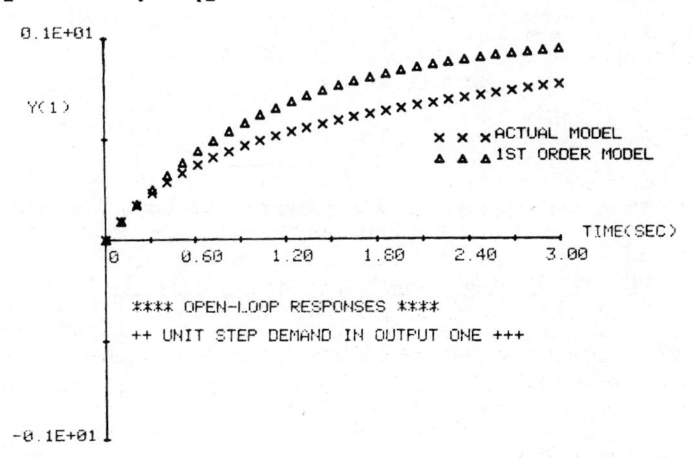

Fig 10.6 **Unit Step Responses of Real and Approximating Liquid Level Models**

s = −2.7. Assuming a sampling interval of h = 0.1 (which intuitively corresponds to a fairly fast sampling rate for the system), then analysis of the response to a unit step input from zero initial conditions following the procedure suggested in section 10.4.1 leads to the data $B_0 = 11.02$ and $B_1 = 1.00$ defining the approximating first order plant model. The unit step responses of real plant and first order model are compared in Fig 10.6. They clearly differ but the first order model captures the 'essential' features of the plant response.

The design of the two-term controller for the approximate model boils down to (dropping subscripts) the choice of k,c. But k and c can be interpreted as governing the rise and reset characteristics of the approximating feedback system in the sense that (equations (20) and (21)) k can represent the fast pole and c the slow pole of the system. We will therefore choose k = 0.5 and c = 0.95 for illustrative purposes. Theorem 10.1 now tells us that, *provided that the chosen sampling rate is fast enough*, we can expect that the resulting controller (obtained from (18) and the given data) will stabilize the real plant with response characteristics very close to those predicted by the first order model. That this is the case is verified by the responses of real and approximating feedback systems to unit step demands shown in Fig 10.7. The design procedure has clearly been successful in this case.

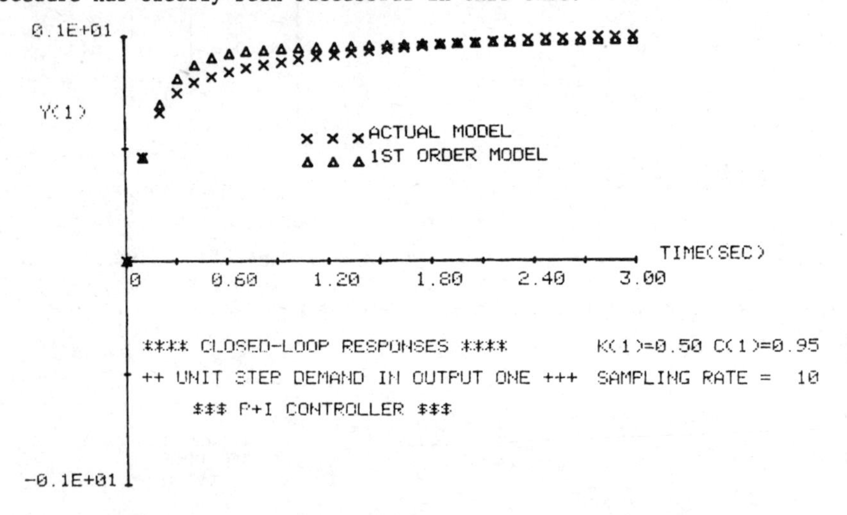

Fig 10.7 Closed-loop Unit Step Responses of Real and Approximating Level Control Configurations

10.5.2 Digital Control of an Open-loop Unstable Multivariable Plant

Consider the digital control of the two-input/two output *unstable* batch process discussed by Rosenbrock (1974) with continuous model defined by the matrices

$$A = \begin{bmatrix} 1.38 & -0.2077 & 6.715 & -5.676 \\ -0.5814 & -4.29 & 0 & 0.675 \\ 1.067 & 4.273 & -6.654 & 5.893 \\ 0.048 & 4.273 & 1.343 & -2.104 \end{bmatrix}$$

$$B = \begin{bmatrix} 0 & 0 \\ 5.679 & 0 \\ 1.136 & -3.146 \\ 1.136 & 0 \end{bmatrix}, \quad C = \begin{bmatrix} 1 & 0 & 1 & -1 \\ 0 & 1 & 0 & 0 \end{bmatrix} \qquad (29)$$

This model is assumed to be known but we will suppose that the designer wishes to attempt controller design using only simple calculations with verifying simulations. It is easily verified that the system is minimum-phase with CB nonsingular and that it is open-loop unstable. Assuming a sampling interval of h = 0.02, simulations of the open-loop responses to unit step demands in each input in the manner described in section 10.4.1 leads to the data (equation (25))

$$B_o = \begin{bmatrix} 0.0018 & 9.17 \\ -16.07 & 0.0057 \end{bmatrix} \qquad (30)$$

but, as the system is open loop-unstable, (26) does not hold and we must use (24) to give

$$B_1 = \begin{bmatrix} 0.141 & 0.296 \\ 0.995 & 2.455 \end{bmatrix} \qquad (31)$$

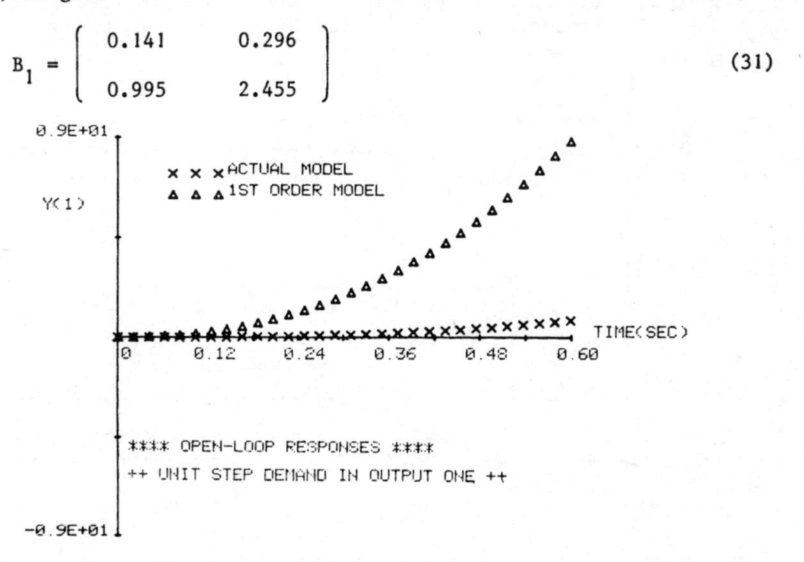

Fig 10.8 Responses of Real and Approximating Batch Process

The responses of the first output of the real and approximate model to a unit step input in the first channel are given in Fig 10.8 for comparative purposes.

The design of the two-term controller for the approximate model relies purely on the choice of tuning parameters. Remembering that k_j and c_j are poles corresponding to rise-time and reset characteristics in the jth loop, choose $k_1 = k_2 = 0.5$ and $c_1 = c_2 = 0.95$ for illustrative purposes and note that the elements of $B_o^{-1}B_1$ are small so that we can

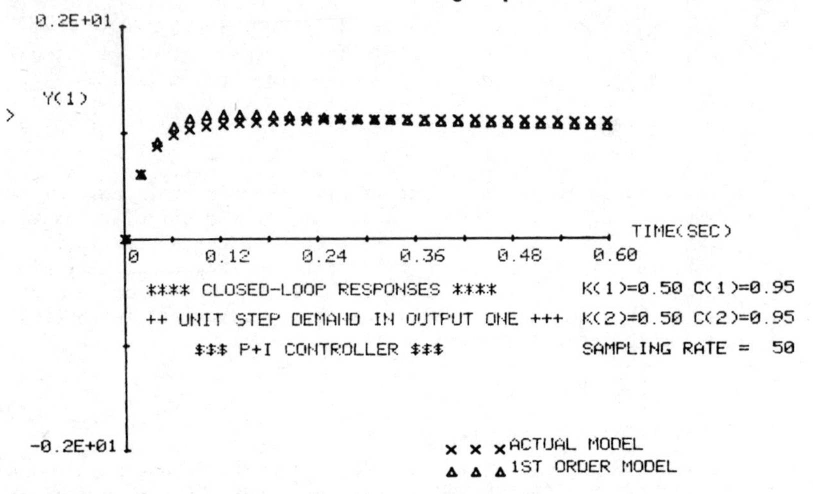

Fig 10.9a Closed-loop Responses of Real and Approximating Batch Processes to a Unit Step Demand in First Output

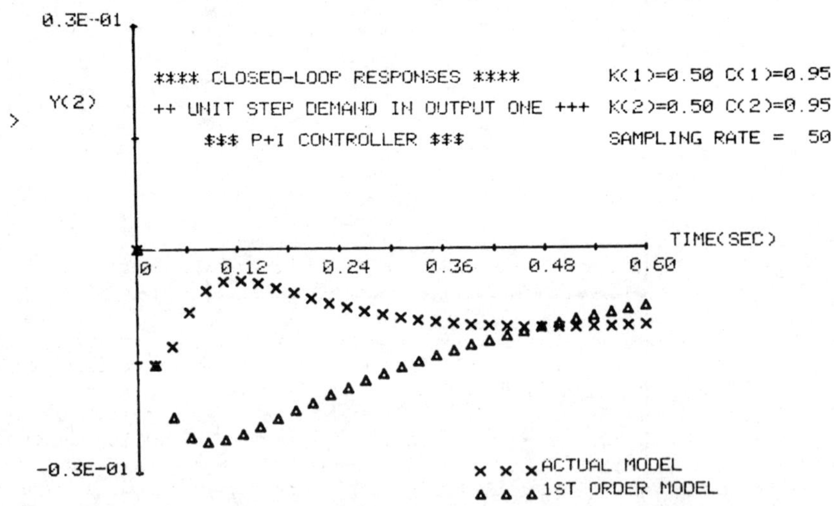

Fig 10.9b Closed-loop Responses of Real and Approximating Batch Processes to a Unit Step Demand in First Output

expect a high performance controller with little interaction between the two loops of the approximating feedback system. Again theorem 10.1 tells us that, if our choice of sample rate is high enough, the resulting controller (obtained from (18) and the given data) will stabilize the real plant with response characteristics close to those predicted by the first order model. This is indeed the case as illustrated by the closed-loop responses of the real and approximating feedback system given in Fig 10.9.

10.6 *Effect of Measurement Nonlinearities*

The assumption of linear plant dynamics is very often justifiable in engineering applications, particularly if the primary function of the controller is to act as a regulator about a fixed operating condition with only occasional and small deviations from this point. Also the assumption of control linearity is reasonable as the control calculations are performed 'inside' a computer. It is frequently the case however that the control actuators and, in particular, output sensors possess simple nonlinear characteristics such as dead-zone or quantization. We will restrict our attention to the case of measurement nonlinearity illustrated in Fig 10.10, where the element N is a time-invariant, memoryless nonlinearity satisfying the constraint (Boland and Owens 1980)

$$\| N\eta - \eta \|_m \leq \frac{q}{2} \tag{32}$$

for all mx1 real vectors η. Here $\| x \|_m = \max\limits_{1\leq j\leq m} |x_j|$ is the normal uniform norm on m-vectors. This class of nonlinearity covers the case of quantization and deadzone and can be extended to cover more general cases (Owens 1981).

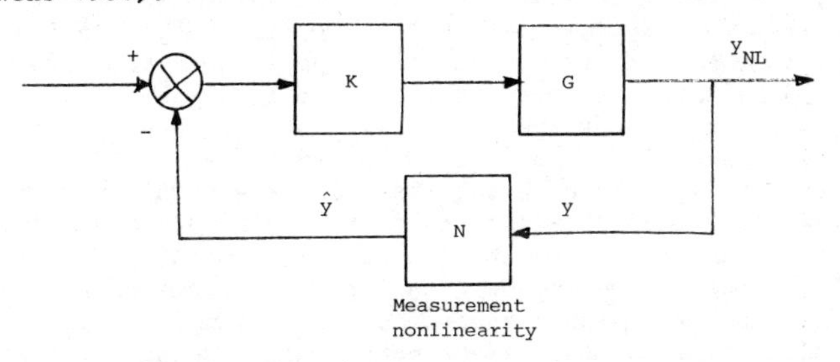

Fig 10.10 Multivariable Feedback System with Measurement Nonlinearity

The precise problem that must be considered if the plant model is unknown and controller design is undertaken in the manner outlined in section 10.4 is the relationship between the dynamics of the approximating *linear* feedback system of Fig 10.1b and those of the real *nonlinear* feedback system of Fig 10.10. The level of ignorance of system dynamics assumed prevents a precise answer to this question that covers all possible conditions. Following the general ideas illustrated in theorem 10.1 it is possible, however, to prove the following nonlinear analogue (Boland and Owens 1980, Owens 1981):

Theorem 10.2: Under the conditions of theorem 10.1, for each reference demand sequence $\{r_k\}_{k\geq 0}$, the responses of the approximating linear feedback system and the real nonlinear feedback system satisfy the relation

$$\lim_{h\to 0+} \| (y_A)_k - (y_{NL})_k \|_m \leq \frac{q}{2} \tag{33}$$

for each k≥0, the limit being uniform on this interval.

The practical interpretation of this abstract result follows from the definition of the uniform norm $\| \cdot \|_m$ i.e., under the stated conditions, the maximum modulus of the transient error involved in the use of the approximating linear feedback system to predict the dynamics of the real nonlinear feedback system is less than the nonlinearity constant q/2 *under fast sampling conditions*. The result can, of course only be applied in an approximate manner in real (finite sampling rate) conditions, but it can be a good working approximation (Boland and Owens 1980) and, at minimum, provides a reassuring guarantee of stability (in the bounded-input/bounded-output sense) and an estimated upper bound on the transient magnitude of any possible limit cycles.

10.7 *Conclusions*

The chapter has reviewed the conceptual basis and some formal mathematical results available for the design of stabilizing two-term controllers for a well-defined class of multivariable process plant for which detailed knowledge of process dynamics is not available and illustrated the application of the results by two non-trivial examples. The power of the techniques lie in the *guaranteed* ability to design high performance, small interaction feedback systems using only a small amount of system data deduced from plant step responses, an assumed knowledge of some elementary structural properties of the underlying continuous system and the implementation of a reasonably fast sampling rate. The ideas do in fact apply in the case of analogue control (Edwards and Owens 1977, Owens 1978, Owens and Chotai 1980a, 1980b, 1981 and Owens 1981) and can be regarded as the natural generalization of the well-known classical notion of fitting a first order model to plant dynamics. In this case, it is a partial generalization (in the delay-free case) of the well-known 'tuning' technique of Ziegler and Nichols (1942) as the designed controller has a simple form, being defined in terms of elementary plant parameters together with a number of 'tuning parameters' that can either be adjusted on-line to produce the required performance or estimated from analysis of the approximate first-order plant model.

The material can be extended in a number of directions to increase its applicability and the information available. Some indication of the effect of nonlinearities has already been obtained, with encouraging results, and the use of more complex approximate models is envisaged to cover cases where a first order model carries insufficient information to make it a useful vehicle for design. This is the topic of present study.

Finally we note that there are clear connections between the ideas expressed here and those expressed in chapter 8 and the work of Davison (1976), Åström (1980), Pentinnen and Koivo (1980) and Owens and Chotai (1981b) which provide distinct and viable approaches to cover other aspects of unknown systems control. All of these techniques do provide information on stability of the final design but the ideas expressed here (when applicable) have the added bonus of providing information on detailed transient performance in both the linear and nonlinear cases.

References

Åström, K J (1980). 'A Robust Sampled Regulator for Stable Systems with Monotone Step Responses', Automatica, 16, 313-315.

Boland, F M, Owens, D H (1980). 'On Quantization and other Bounded Nonlinearities in First-order Discrete Multivariable Control', Proc IEE, Pt D, 127, 2, 38-44.

Davison, E J (1976). 'Multivariable Tuning Regulators: the Feedforward and Robust Control of a General Servomechanism Problem', IEEE Trans, Aut Con, AC-21, 35-47.

Edwards, J B, Owens, D H (1977). 'First-order-type Models for Multivariable Process Control', Proc IEE, 124, 11, 1083-1088.

Eykhoff, P (1974). 'System Identification', Wiley.

Harris, C J, Owens, D H (1979). 'Multivariable Control Systems', Special Issue of the Control and Science Record, Proc IEE, 126, 6, 537-648.

MacFarlane, A G J (1980). 'Frequency Response Methods in Control Systems', IEEE Press.

Owens, D H (1978). 'Feedback and Multivariable Systems', Peter Peregrinus.

Owens, D H (1979). 'Discrete First-order-models for Multivariable Process Control', Proc IEE, 126, 6, 525-530.

Owens, D H (1981a). 'On the Effect of Nonlinearities in Multivariable First Order Process Control', Proceedings of the IEE Int Conf 'Control and Its Applications', Univ Warwick, UK, March 23-25, 240-244.

Owens, D H (1981b). 'Multivariable and Optimal Systems', Academic Press.

Owens, D H, Chotai, A (1980a). 'Robust Control of Unknown or Large-scale Systems using Transient Data Only', Research Report No 134, Dept Control Eng, Univ Sheffield, UK.

Owens, D H, Chotai, A (1980b). 'High Performance Controllers for Unknown Multivariable Systems', Research Report No 130, Dept Control Eng, Univ Sheffield, UK.

Owens, D H, Chotai, A (1981). 'Robust Control of Unknown Discrete Multivariable Systems', Research Report No 135, Dept Control Eng, Univ Sheffield, UK.

Owens, D H, Chotai, A (1981b). 'Robust Control of Unknown Multivariable Systems using Monotone Modelling Errors', Research report No 144, Dept Control Eng, Univ Sheffield, UK.

Pentinnen, J, Koivo, H N (1980). 'Multivariable Tuning Regulators for Unknown Systems', Automatica, 16, 393-398.

Raven, F H (1978). 'Automatic Control Engineering', McGraw-Hill.

Rosenbrock, H H (1974). 'Computer-aided-design of Control Systems', Academic Press.

Ziegler, J G, Nichols, J G (1942). 'Optimum Settings for Automatic Controllers', Trans ASME, 64, 75-91.

Chapter Eleven

Single and multivariable application of self-tuning controllers

A J MORRIS, Y NAZER and R K WOOD

11.1 *Introduction*

The work of Peteraka (1970), Åström and Wittenmark (1973), Clarke and Gawthrop (1975, 1979) and Gawthrop (1977) on self tuning regulators and controllers has given rise to serious consideration being given to their implementation on large scale plant. However, the simple two or three term regulator (PI or PID) unquestionably remains the most common regulator in industrial process control due to its inherent robustness, ease of tuning, tolerance of process changes, and lack of a detailed process knowledge required during the initial design stage. In spite of this there are cases where detuning is necessary, to ensure non-oscillatory behaviour and stability over a wide range of operating conditions, giving rise to sloppy control.

Chemical process control systems can in many respects pose more complex problems than those usually met in general automatic control. This is because they are characterised by large time lags, in some cases time varying and sometimes of the non-minimum phase type, they can exhibit large and time varying transportation delays (dead times) can be grossly non-linear and interactive and in many cases although subjected to some stochastic disturbances can be almost deterministic in operation. In such processes the main aim is plant output regulation rather than servo-type control, although a fast stable response to set point demand changes and sudden large load disturbances is important. In general there are three conflicting requirements for good chemical process control: optimum long term steady state regulation; good set point following properties; and the need to ensure that any control actions applied or transducer failure experienced, and the effects they may produce on the plant, do not in any way endanger plant safety. It is not only by rapid and sustained control valve movement that dangerous conditions can be induced but also unconstrained response to rapid rates of change in the measured process variables could cause serious degradation of the product or even damage to the plant itself. Although during normal operation plant safety and product quality requirements are rarely violated, in the presence of large errors between measured and desired process output they may become of overriding importance.

A J Morris and Y Nazer are with the Department of Chemical Engineering, University of Newcastle upon Tyne.

R K Wood is with the Department of Chemical Engineering, University of Alberta, Canada.

Because of these real industrial problems work has been directed towards the application of robust versions of the self tuning controller. This has involved extensions of the basic minimum variance control law by (see chapter two) Clarke and Gawthrop (1979, 1977), Morris and Nazer (1977, 1979), Morris, Fenton and Nazer (1977), Lieuson, Morris, Nazer and Wood (1980) and pole assignment by Wellstead (see chapter three), Zanker and Edmunds (1978). Indeed the volume of literature devoted to single-input single-output self tuning controllers is now considerable. However, in the multivariable case the self tuning controller has not been extensively considered (see chapter three). Some attention though has been focused on such systems, in particular Borrison (1979) has extended the minimum variance regulator of Åström to deal with multi-input multi-output systems with equal numbers of inputs and outputs, and some improvements have been suggested by Keviczky and Hatthessy (1977). A practical application of this controller to a cement blending process has been described by Keviczky et al (1978). Minimum variance decoupling control has been investigated in Sinha (1977), who used cascaded sets of single-input single-output self tuners onto decoupled subsystems of the overall multivariable process. Here the "decoupler" can be designed by any standard technique such as that suggested by Rosenbrock (1974). In all of the above versions of the minimum variance multi-input multi-output self tuning controller, restrictions must be placed on the equality of the individual loop time delays and all suffer from the disadvantages common to the single variant cases from which they were developed. A particularly severe limitation being their inability to control non-minimum phase systems which can cause problems in digital control.

This contribution outlines the development of both single and multivariable self tuning controllers based on the single variable controller of Clarke and Gawthrop (1975, 1977) (see chapter two). The controller can deal with non-minimum phase systems, does not have the limitation that all loop time delays must be equal and is amenable to pole location specification for pre-design, or even on-line design, of the desired system transient behaviour. The proposed self tuning controller uses a recursive factorisation parameter estimator based on the method of least squares. Both set point following and feedforward control actions are included in the tuning strategy. Particular attention is paid to processes whose dynamics can be adequately represented by multivariable transfer functions having time delays. Such systems are found widely in the process industries. Since one of the most important and interactive in this category is distillation column product composition control, the distillation process is used for the purpose of control strategy evaluation. The problems associated with multivariable control of such a complex process have been discussed widely in the literature and the conclusions reached would appear to suggest that satisfactory control is only possible for moderate disturbances, McGinnis and Wood (1974). The difficulties met in attempting to apply linear state space based optimal control theory usually arises from the severe non-linear behaviour of the column and the large number of state variables required to adequately describe its dynamic behaviour, Hu and Ramierez (1972). Control schemes which require only terminal measurements of top and bottom product compositions are therefore attractive, although in-line measurement of these using gas chromatographs, etc, is sometimes so time consuming that direct feedback control becomes extremely difficult. Furthermore, due to the heavy non-linearities in the process some form of adaptive control scheme is desirable if the operating regions are to be changed, or process characteristics change, over periods of time.

Feedforward control of product compositions, Luyben (1970, 1975), Nisenfield and Miyasaki (1973), not only requires measurement of process disturbances which may only be possible at large cost but also may require feedforward controller parameters that change to adapt to the column non-linear behaviour. In addition feedforward controller can be notoriously difficult to tune especially if the process dynamics are inadequately known and if disturbances only occur at infrequent intervals of time. In this contribution the robust and generalised nature of the controllers developed is highlighted and their performance evaluated by application to the separate and simultaneous regulation of distillation column terminal compositions. The results of both comprehensive computer simulation studies and pilot plant tests are presented. These are compared with the controlled performances achieved when using conventional regulators and regulators based upon steady state gain decoupling techniques. The main control objective is to regulate the top and bottom product compositions at their desired values despite feed flow rate disturbances of magnitude of 20-25% of nominal steady state conditions.

11.2 *Self Tuning Control*

11.2.1 Multi-input Single-output Systems

A general discrete time model of the process being controlled is assumed to take the form:

$$A(z^{-1})y(t) = z^{-k}\,B(z^{-1})u(t) + C(z^{-1})\xi(t) + z^{-k_\ell}\,L(z^{-1})v(t) \quad (1)$$

Where $A(z^{-1})$, $B(z^{-1})$, $C(z^{-1})$ and $L(z^{-1})$ are z-transform polynomials with $a(0) = 1$, $b(0) \neq 0$, and $c(0) = 1$. $C^{-1}(z^{-1})$ is assumed a stable transfer function with a zero mean uncorrelated random input $\xi(t)$. The system time delay is denoted by k sample intervals ($k \geq 1$). The term v(t) allows for the inclusion of other control loop inputs in a feed-forward manner, e.g. load disturbances, with the associated time delay denoted by k_ℓ sample intervals. The controller design is based upon the minimisation of the general performance function (see chapter two):

$$J = E\{[P(z^{-1})y(t+k) - R(z^{-1})w(t)]^2 + [Q'(z^{-1})u(t)]^2\} \quad (2)$$

Where $P(z^{-1})$, $R(z^{-1})$ and $Q'(z^{-1})$ are transfer functions in (z^{-1}) placing general weightings on system output y(t), set point w(t) and control effort u(t) respectively and are expressed as polynomial ratios:

$$P(z^{-1}) = \frac{P_n(z^{-1})}{P_d(z^{-1})}\;;\quad R(z^{-1}) = \frac{R_n(z^{-1})}{R_d(z^{-1})}\;;\quad Q'(z^{-1}) = \frac{Q_n'(z^{-1})}{Q_d'(z^{-1})} \quad (3)$$

The performance index is minimised by selecting a single step control action which at each sample interval attempts to drive a k-step ahead prediction of the deviation of the weighted output of the process, with respect to the weighted set point, to zero. Recalling equation (1) and rewriting it in terms of the weighted output of the process gives:

$$Py(t+k) = \frac{P\ B}{A} u(t) + z^k \frac{P\ C}{A} \xi(t) + z^{k-k_\ell} \frac{P\ L}{A} v(t) \qquad (4)$$

where for the sake of simplicity the term (z^{-1}) has been dropped from the z-transfer function polynomials. In this equation the terms associated with the noise process $\xi(t)$ can be expanded as:

$$z^k \frac{P_n\ C}{P_d\ A} \xi(t) = z^k\ E\ \xi(t) + \frac{F}{P_d\ A} \xi(t) \qquad (5)$$

and is expressed in terms of the sum of future disturbances and those up to and including time t. Here

$$E(z^{-1}) = e(0) + e(1)\ z^{-1} + \ldots + e(k-1)\ z^{-k+1}$$

$$F(z^{-1}) = f(0) + f(1)\ z^{-1} + \ldots + f(n_F)\ z^{-n_F}$$

The weighted system output given by equation (4) can now be written in terms of the known sequences of y(t), u(t), v(t) and the unknown future disturbances $\xi(t+k)$ by combining equations (1), (4) and (5) to give:

$$C\ Py(t+k) = E\ \xi(t) + \frac{F}{P_d}\ y(t) + E\ B\ u(t) + z^{k-k_\ell}\ E\ L\ v(t) \qquad (6)$$

Now by defining two new polynomials as:

$$G(z^{-1}) = E(z^{-1})\ B(z^{-1}) = g(0) + g(1)\ z^{-1} + \ldots + g(n_G)\ z^{-n_G}$$

$$D(z^{-1}) = E(z^{-1})\ L(z^{-1}) = d(0) + d(1)\ z^{-1} + \ldots + d(n_D)\ z^{-n_D}$$

and by defining a predicted weighted output of the process, using data up to and including time t, as:

$$[Py(t+k|t)]^* = Py(t+k) - E\xi(t+k) \qquad (7)$$

a k-step ahead output prediction is given by:

$$C\ [Py(t+k|t)]^* = \frac{F}{P_d}\ y(t) + G\ u(t) + z^{k-k_\ell}\ D\ v(t) \qquad (8)$$

Thus at any time t, since c(0) was assumed equal to 1, the weighted predicted output of the process becomes:

$$[Py(t+k|t)]^* = \frac{F}{P_d}\ y(t) + G\ u(t) + z^{k-k_\ell}\ D\ v(t) - \sum_{i=1}^{n_C} c(i)\ [Py(t+k-i|t-i]^* \qquad (9)$$

By defining

$$H(z^{-1}) = z - z\ C(z^{-1})$$

equation (9) becomes:

$$[Py(t+k|t)]^* = \frac{F}{P_d} y(t) + G\, u(t) + z^{k-k_\ell} Dv(t) + H\, [Py(t+k-1|t-1)]^*$$

Since the value of $[Py(t+k|t]^*$ is now known, the performance function J can be rewritten by substituting for the weighted process output using equation (7) to give

$$J = E\{([Py(t+k|t)]^* - E\,\xi(t+k) - R\,w(t))^2 + (Q'\,u(t))^2\} \qquad (10)$$

Assuming $\xi(t+k)$ is uncorrelated with $u(t-i)$ and $y(t-i)$ for all $i \geq 0$, and since $\xi(t)$ represents a noise process with zero mean the performance function is minimised, $(dJ/du(t) = 0)$, when:

$$[Py(t+k|t)]^* - R\,w(t) + Q\,u(t) = 0 \qquad (11)$$

where Q is a new control effort weighting function given by:

$$Q = Q' \frac{q'(0)c(0)}{e(0)b(0)}$$

Now by defining a measurable scalar output function as:

$$\phi(t+k|t) = Py(t+k) - R\,w(t) + Q\,u(t) \qquad (12)$$

together with a corresponding predictable output function as:

$$\phi^*(t+k|t) = [Py(t+k|t)]^* - R\,w(t) + Q\,u(t) \qquad (13)$$

the control strategy is obtained by setting the prediction $\phi^*(t+k|t)$ to zero at each step of the control policy calculation to give:

$$u(t) = \frac{1}{Q}\,(T\,w(t) - [Py(t+k|t]^*) \qquad (14)$$

This control law represents the inclusion of a k-step prediction of the weighted process output in the feedback path, with series compensation Q^{-1} in the forward path.

The control strategy developed for known constant parameter systems can be extended to include those systems with unknown but constant parameters and also those with slowly time varying parameters. This is achieved by using on-line parameter estimation to identify the co-efficients of the controller polynomials as $\hat{F}$, $\hat{G}$, $\hat{C}$ and $\hat{D}$. Fig 11.1 shows schematic diagram illustrating the structure of the self tuning control law.

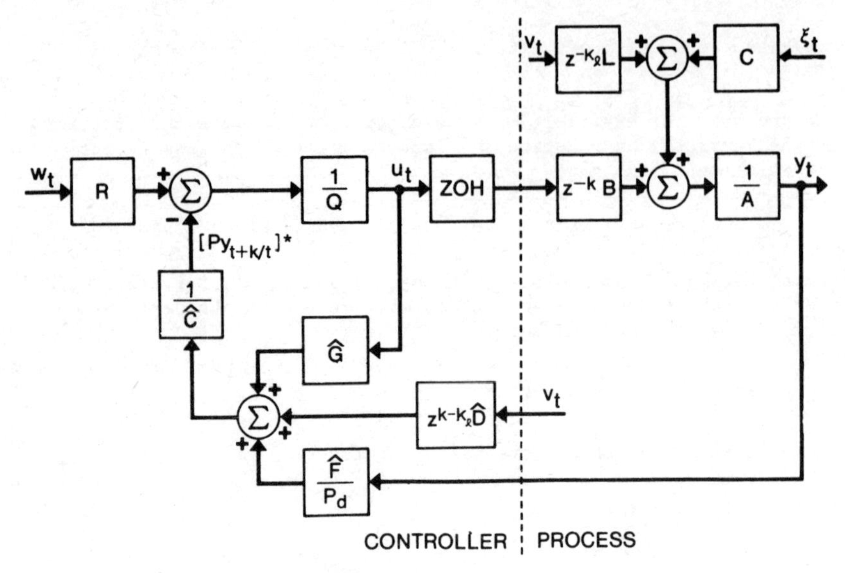

Fig 11.1 Schematic Diagram of the Multi Input Single Output Self Tuning Controller.

11.2.2 Parameter Estimation

A general overview of several methods of recursive parameter estimation has been given by Söderström et al (1974). Although no single recursive parameter estimator can be said to be uniformly best, studies by Morris and Nazer (1977) have indicated that those based on covariance matrix factorisation techniques, Peterka (1975) and Bierman (1975), do tend to provide a far superior performance than conventional recursive least squares methods. A recursive upper diagonal factorisation method is used in this work, Bierman (1975).

Given the known constant parameter control law developed above, the extension to unknown parameter systems can be achieved using a simple recursive least squares technique. Introducing a vector of parameter estimates:

$$\theta^T(t) = [g(0),\ g(1),\ \ldots,\ g(n_G)\ ;\ f(0),\ f(1),\ \ldots,\ f(n_F)\ ;\ d(0),\ d(1),\ \ldots,\ d(n_D)\ ;\ h(0),\ h(1),\ \ldots,\ h(n_H)] \quad (15)$$

and a vector of regressors:

$$\begin{aligned} x(t) = [u(t) &, u(t-1) &&, \ldots, u(t-n_G) \\ y(t)/P_d &, y(t-1)/P_d &&, \ldots, y(t-n_F)/P_d \\ v(t+k-k_\ell) &, v(t+k-k_\ell-1) &&, \ldots, v(t+k-k_\ell-n_D) \\ [Py(t+k-1|t-1)]^* &, [Py(t+k-2|t-2)]^* &&, \ldots, [Py(t+k-n_H|t-n_H]^* \end{aligned} \quad (16)$$

The predicted weighted output of the process can then be written as:

$$[Py(t+k|t)]^* = x(t)\ \theta(t) \quad \text{or} \quad [Py(t|t-k)]^* = x(t-k)\ \theta(t) \tag{17}$$

Now if at time t the elements of x(t-k) contain known sequences, and if E ξ(t) were white it would be possible to estimate the parameter vector θ using normal linear least squares. However, when k>1 E ξ(t) will not be white and ordinary least squares cannot be applied. The approach in self tuning control is therefore to cascade the control strategy with a parameter estimator to replace $[Py(t|t-k)]^*$ by an estimated value $[P\hat{y}(t|t-k)]^*$ and use a linear least squares based method as though $[P\hat{y}(t|t-k)]^*$ was correct and the prediction error was white, ε(t).

i.e.

$$[P\hat{y}(t|t-k)]^* + \varepsilon(t) = [Py(t|t-k)]^* + E\ \xi(t) \tag{18}$$

and

$$Py(t) = [P\hat{y}(t|t-k)]^* + \varepsilon(t) = \hat{x}(t-k)\ \theta(t) + \varepsilon(t) \tag{19}$$

Here $\hat{x}(t-k)$ is equivalent to x(t-k) except that the elements $[Py(t-i|t-k-i)]^*$ are replaced by their estimated values $[P\hat{y}(t-i|t-k-i)]^*$, $1 \le i \le n_H$. The recursive least squares estimate of θ(t) is then given by

$$\left.\begin{aligned} r &= \rho_1 + \hat{x}^T(t-k).P_c(t).\hat{x}(t-k) \\ K_g(t+1) &= P_c(t).\hat{x}^T(t-k)/r \\ P_c(t+1) &= (1/\rho_2)\ [P_c(t) - K_g(t+1).K_g{}^T(t+1).r] \\ \hat{\theta}(t+1) &= \hat{\theta}(t) + K_g(t+1).[Py(t+1) - \hat{x}(t-k).\hat{\theta}(t)] \end{aligned}\right\} \tag{20}$$

where $P_c(t)$ represents the covariance matrix of the estimated parameters and $K_g(t)$ is the estimator or Kalman gain vector. ρ_1 represents the standard deviation of the process noise sequence and ρ_2 (≤ 1) is introduced to make the estimator adaptive and allow tracking of slowly time varying parameters. The k-step ahead estimated weighted process output can then be calculated at time t by:

$$[P\hat{y}(t+k|t)]^* = \hat{x}(t)\ \hat{\theta}(t) \tag{21}$$

The forgetting factor ρ_2 in equation (20) is introduced to discount past data when performing the least squares procedure, Clarke (chapter six). In many cases during process regulation, the estimation algorithm is continually excited by process disturbances (noise) which are normally relatively uniform in time. In these circumstances a value for ρ_2 between 0.95 and 0.99 usually works satisfactorily. However, there are situations, especially in the process industries, where process and measurement noise is so small that the major excitation comes from deterministic load disturbances and set point changes. Such disturbances or changes are usually irregular resulting in insufficient excitation of the estimator leading to "blow up" or "bursting" of the system outputs and inputs when ρ_2 is chosen to be less than unity, Morris et al (1977) Clarke, (chapter six). This effect can be intuitively understood by considering the behaviour of the covariance matrix update equation where the negative term reduces parameter estimate uncertainty from the last measurement. When the system is unexcited the vector $P_c(t)\,\hat{x}^T(t-k)$ tends to zero and the covariance matrix is updated as:

$$P_c(t+1) = \frac{1}{\rho_2} P_c(t) \tag{22}$$

causing an exponential inflation of P_c when ρ_2 is chosen less than unity. Over long periods of unexcited operation P_c can become very large, as indeed can the values of the parameter estimates, due to the lack of confidence being expressed. This inflation can lead to both numerical problems in the estimator as well as large erratic parameter and process behaviour when the system is eventually disturbed. There are several ad-hoc methods of eliminating blow up, ranging from modification of the forgetting factor to prevent over inflation of P_c when the system becomes under excited, to ensuring P_c remains bounded by preventing updating taking place when the prediction error is small. In this work the covariance matrix is uniquely factorised as:

$$P_c(t) = U(t)\, D(t)\, U^T(t) \tag{23}$$

where U(t) and D(t) are unit upper triangular and diagonal matrices respectively. $P_c(t)$ is replaced by its U-D factors in the updating procedure of equations (20) to give:

$$U(t+1)\, D(t+1)\, U^T(t+1) =$$

$$\frac{1}{\rho_2} U(t)[D(t) - \frac{1}{r}(D(t)U^T(t)\hat{x}^T(t-k))\,(D(t)U^T(t)\hat{x}^T(t-k))^T]U^T(t)$$

and it becomes straightforward to arrange for the factors of the covariance matrix to be recursively updated and the controller parameters estimated at any desired interval of time, Nazer (1981).

11.2.3 Closed Loop Properties

It has been found useful at this point to define the inverse of the control effort weighting, Q, as being a dynamic compensating function, G_c. The substitution $G_c = Q^{-1}$ is advantageous, in that it allows the system closed loop response to be modified by suitable design of G_c using conventional compensator design techniques. The system closed loop behaviour is then given by:

$$y(t) = \frac{(G_c\, EB + C)\, \xi(t) + z^{-k} G_c\, BR\, w(t) + z^{-k}\ell\, L\, v(t)}{(G_c\, PB + A)} \tag{25}$$

Stability of the optimally controlled system is thus dependent upon the eigenvalues of the characteristic equation:

$$G_c\ P\ B + A = 0 \quad \text{or} \quad P\ B + Q\ A = 0 \tag{26}$$

Under conditions of minimum variance control ($P = 1$, $R = 1$ and $Q = 0$ in performance function J), the characteristic equation becomes $B = 0$. It can be seen from equation (26) that the system dead time has been eliminated from the characteristic equation and that both the weighting functions $G_c = Q^{-1}$ and P allow modification of the system closed loop eigenvalues, subject to nonviolation of system controllability constraints. An analogy can be drawn with Smith and Analytical Predictor techniques, Smith (1959), Moore et al (1970), and Meyer et al (1977, 1979). Prespecification of the desired system response is now possible by designing G_c and/or P, as though they were tunable compensators. Well known controller tuning techniques can be used, for example those due to Lopez, Moore, Smith and Murrill, and Ziegler-Nichols, all of which are usefully summarised in Smith (1972). For example if G_c were restricted to being a three term compensator, settings could be designed for proportional band (%PB), integral time (T_I) and derivative time (T_D).

Pole assignment methods are equally applicable by designing G_c and/or P to satisfy a pre-specified system performance requirement by:

$$G_c\ P\ B + A = \delta\Delta \tag{27}$$

where δ is a scalar constant and Δ specifies the desired closed loop response. However, care must be exercised to ensure that the system polynomials A and B do not have common factors which cancel. This would preclude the solution of equation (27). In the self tuning controller the system parameters A and B are unknown and the system characteristic equation for pole assignment must be written in terms of the estimated controller polynomials as:

$$G_c\ P\ \hat{G} + [P\ \hat{C} - z^{-k}\frac{\hat{F}}{Pd}] = 0 \tag{28}$$

where $\hat{G}$, $\hat{C}$ and $\hat{F}$ are identified recursively. Solution of this equation is then possible to provide continuous, or periodic if so desired, tuning of G_c and/or P quite independently of the values of the estimated polynomial coefficients. Pole assignment can then be achieved by solving equation (28) with the right hand side equal to $G_c\delta\hat{E}\Delta$.

Some particular operational problems can result when the self tuning controller is required to regulate against unmeasurable, non zero mean, deterministic process load disturbances. In this case the closed loop performance is given by:

$$y(t) = \frac{(EB + CQ)\ \xi(t) + z^{-k}\ BR\ w(t) + z^{-k_\ell}\ (EB + CQ)\ L/C\ v(t)}{(PB + QA)} \tag{29}$$

and the effect of the "unknown" disturbance $v(t-k_\ell)$ is similar to that of process noise with a non-zero mean. The only way that the steady state effects of unknown load disturbances can be eliminated, even with minimum variance control or incremental control effort weighting where Q is replaced by $(1-z^{-1})Q$, is by the controller polynomial coefficients becoming biased to treat the load disturbance as though it were a change in plant parameters. There is no guarantee though that such a movement in the coefficient values will lead to an acceptable controlled performance. However, the predictive nature of the controller can be used to estimate the long term effects of an unmeasurable non-zero mean load disturbance as $v^*(t)$:

$$v^*(t) = v^*(t-1) + \lambda[Py(t) - [Py(t|t-k]^*] \tag{30}$$

where λ is a scaling factor. In this way $v^*(t)$ can be used as if it were measured and the coefficients of the D polynomial estimated accordingly. The closed loop equation (25) applies with $v^*(t)$ replacing v(t).

11.2.4 Multi-input Multi-output Systems

The structure of the optimal control strategy developed in section 11.2.1 enables the control of a class of multivariable systems to be treated essentially as a number of individual multi-input single-output systems, Morris and Nazer (1979). The analysis presented here is restricted to systems with equal numbers of inputs and outputs, m, operating in such a way that the forward loop time delay corresponding to any particular control variable is always less than or equal to the time delays associated with other interaction causing variables. Consider a general multi-input multi-output discrete time model of a process as:

$$A(z^{-1})Y(t) = z^{-K_{ij}} B(z^{-1})U(t) + z^{-K_{\ell ij}} L(z^{-1})V(t) + C(z^{-1})\,\Xi(t) \tag{31}$$

where A, B, L and C are polynomial matrices in z^{-1}, with Y(t) and U(t) 1xm column vectors of m measurable plant outputs and m control inputs respectively. V(t) denotes an 1xn vector of measurable disturbances and $\Xi(t)$ is 1xm vector of m independent distributed random variables with zero mean and covariance $E\{\Xi(t)\Xi(t)^T\} = \Omega$. Ω is assumed to be equal to an mxm diagonal matrix. K_{ij} and $K_{\ell ij}$ represent the time delays in terms of integer numbers of the sampling interval, T_s, for the (ij)-th elements of $B(z^{-1})$ and $L(z^{-1})$ polynomial matrices respectively. Without loss of generality the polynomial matrices $A(z^{-1})$ and $C(z^{-1})$ can be assumed to be diagonal polynomial matrices with A(0) and C(0) equivalent to identity matrices. The multi-input multi-output optimal stochastic controller design is based upon the single step minimisation of a performance index J, $(\partial J)/\partial U(t) = 0$. The performance index can be defined as:

$$J = E\{[P(z^{-1})Y(t+K_{ii}) - R(z^{-1})W(t)]^T [P(z^{-1})Y(t+K_{ii}) - R(z^{-1})W(t)] + [Q'(z^{-1})U(t)]^T [Q'(z^{-1})U(t)]\} \tag{32}$$

where W(t) is a 1xm vector of set-points and $P(z^{-1})$, $R(z^{-1})$ and $Q'(z^{-1})$ are mxm diagonal transfer function matrices enabling general weightings to be placed upon the system outputs, set points and inputs respectively.

Now using the definitions:

$$P(z^{-1}) = \frac{P_n(z^{-1})}{P_d(z^{-1})} \; ; \; R(z^{-1}) = \frac{R_n(z^{-1})}{R_d(z^{-1})} \; ; \; Q'(z^{-1}) = \frac{Q'_n(z^{-1})}{Q'_d(z^{-1})}$$

and

$$P(z^{-1})C(z^{-1})\Xi(t) = A(z^{-1})\, E(z^{-1})\Xi(t) + z^{-K_{ii}}[P_d(z^{-1})]^{-1}\, F(z^{-1})\Xi(t) \quad (33)$$

and following the analysis for the single variable case, the K_{ii}-step ahead weighted output predictor can be defined as:

$$[PY(t+K_{ii}|t)]^* = PY(t+K_{ii}) - E\Xi(t+K_{ii})$$

or

$$[PY(t+K_{ii}|t)]^* = PY(t+K_{ii}) - \varepsilon(t+K_{ii}) \quad (34)$$

Here the (z^{-1}) term has been dropped for simplicity. Substitution of equation (34) into equation (32) and minimising with respect to the control vector U(t), gives:

$$Q\, U(t) + [PY(t+K_{ii}|t)]^* - R\, W(t) = 0 \quad (35)$$

where a new control weighting, an mxm diagonal transfer function matrix Q, is defined as:

$$Q = [B(0)^T]^{-1}\, [E(0)^T]^{-1}\, C(0)\, Q'^T(0)\, Q' \quad (36)$$

The K_{ii}-step ahead weighted output predictor is given by:

$$[PY(t+K_{ii}|t]^* = [C]^{-1}\, \{F\, Y_{Pd}(t) + z^{K_{ii}-K_{ij}}\, G\, U(t) + z^{K_{ii}-K_{\ell ij}}\, D\, V(t)\}$$

Here F, G and D represent controller polynomial matrices and the (i)-th elements of the 1xm column vector $Y_{Pd}(t)$ are defined as:

$$y_{P_{d_i}}(t) = \frac{1}{P_{d_{ii}}}\, y_i(t) \qquad \text{for } i=1,2,\ldots,m$$

The control vector U(t) chosen to minimise the objective function then gives the control law:

$$C[\phi(t+K_{ii}|t)]^* = F\, Y_{P_d}(t) + z^{K_{ii}-K_{ij}}\, G\, U(t) + C\, Q\, U(t) - C\, R\, W(t) = 0$$

(37)

where $[\phi(t+K_{ii}|t)]^*$ is a predicted output vector function related to a measurable vector of output function as:

$$\phi(t+K_{ii}|t) = P\ Y\ (t+K_{ii}) - R\ W(t) + Q\ U(t)$$

with

$$[\phi(t+K_{ii}|t)]^* = [PY(t+K_{ii}|t]^* - R\ W(t) + Q\ U(t) \tag{38}$$

Following the techniques outlined in section 11.2.2 the control law is made to self-tune by fixing the controller structure and allowing the polynomial coefficients of the control law to be identified recursively. The control law is cascaded with an on-line recursive parameter estimator allowing $[\phi(t+K_{ii}|t)]^*$ to be set equal to zero at each stage of control calculation to give:

$$\{z^{K_{ii}-K_{ij}}\ \hat{G} + \hat{C}\ Q\}\ U(t) = -\ \{\hat{F}\ Y_{Pd} + z^{K_{ii}-K_{\ell ij}}\ \hat{D}\ V(t) - \hat{C}\ R\ W(t)\} \tag{39}$$

where $\hat{F}$, $\hat{G}$, $\hat{C}$ and $\hat{D}$ are estimated controller polynomial matrices.

11.3 *Distillation Column Pilot Plant and Mathematical Models*

The pilot plant column, Sastry et al (1977), Meyer et al (1977, 1979) and Wood and Berry (1973), which was operated with a 50 weight per cent methanol and water feed at a rate of 18.86 g/s, is interfaced to an HP-1000 based distributed computer system. The 22.5 cm diameter column contains eight trays fitted with four bubble caps per tray at a tray spacing of 30.48 cm. Top composition is measured continuously by an in-line capacitance probe whilst bottom composition is measured using an HP-5720A gas chromatograph under computer controlled sampling and analysis of the chromatogram. In line sampling on a three minute cycle time is employed. Control is applied to the column using reflux and steam flow rates as manipulated variables for the top and bottom composition control loops respectively. A schematic diagram of the column is shows in figure 11.2 and typical steady state operating conditions are given in table 11.1.

Table 11.1. Steady state operating conditions.

Feed Flow Rate	18.06	(g/s)
Feed Temperature	72.0	(°C)
Feed Composition	50.0	(wt%)
Reflux Flow Rate	10.22	(g/s)
Reflux Temperature	63.9	(°C)
Steam Flow Rate	13.93	(g/s)
Top Product Composition	95.00	(wt%)
Top Product Flow Rate	8.92	(g/s)
Bottom Product Composition	5.00	(wt%)
Bottom Product Flow Rate	9.14	(g/s)

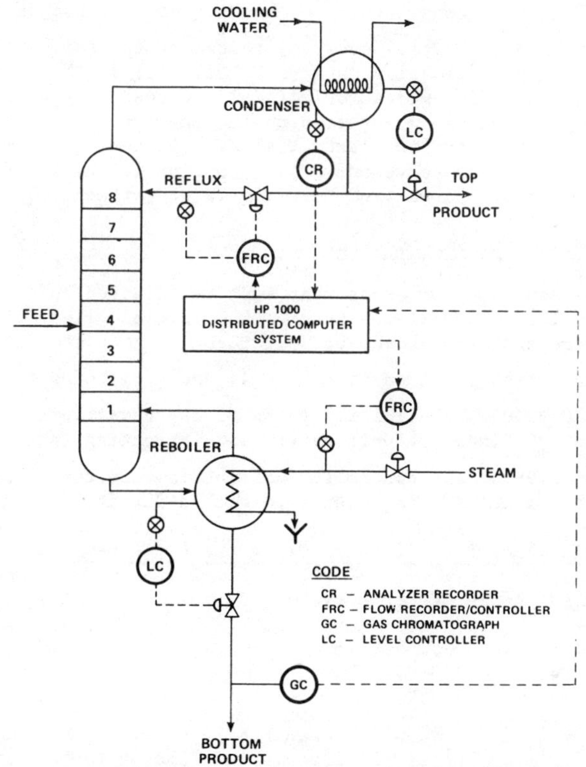

Fig 11.2 Schematic Diagram of the Pilot Plant Distillation Column

A nonlinear differential equation model based on individual tray, reboiler and condenser mass and energy balance relationships is used for simulating the dynamic behaviour of the column, Simonsmeier (1977). Initial analysis results in 30 differential equations with 140 variables. These can however be reduced to 20 differential equations by making the assumptions of constant tray head loss, constant tray efficiency and constant stage liquid hold up. Extensive testing of the dynamic behaviour of the pilot plant column, using both step and pulse analysis indicated that the behaviour of the pilot plant column could be simply modelled, for control purposes, by first order plus time delay transfer functions. Although the parameters of these simple models are found to be very dependent upon the direction and magnitude of the corresponding testing signals, they do enable classical and decoupling controllers to be designed for normal operating conditions. The linear column dynamics are represented by:

$$\begin{vmatrix} x_D(s) \\ x_B(s) \end{vmatrix} = \begin{vmatrix} \dfrac{1.0081\ e^{-0.35s}}{(4.36\ s + 1)} & \dfrac{-1.1161\ e^{-0.925s}}{(5.83\ s + 1)} & \dfrac{0.0928\ e^{-8.247s}}{(20.71\ s + 1)} \\ \dfrac{2.3375\ e^{-0.780s}}{(10.18\ s + 1)} & \dfrac{-5.0712\ e^{-3.480s}}{(10.27\ s + 1)} & \dfrac{2.3245\ e^{-1.725s}}{(10.43\ s + 1)} \end{vmatrix} \cdot \begin{vmatrix} R(s) \\ St(s) \\ F(s) \end{vmatrix}$$

where x_D and x_B represent the top of bottom terminal compositions respectively, and R(s) St(s) and F(s) represent the reflux flow rate, steam flow rate and feed flow rate respectively. In the model all the process gains have dimensions (wt%)/(g/s) and the time constants and time delays are given in minutes. In studying the controlled performance of the multivariable system a rate limit was placed on the maximum allowable control valve movement at any control interval. This limit was fixed at 20% of the maximum possible valve movement.

11.4 *Simulation and Experimental Results*

There are several parameters that must be selected before implementing the ST controller. In all the studies the initial values of the controller parameters were chosen to be zero except for the leading $\hat{C}$ coefficient which is fixed as $\hat{c}(0) = 1$, and $\hat{g}(0)$ being initially set such that $\hat{g}(0) \geq g(0)$. The initial value of the covariance matrix was chosen to be 100 times the unit matrix and forgetting factors ρ_1 and ρ_2 set equal to 0.998. The controller parameters were then allowed to tune in whilst the "process" was controlled with a digital PI regulator.

11.4.1 Performance Specification Using the G_c Compensator

A simulated continuous time system

$$G(s) = \frac{5.0\, e^{-2.0\, s}}{(1 + 12.0\, s)\,(1 + 20.0\, s)} \tag{40}$$

was subjected to a square wave reference input with amplitude 10 and period 50 minutes. The sample interval was chosen to be 2 minutes. Figure 3 shows the controller performance for different G_c compensator designs. Response (a) results from the application of minimum variance control achieved by setting Q = 0. The oscillations observed here are due to the location of the poles of the discrete time system. Responses (b) and (c) result from designing G_c as a PI regulator according to the Ziegler-Nichols procedure (PB = 250%), and (PB - 111.11%, T_I = 13.32 min) respectively. In the case of purely proportional compensation, response (b), where the effective Q filter value is equal to 2.5, an offset results which can only be removed by introducing a weighting on incremental changes in control effort. This is achieved by designing G_c to include integral action as demonstrated by response (c).

The predictive feedback properties of the controller can be demonstrated by comparing the set point following performance of the ST controller, with G_c compensation, to that resulting from conventional PI regulation. Fig 11.4 shows the responses obtained when the system time delay is chosen to be 2,4 and 8 minutes respectively. Studies (a), (b) and (c) correspond to ST control and (d), (e) and (f) to PI control. Here the digital PI regulator parameters were tuned using the techniques suggested by Moore (Smith, 1972) and the ST controller G_c parameters designed in a similar manner, but independently of the system time delay value, to give (PB = 52.5% and T_I = 6.46 min). The Moore design for the PI regulators gave values of (PB = 103.94%, T_I = 10.54 min), (PB = 155.03%,

T_I = 14.036 min) and (PB = 256.54%, T_I = 20.142 min) respectively for system time delays of 2, 4 and 8 minutes. In the ST studies the system time delay was assumed to be known exactly. It can be seen from responses (a), (b) and (c) that the resulting transient performance under ST control is independent of the system time delay, apart from the obvious initial dead period. However under conventional control, responses (d), (e) and (f), the controlled performance becomes progressively degraded as the system time delay to time constant ratio increases. The ST controller behaves in a very similar manner to the Smith and Analytical predictors except that here the predictive model is continuously being updated to meet changing plant conditions. However, when the time delay is less than about 1/10th of the dominant

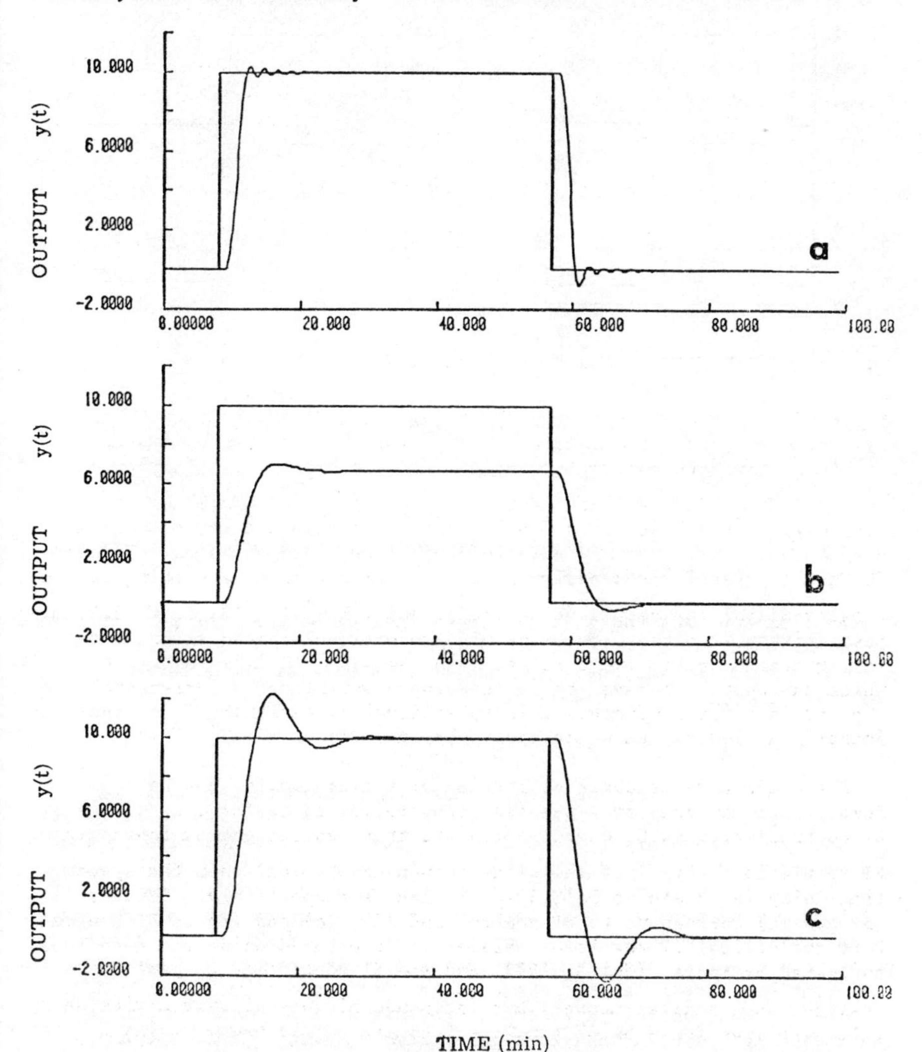

Fig 11.3 ST Performance Specification using the G_c Compensator.

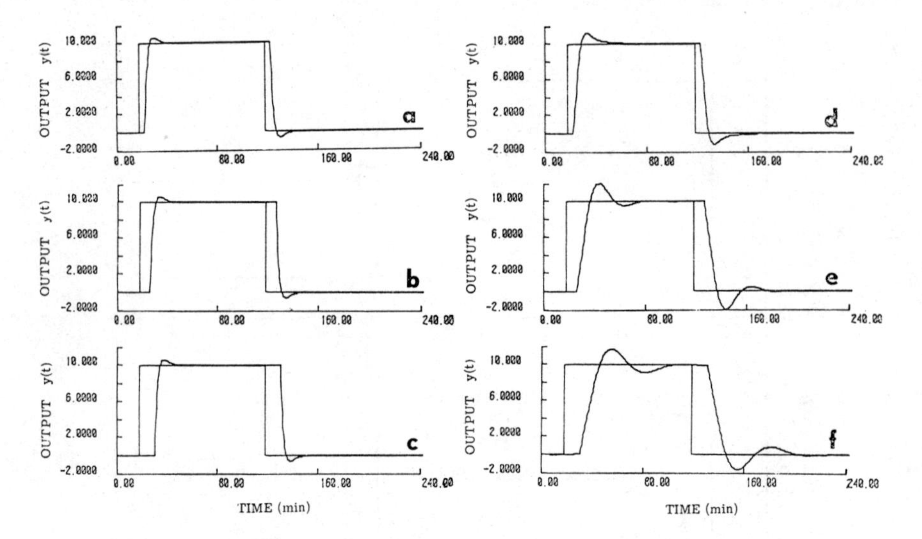

Fig 11.4 Comparison of PI and ST Controller Performance in Systems with Time Delays.

system time constant there is little difference between the performances to be achieved using tuned ST or PI controllers. This is demonstrated both by simulation responses (a) and (c), as well as the experimental tests shown in Figs 11.7 and 11.8.

11.4.2 Load Disturbance Rejection

Fig 11.5 demonstrates one very undesirable characteristic of the minimum variance (MV) ST controller. Here the original system is subjected to "unknown" deterministic load disturbances which are additive to the system output via an "unknown" transfer function $1/(1 + 10.0\ s)$. The system time delay is taken to be 2 minutes and the sample interval equal to 2 minutes. The load disturbance was a square wave of amplitude 10 and period 50 minutes. The unpredictable variations of the controller polynomial coefficients occurs since the adaptive controller cannot differentiate between a distrubance or actual process parameter changes as only process control input and output are available to the controller. The closed loop system behaviour is extremely degraded as the $\hat{G}$ co-efficients become numerically very small. However, once the disturbance

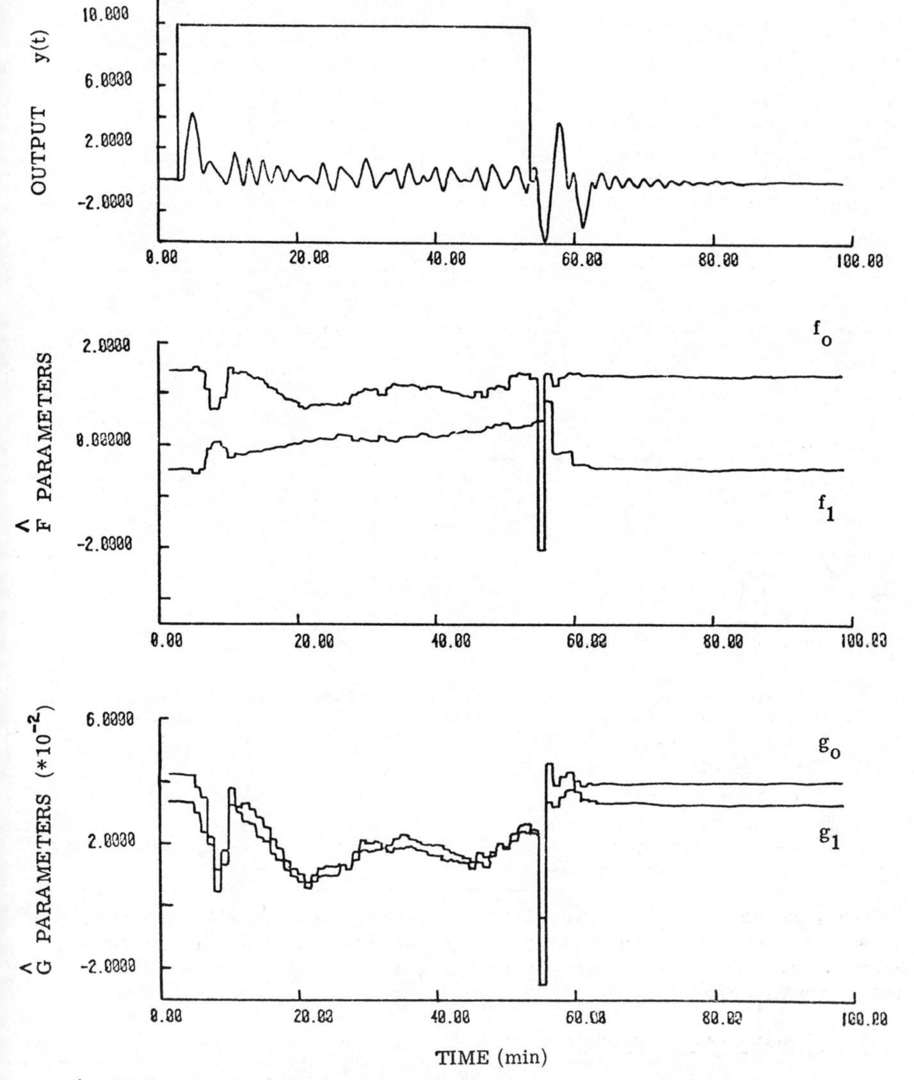

Fig 11.5 Load Disturbance Performance of the Minimum Variance ST Controller.

is removed the controller coefficients reconverge to their MV optimal values. If the controller coefficients remain fixed at their optimum values, for the system being controlled, then an output offset will always result from an unknown load disturbance. This is shown by Fig 11.6a. Clearly an acceptable closed loop performance can only be achieved by providing the ST controller with sufficient information to allow a distinction to be made between disturbance effects and the effects of system parameter changes. Fig 11.6b shows the improved performance of the minimum variance ST controller, over that shown in

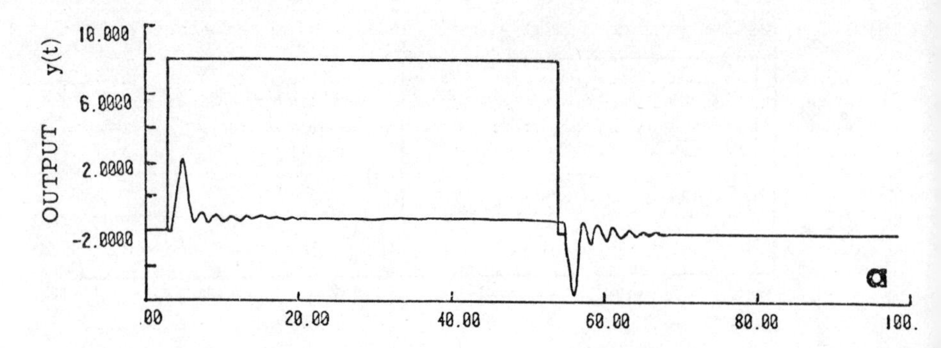

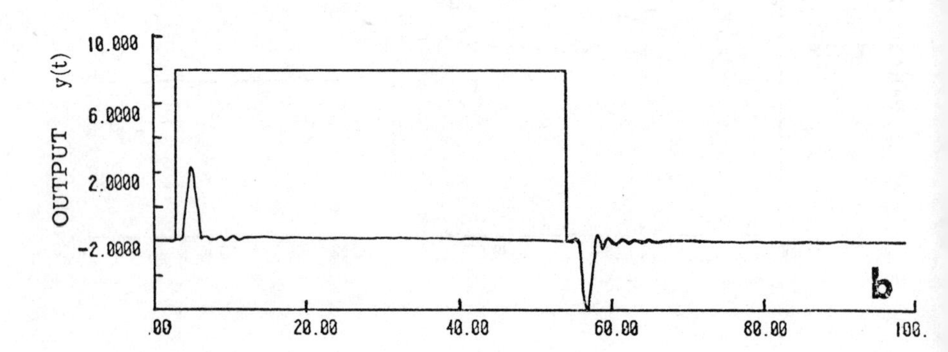

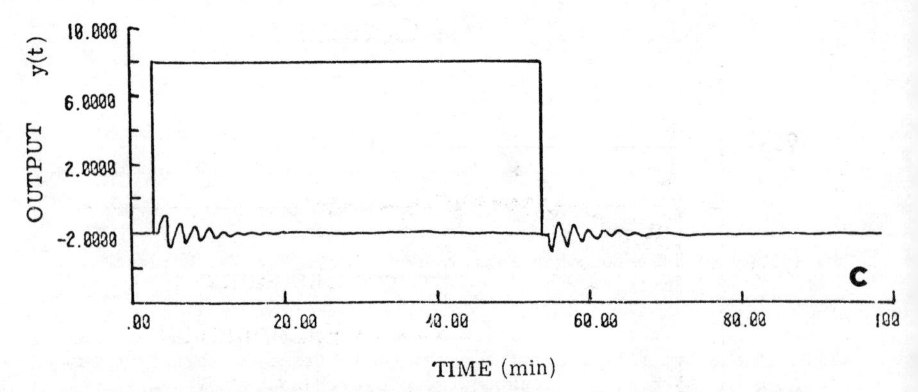

Fig 11.6 Comparison of the Load Disturbance Performances of the Minimum Variance Controller.

Fig 11.5 when the disturbance is predicted using equation (30) and the coefficients of the feedforward polynomials, $\hat{D}$, identified accordingly. The first peak overshoot is still present since the controller relies on feedback of output information. If the disturbance is measurable it can be used as a feedforward control signal, v(t), giving a further improvement in controlled performance as shown in Fig 11.6c. In both studies (b) and (c) the controller polynomial coefficients varied only slightly from their correct values. The intersample oscillatory behaviour of the system is easily modified by introducing compensation as discussed in section 11.4.1.

11.4.3 Distillation Column Control

Further evaluation of the single variant ST controller was achieved by studying the control of single product composition of a binary distillation column. In all the studies it will be seen that the initial composition deviation, as a result of a sudden change in feed flow rate, is always present since the distrubance occurs immediately after sampling. This represents the most severe case since the system

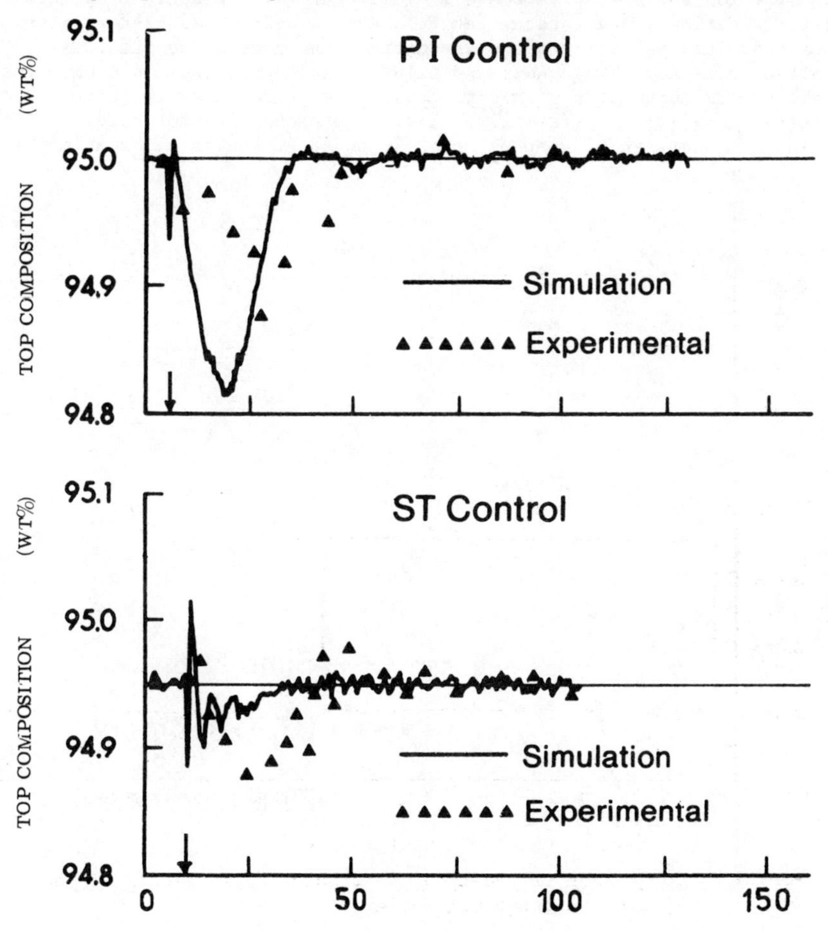

Fig 11.7 Comparison of PI and ST Control of Top Product Composition for a 25% decrease in Feed Flow Rate.

remains "uncontrolled" for the duration of one sample interval. Fig 11.7 shows simulated and experimental top composition responses, with the column operated under PI and ST control, for a 25% step decrease in feed rate (↓). In comparing the experimental responses it can be seen that the maximum deviation of composition from its set value is smaller and faster under ST control. The same conclusion applies to the simulated responses which show a faster response than those obtained experimentally. An important difference between the simulation and the pilot plant is that although the pilot column was uninsulated a constant heat loss was assumed in the mathematical model. Further simulations are under way aimed at investigating this discrepancy using a variable heat loss representation of the column.

A comparison of ST and PI control of bottom composition is shown in Fig 11.8 for a 25% step decrease in feed rate (↓). Clearly ST control is far superior to that resulting from even a well tuned PI regulator. These results substantiate and demonstrate the general conclusions reached from many simulations and pilot plant experiments that when the ratio of the dominant system time constant to time delay is large (greater than 10:1) there will be little to choose in controlled performance between ST control and well tuned PI control. Experimental

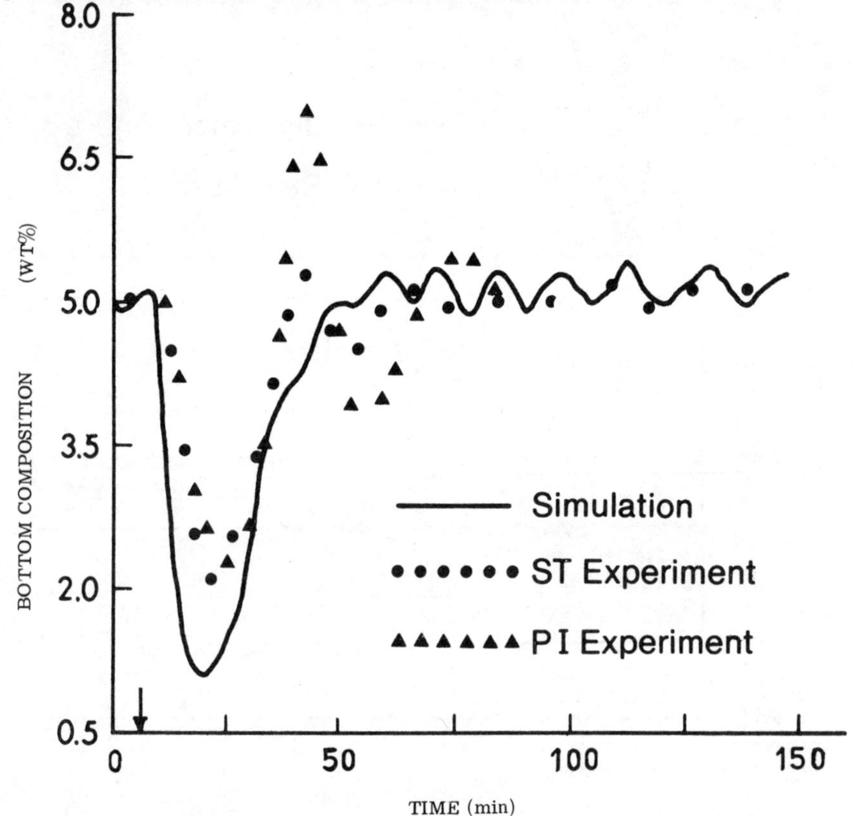

Fig 11.8 Comparison of PI and ST Control of Bottom Product Composition for a 25% decrease in Feed Flow Rate.

dynamic testing of the column shows the dominant time constant to be of the order of 4 min with a time delay of 0.4 min for top composition, while the corresponding values for bottom composition are 10 min and 3.5 min respectively. As would be expected in these circumstances ST control shows superior performance in the control of bottom composition.

Evaluation of the extension of the single variant controller to a class of interactive multivariant systems was first studied by investigating the simultaneous control of both product streams of the distillation column using the nonlinear simulation package. In all the simulation studies a step decrease in feed flow rate of magnitude 20% of nominal flow, followed by a return to steady state, was used as the load disturbance; the point of application being indicated on the figures as (↓,↑). Fig 11.9 shows the performance of the distillation column under simultaneous terminal composition control using discrete multiloop PI regulators. The regulator parameters were tuned to minimise an integral of absolute error (IAE) criteria resulting in top composition loop settings of (PB = 200%, T_I = 1.66 min) and bottom

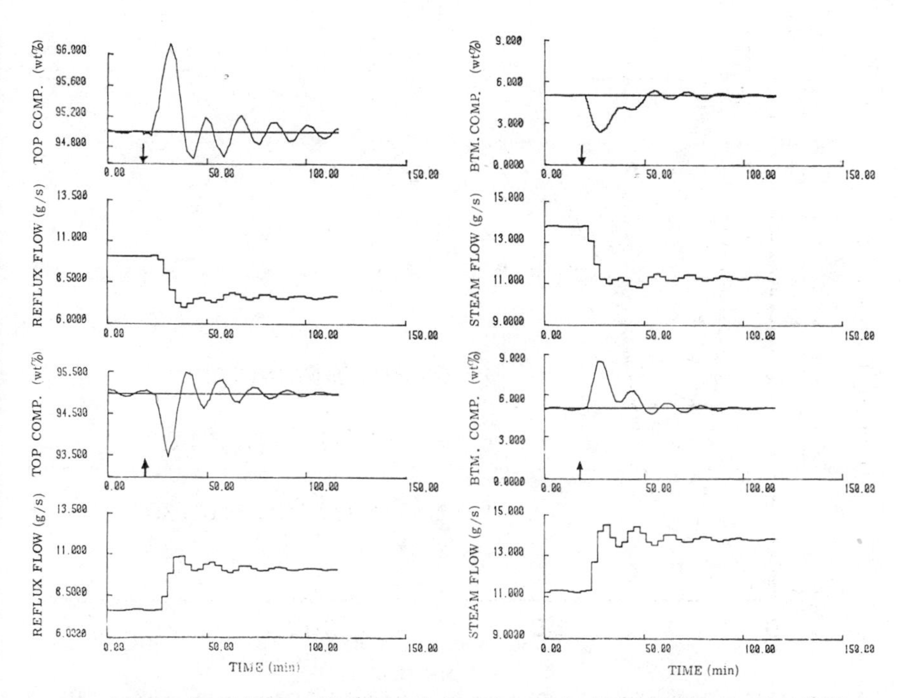

Fig 11.9 Multiloop PI Composition Control (Nonlinear Simulation).

composition loop settings of (PB = 153.4%, T_I = 10.14 min). The oscillatory controlled behaviour is similar to that reported by Wood and Berry (1973) and can only be eliminated by degaining the controllers. The manipulated variables, reflux flow and steam flow, are also shown.

An improvement in overall system performance can be achieved by using steady state gain decoupling, Luyben (1970). Using a decoupling matrix:

$$\begin{vmatrix} 1.000 & 1.107 \\ 0.461 & 1.000 \end{vmatrix}$$

with the same PI controller settings as above, results in the controlled performance shown in Fig 11.10. It is interesting to observe here that the first peak overshoot in top composition is in the opposite direction to that found under multiloop PI control, Fig 11.9 (note also the different ordinate scales for top composition).

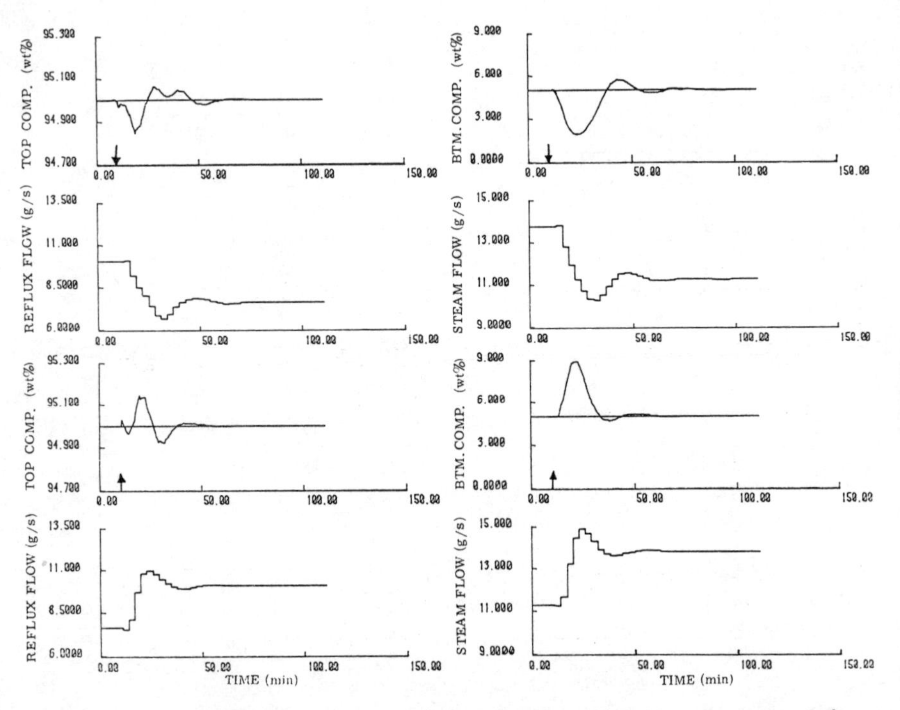

Fig 11.10 Multiloop PI Composition Control with Steady State Gain Decoupling (Nonlinear Simulation).

Distillation column behaviour is such that for decreasing feed flow changes, and for decreasing reflux flow changes, both top and bottom compositions will decrease. However, for decreasing steam flow changes, both top and bottom compositions will increase. In this particular column the effect of steam flow rate changes on top composition are very severe resulting in the inverse response effects observed under multiloop PI control. With steady state gain decoupling, changes in steam flow rate are reflected in reflux flow changes and these inverse properties are eliminated. Comparison of the manipulated variables just after the load disturbance occurs, Fig 11.9 and 11.10 highlights this behaviour.

Application of the multivariable minimum variance ST controller, with unmeasurable but "predicted" feed flow changes, is shown in Fig 11.11. Clearly, although the bottom product response shows improvement, the behaviour of the top composition loop is far from being acceptable as a result of the rapid control action responses. However, the decoupling nature of the ST controller becomes apparent by comparing these responses with those of Fig 11.10.
A further problem hindering good top composition control here, is the sample time of 3 minutes chosen to satisfy bottom product analysis restrictions. A one minute sample interval would be more appropriate for good top composition control.

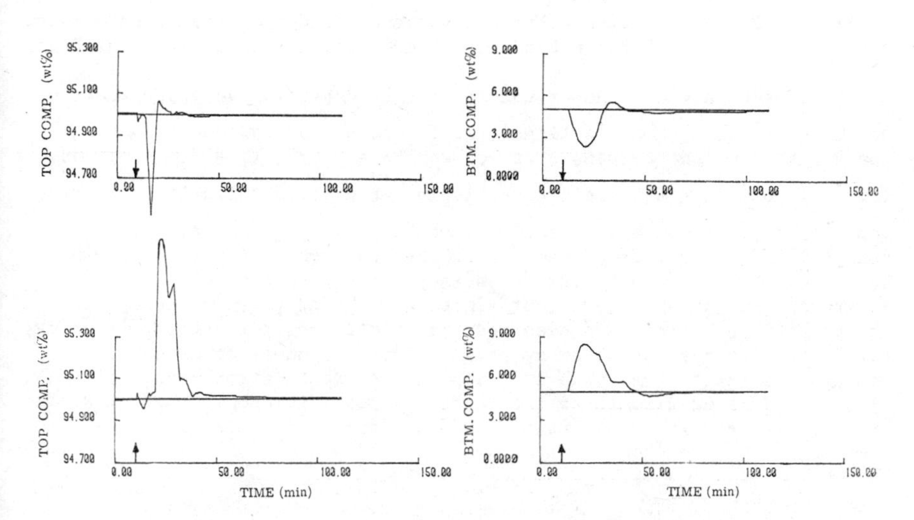

Fig 11.11 Multivariable Minimum Variance ST Composition Control with Predicted Feed Flow Disturbance (Nonlinear Simulation).

Implementation of the multivariable minimum variance ST controller with feed-forward measurement of feed flow changes is shown in Fig 11.12. A further improvement in controlled performance of bottom product is achieved whilst the top product behaviour again suffers from the incompatibility of minimum variance control with the fast steam and slow reflux interactive dynamics of this particular process.

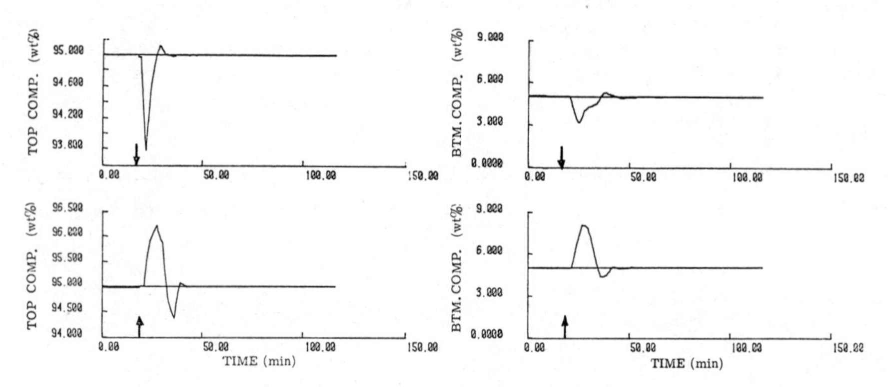

Fig 11.12 Multivariable Minimum Variance ST Composition Control with Feed Forward Feed Flow Compensation (Nonlinear Simulation).

The control actions can be calmed down by designing an appropriate G_c compensator, using Moore's technique, for each product loop resulting in top and bottom loop compensators set to (PB = 35.7%, T_I = 3.4 min) and (PB = -76.6%, T_I = 4.3 min) respectively. Fig 11.13 shows the resulting controlled performances. Clearly a much improved bottom product composition response is obtained with the top composition loop showing no improvement over that with the steady state gain decoupler. It is interesting to observe here that since the initial rates of change in the control variables have been reduced, due to control effort filtering, the first peak overshoot in top product composition is reversed and becomes similar to that with multiloop PI control. Clearly the reflux flow rate must be allowed to change more rapidly to compensate for steam flow control changes and thus provide more decoupling action.

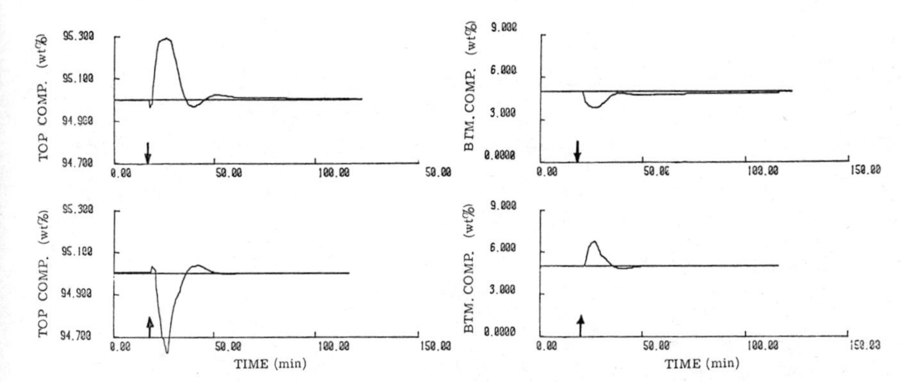

Fig 11.13 Multivariable ST Composition Control with G_c Compensation in both Top and Bottom Control Loops and Feed Forward Feed Flow Compensation (Nonlinear Simulation).

By allowing the top composition loop to be controlled without any weighting on control effort (Q = 0) and using the same bottom composition loop compensator settings, the controlled performance shown in Fig 11.14 is achieved. This represents the best performance that can be obtained with this column, giving good decoupling and elimination of the inverse response effects caused by steam flow variations. The manipulated variable responses are very acceptable and would not cause any control valve operational problems. It is interesting to note here that an almost identical controlled behaviour can be achieved by setting the bottom loop control effort weighting to zero and using a model following approach with the bottom loop P filter set as $P(s) = (1+20.0\ s)$. This would imply a specification to follow a model $1/(1+20.0\ s)$.

Application of multi-loop PI control to the pilot plant column results in the controlled responses shown in Fig 11.15. A 25% step decrease in feed flow rate, followed by a return to steady state, was used as a system disturbance. The composition responses shown here should be compared with those of Fig 11.9, obtained from the non-linear simulation. Considering the complexity of the pilot plant being used, an acceptable agreement between experimental and simulated plant behaviour is achieved.

Replacement of the multiloop PI regulators by two single-input single-output ST controllers, multiloop ST control, results in the plant responses shown in Fig 11.16. Although the performance of the bottom composition control loop has been improved, as a direct result of the predictive nature of the ST controller, the behaviour of the top composition loop does not show any real improvement over that achieved with the multiloop PI control strategy. The nonlinear interactive

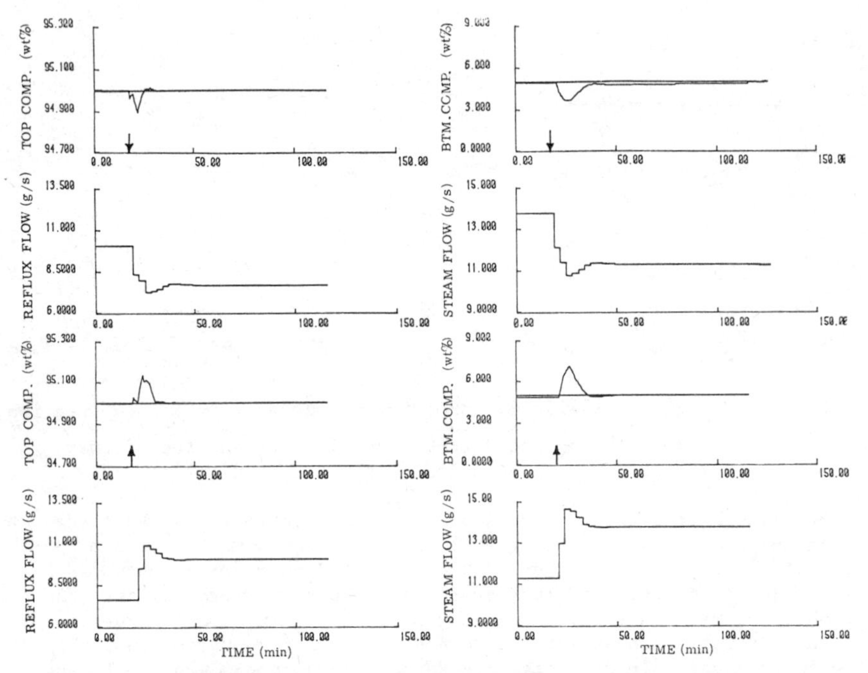

Fig 11.14 Multivariable ST Composition Control with Minimum Variance Top Composition Control, G_c Compensated Bottom Composition Control and Feed Forward Feed Flow Compensation (Nonlinear Simulation).

behaviour of the column cannot be completely accounted for by using simple multiloop control strategies.

By comparison, application of the multivariable self tuning controller is shown in Fig 11.17. Clearly a much improved terminal composition controlled performance has been achieved in spite of the non-linear behaviour and severe loop interaction. When, however, the load disturbance was measured using a conventional orifice plate/differential pressure cell flow measurement, the transient response of bottom composition loop is significantly improved. This is shown in Fig 11.18. Here the initial excursion in bottom product composition is almost eliminated completely. In comparison, due to the rapid transient effects of steam flow manipulation on top product, there was little improvement in top composition controlled performance. It is important

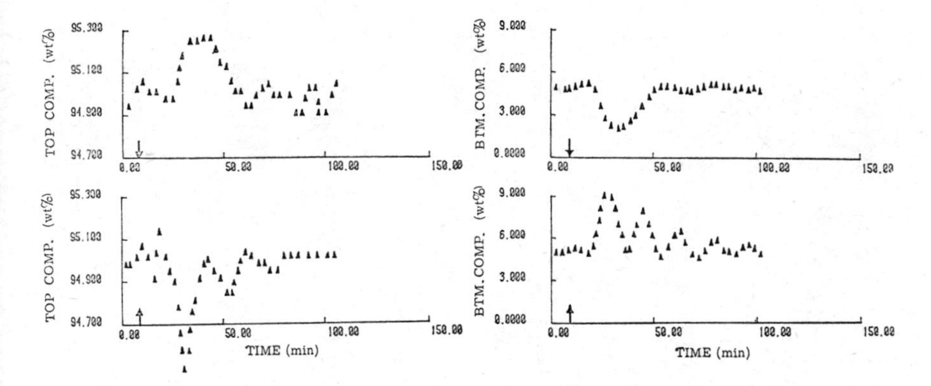

Fig 11.15 Multiloop PI Composition Control (Pilot Plant Tests).

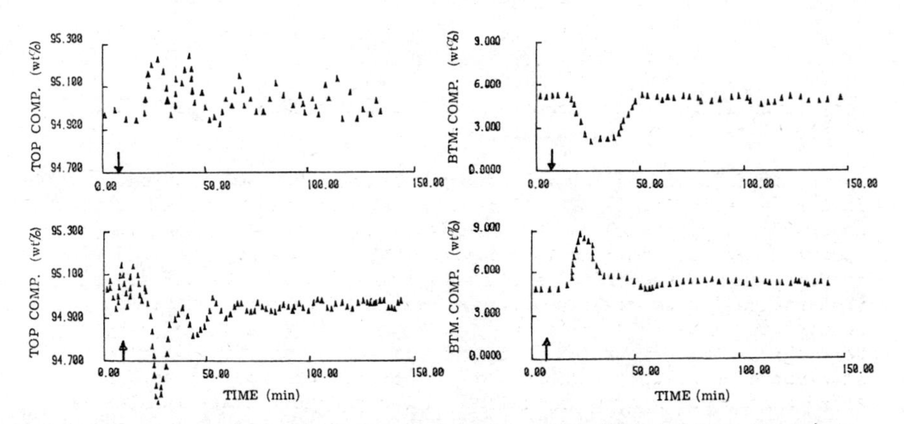

Fig 11.16 Multiloop ST Composition Control (Pilot Plant Tests).

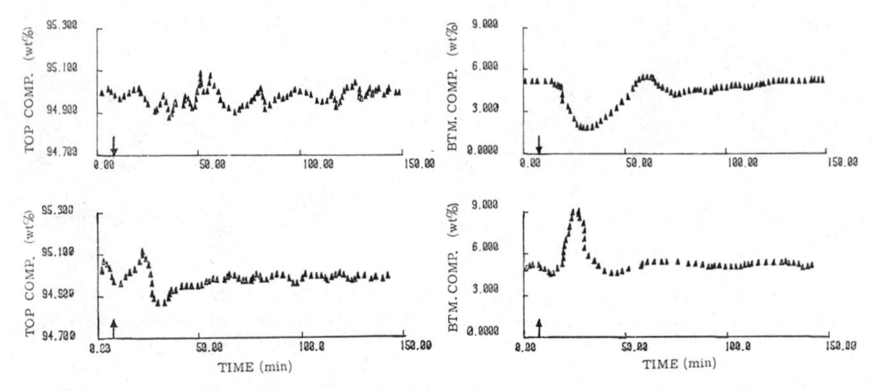

Fig 11.17 Multivariable ST Composition Control with G_c Compensation in both Top and Bottom Control Loops (Pilot Plant Tests).

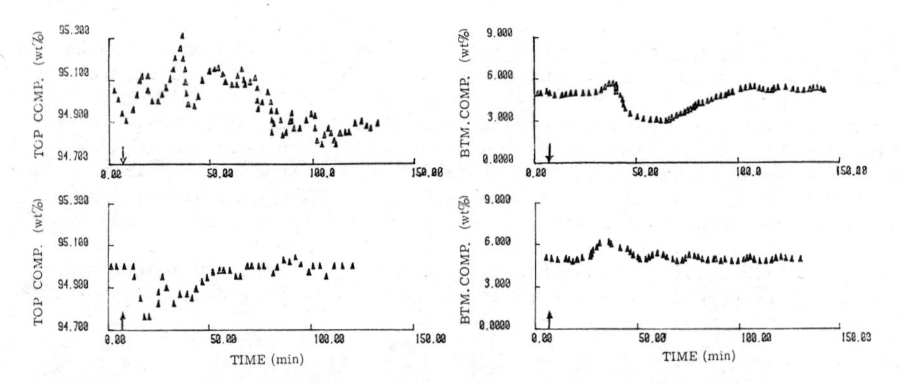

Fig 11.18 Multivariable ST Composition Control with G_c Compensation in both Top and Bottom Control Loops and Feed Forward Feed Flow Compensation (Pilot Plant Tests).

to note here that during the first disturbance, decreasing feed flow rate, the $\hat{D}$ polynomial coefficients were being estimated from zero initial values such that when the second disturbance occured, return to steady state feed flow conditions, good feed forward control was achieved. This demonstrates a very useful feature of the ST controller, the automatic tuning of feed-forward control actions.

There is no doubt however, that even much better regulation would result if the limit on the maximum allowable rate of control valve movement were relaxed and/or the control interval shortened by faster product composition analysis. Further experimental work is reported in Lieuson (1980) and Nazer (1981).

11.4.4 Multivariable Multi-Rate Self Tuning Control

In multivariable control the overall sample time chosen is usually a compromise between the ideal sample intervals required for the individual loops making up the system. However, there are some systems where such a choice may not be satisfactory. For example the distillation column used in this work, where the sample interval is chosen to be equal to the chromatograph cycle time and is too large for good control of top composition. An improvement in controlled performance might be expected therefore if different sample intervals, $T_s(i)$, could be used to satisfy the dynamic requirements of each (i-th) individual control loop. This choice is possible with the multivariable ST controller and arises from the structure, and decoupling nature, of the control strategy developed. Here each individual sub-system of the multivariable process is treated as a multi-input single output process.

The implementation of the multirate self-tuning controller assumes that all the sampling intervals of the process are integer multiples of the smallest, or base, sample interval T_b. In addition all the control signals acting as measurable disturbances $(u_j(t);j=1,2...,m;j\neq i)$ on the i-th loop are accessible in the time sequence $T_s(i)$, $2T_s(i)$, $3T_s(i)$... etc, Nazer (1981).

Fig 11.19 compares the simulated controlled performance of the multivariable ST controller with a common sample interval of 3 minutes, to that of the multirate ST controller where the top composition loop is sampled at 1 minute, with the bottom loop still sampled at 3 minute intervals to coincide with the chromatograph cycle time. In this study process and measurement noise was added to more accurately represent the pilot plant behaviour, and a 25% step decrease in feed flow rate considered as an "unmeasurable" disturbance. The faster response of the reflux flow control variable in the case of multirate sampling is evident, resulting in a significant improvement in top composition dynamic performance. A marginal improvement in bottom composition transient behaviour is also achieved. Clearly multirate sampling is attractive and further studies are in progress, particularly with respect to pilot plant experiments.

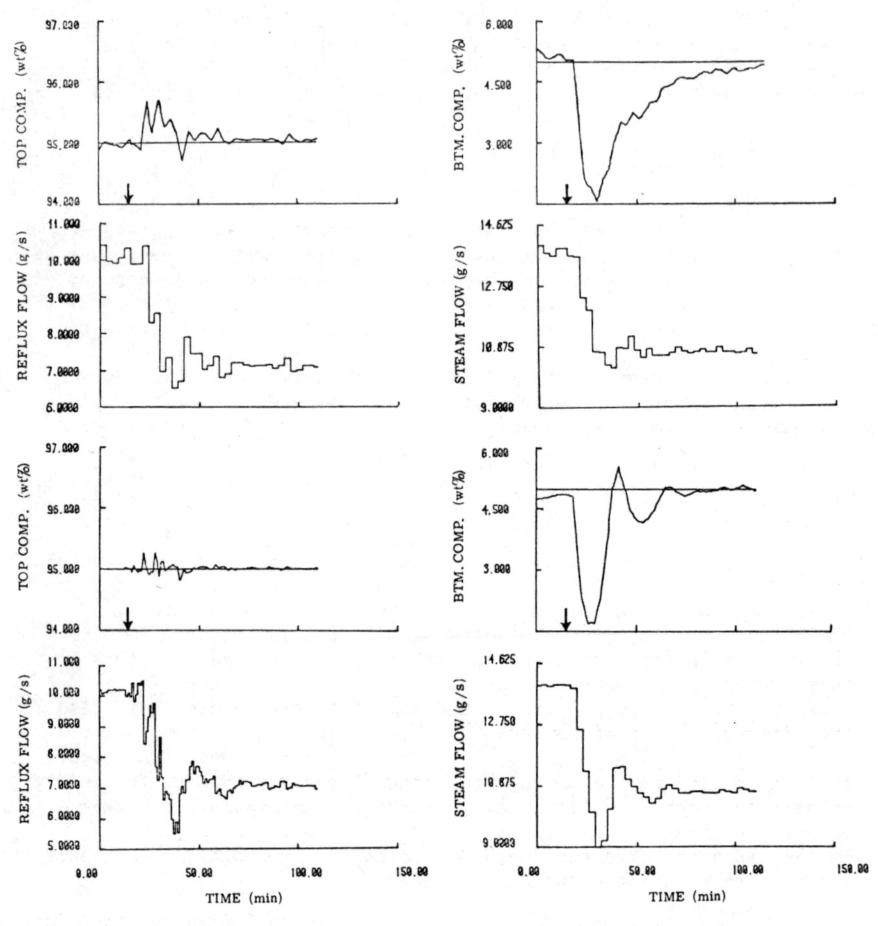

Fig 11.19 Multivariable-multirate ST Composition Control.

11.5 *Conclusions*

The theoretical development of a general and robust self tuning controller to include closed loop performance specification, feed-forward regulatory control and unknown load disturbance rejection, has been presented. Although specification of the controller performance by design of G_c and/or P, or even pole assignment is achieved at the expense of closed loop optimality, this is not necessarily a disadvantage since in most process control applications ideal optimal control is not the main objective. Also the provision of well known "tuning parameters", G_c, for use by the process control engineer is considered to be very beneficial to plant operation. A new multivariable self tuning controller has been developed which exhibits several advantages over existing minimum variance controllers in that it is applicable to non-minimum phase systems, systems in which the individual loops have different time delays and systems where multirate sampling might be advantageous. In common with its single variant counterpart it is found to be very robust. However, the simulation and experimental studies carried out so far do confirm that careful thought must be given to the initial overall system design taking into account all the plant physical constraints and non-linearities. Since this should be common practice in the design of process control systems no additional system knowledge should be required. In fact once the controller has been commissioned on the plant it will require less attention than conventional control systems in that adaption of the predictor parameters is automatic, and tuning of the compensating function (G_c) parameters can also be made automatic if desired.

Acknowledgments

The authors gratefully acknowledge the financial support of the Natural Sciences and Engineering Research Council of Canada under grant A-1944, and use of the facilities provided by the Departments of Chemical Engineering at the Universities of Alberta and Newcastle upon Tyne. The contribution of H Lieuson towards the experimental work is also gratefully acknowledged.

References

Åström K J and Wittenmark B (1973). 'On Self Tuning Regulators'. Automatica, 9, 185-199.

Bierman G J (1975). 'Measurement Updating using the U-D Factorization'. Proc IEEE Conf on Decision and Control.

Borisson U (1979). 'Self Tuning Regulator for a Class of Multivariable Systems'. Automatica, 15, 209-215.

Clarke D W and Gawthrop P J (1975). 'Self Tuning Controller'. Proc IEE, 122, 929-934

Clarke D W and Gawthrop P J (1979). 'Self-tuning Control'. Proc IEE, 126, 633-640.

Gawthrop P J (1977). 'Some Interpretations of the Self Tuning Controller'. Proc. IEE, 124, 889-894.

Hu Y C and Ramierez W F (1972). 'Application of Modern Control Theory to Distillation Columns'. A J Ch E, 18, 479-486.

Keviczky L and Hetthessy J (1977). 'Self Tuning Minimum Variance Control of MIMO Discrete Time Systems'. Automatic Control Theory and Application, 14, 525-532.

Keviczky L, Hetthessy, J Hilger M and Kolostory J (1978). 'Self Tuning Adaptive Control of Cement Raw Material Blending'. Automatica, 14, 252-532.

Lieuson H (1980). 'Evaluation of Self Tuning Control of a Binary Distillation Column'. MSc Thesis, Univ Alberta, Canada.

Lieuson H, Morris A J, Nazer Y and Wood R K (1980). 'Experimental Evaluation of Self Tuning Controllers Applied to Pilot Plant Units'. Methods and Applications in Adaptive Control, Vol 24, Ed Unbehauen, Springer Verlag Press, 301-309.

Luyben W L (1975). 'Feed-forward Control of Distillation Columns'. Chem Eng Proc, 61, 74-78.

McGinnis R G and Wood R K (1974). 'Control of a Binary Distillation Column Utilizing a Simple Control Law'. Can J Chem Eng, 52, 806-809.

Meyer C, Seborg D E and Wood R K (1977). 'An Experimental Application of Time Delay Compensation Techniques to Distillation Column'. 5th IFAC/IFIP International Conf on Digital Computer Applications to Process Control, The Hague, Netherlands 439-446.

Meyer C. Wood R K and Seborg D E (1979). 'Experimental Evaluation of Analytical and Smith Predictors for Distillation Column Control', A I Ch E Journal, 25, 24-32.

Moore C F, Smith C L and Murrill P W (1970). 'Improved Algorithm for Direct Digital Control'. Ints and Control Systems, Vol 43, No 1, pp 70-74.

Morris A J and Nazer Y (1977). 'Self-tuning Control of a Binary Distillation Column'. Dept Chem Eng Report, University of Newcastle Upon Tyne, England.

Morris A J, Fenton T P and Nazer Y (1977). 'Application of Self Tuning Regulator to the Control of Chemical Process'. 5th IFAC/IFIP International Conf on Digital Computer Applications to Process Control, The Hague, Netherlands, 447-455.

Morris A J and Nazer Y (1979). 'Self Tuning Process Controller for Single and Multivariable Systems'. Submitted for publication in Automatica.

Morris A J, Nazer Y, Wood R K and Lieuson H (1980). 'Evaluation of Self Tuning Controllers for Distillation Column Control'. 6th IFAC/IFIP International Conf on Digital Computer Application to Process Control, Dusseldorf, Germany.

Nazer Y (1981). 'Single and Multivariable Self Tuning Controllers'. PhD Thesis, University of Newcastle Upon Tyne, England.

Nisenfield A E and Miyasaki R K (1973). 'Application of Feed-forward Control of Distillation Columns'. Automatica, 9, 319-327.

Peterka V (1970). 'Adaptive Digital Regulation of Noisy Systems'. Proceeding of 2nd IFAC Symp on Identification and Process Parameter Estimation, Prague, Paper 6.2.

Peterka V (1975). 'A Square Root Filter for Real-time Multivariable Regression'. Kybernetika, 11, 53-67.

Rosenbrock H H (1974). 'Computer Aided Control System Design'. Academic Press.

Sastry, V A, Seborg D E and Wood R K (1977). 'A Self Tuning Regulator Applied to a Binary Distillation Column'. Automatica, 13, 417-424.

Simonsmeir U F (1977). 'Nonlinear Binary Distillation Column Models'. MSc Thesis, Univ of Alberta, Canada.

Sinha A K (1977). 'Minimum Interaction Minimum Variance Controller for Multivariable Systems'. IEEE Trans on Automatic Control, 22, 274-275.

Smith C L (1972). 'Digital Computer Control', Intext Ed Publishers.

Smith O J M (1959). 'A Controller to Overcome Dead Time'. ISA Journal, Vol 6, No 2, pp 28-33.

Soderstrom T, Ljung L and Gustavsson I (1974). 'A Comparative Study of Recursive Identification Methods'. Rpt TFRT-3085, Dept Auto Control, Lund Inst of Tech, Sweden.

Wellstead P E, Zanker P and Edmunds J M (1978). 'Self-tuning Pole/Zero Assignment Regulators'. Control Centre Report 404, UMIST, Manchester, England.

Wood R K and Berry M W (1973). 'Terminal Composition Control of a Binary Distillation Column'. Chemical Engineering Science, 28, 1707-1717.

Chapter Twelve

Application of self-tuning to engine control

P E WELLSTEAD and P M ZANKER

12.1 *Introduction*

It has been clear for many years that the next generation of internal combustion engines will be equipped with extensive micro-computer based management systems. Indeed, certain currently available vehicles (military and civilian) involve a limited degree of digital engine management. As technology develops these will inevitably extend beyond elementary monitoring to general surveillance and control activities. Moreover, the latter will involve exercising closed-loop or quasi-closed loop control over certain engine variables, such as speed, ignition, exhaust, gear change, fuel delivery, efficiency and the like.

It was with developments of this nature in mind that a group of researchers, drawn from the Control Systems Centre and the Thermodynamics Division at the University of Manchester Institute of Science and Technology, addressed the problems associated with direct digital engine control. Because of the diversity of control problems associated with engine systems, a single "theme-problem" was selected - to be specific, the adaptive regulation of diesel engine speed. This topic was chosen principally because it combined the existing expertise at UMIST in the areas of diesel engines and adaptive control. From a broader viewpoint, however, the diesel engine is important as a medium term solution to fuel economy.

By the same taken, but at a more technical level, the diesel engine presents an interesting control problem because of its intrinsic instabilities. Normally, stabilization is provided by a mechanical centrifugal governor or (indirectly) by the non-linear characteristic of frictional loading. Our aim was to demonstrate the feasibility of direct digital speed regulation by placing a turbo-charged diesel engine under adaptive, or in this instance, self-tuning control. Indeed, the speed loop of a diesel engine is particularly appropriate for adaptive control because it involves significant non-linearities, which can be interpreted as time variations in the system parameters.

This then is the background against which the technical studies outlined below should be set. We begin (section 12.2) by outlining in

P E Wellstead is with the Control Systems Centre, University of Manchester Institute of Science and Technology.

P M Zanker is with Marconi Space and Defence Systems.

their basic form, the control structures which are commonly encountered in engine systems, together with their self-tuning equivalents. We then begin to specialize the arguments (section 12.3) by introducing the diesel engine speed loop problem and control system models for turbo-charged diesel engines and a discussion of the self-tuning controller structure as it relates directly to the diesel engine system. This then links naturally to a physical description of the test bed layout (section 12.4) and a detailed discussion of the self-tuning trials carried out on the engine.

The article concludes with a brief appraisal of the specific exercise and an evaluation of future directions for self-adaptive engine control.

For a detailed description of the diesel regulation problem see (Zanker 1980; Zanker and Wellstead 1978). Further practical insights into self-tuning are given in Wellstead and Zanker (1979), Oates (1980) and chapters 6, 11, 13 and 14.

12.2 *Control Structures*

Here we make some general points concerning traditional control structures which, while commonplace, are of renewed interest in the context of direct digital control. In particular, these relate to the

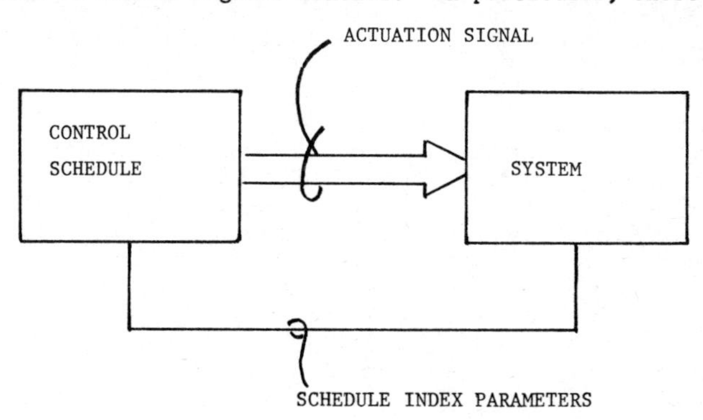

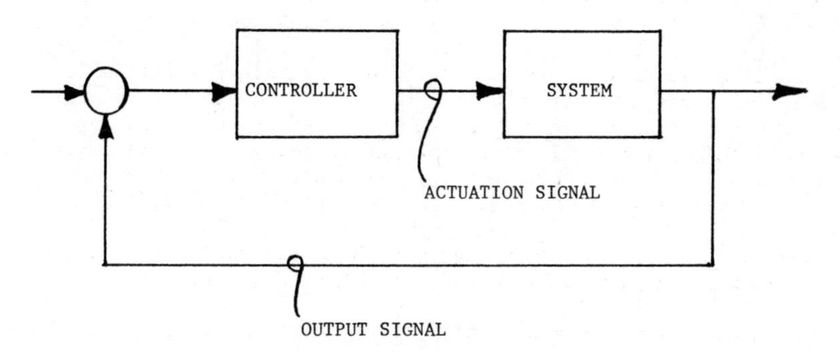

Fig 12.1 Scheduled and feedback control schemes.

essential similarity of open-loop schedule control and closed-loop feedback control. Because these techniques are superficially different, there has been a natural tendency to treat the two as distinct techniques, particularly in respect to engine management systems where much of the computational burden is associated with scheduled control (Laurance 1978). By the same token, the need to stress the distinctions (and links) between open-loop scheduling and feedback control has, until the wide scale implementation of digital control (with their concomitant ability to store the large look-up tables associated with schedules), been largely unnecessary. With these points in mind then we consider afresh open-loop scheduling and closed-loop control.

Fig 12.1 illustrates the basic forms of open-loop scheduled control and conventional control. Notice that (in Fig 12.1a) the open-loop scheduler in fact contains feedback in terms of system variables which are used to index through the control schedule. For example, petrol engine ignition timing schedules are of this form, with speed and manifold vacuum forming the schedule parameters which are used to index

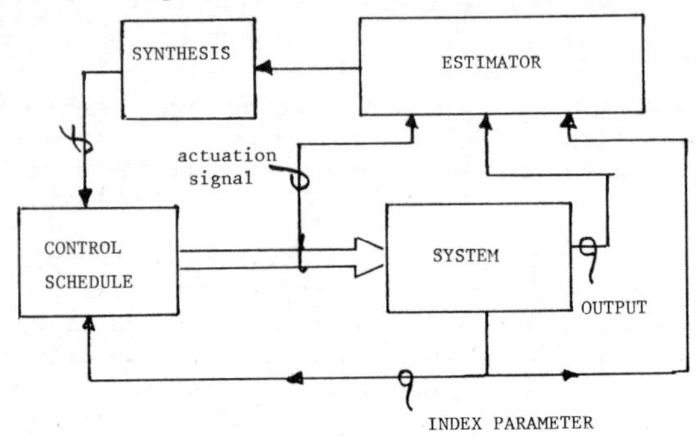

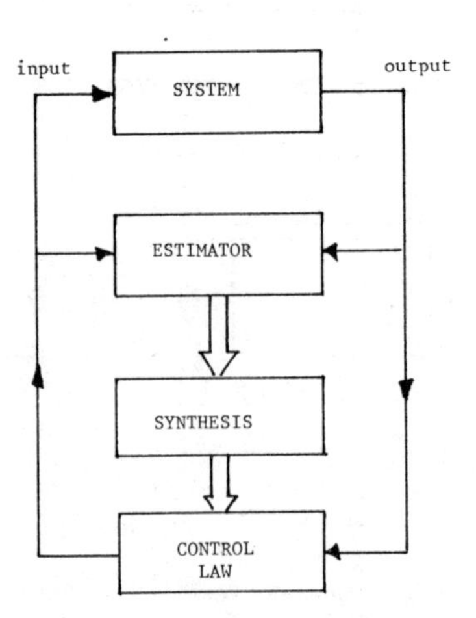

Fig 12.2 Self-tuning versions of schedules and feedback controls.

through a matrix to produce the actuation signal (in this case the spark ignition angle). Moreover, such scheduled control systems can manifest instabilities, which are due to the fedback index parameters, in the same way that the traditional feedback system (Fig 12.1b) will respond to the feedback output signal. The self-induced limit cycling between indices in the control schedule is a well-known example of this phenomenon.

In many ways, therefore, "open loop" scheduling and closed loop control share feedback system features, with the essential difference that scheduling involves no explicit tracking task. Both can be posed in a self-timing format as indicated in Fig 12.2, and both can exhibit the characteristic features of self-tuning systems (cf. Wellstead and Zanker, 1978).

12.3 *Diesel Engine Regulation - System Models*

There is then little operational difference between scheduling and closed loop control in that both are amenable to self-tuning. With this thought in mind we turn to the theme-problem of closed loop diesel engine regulation. The traditional way of regulating the speed of stationary diesel engines is by mechanical centrifugal governor. Incidentally, the more sophisticated of these involve non-linear spring mechanisms which offer a schedule of control action which is analogous to the scheduling of control parameters mentioned previously.

Scheduling and regulation are necessary because of the non-linear, open loop unstable, nature of the diesel engine. To see this, consider the simple block diagram of Fig 12.3 which indicates the essentially positive feedback action of turbo-charging from the exhaust gases and fuel pumping from the engine output shaft. The joint influence of these positive feedback loops makes the engine unstable at certain speeds. The non-linear nature of diesel engine behaviour is to a large extent due to the dissipative loading due to frictional losses. These increase (approximately) with the square of engine speed and tend to counteract

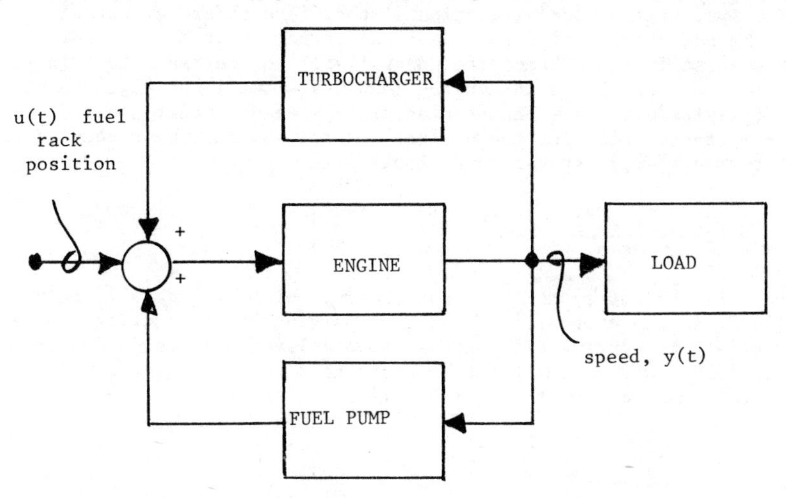

Fig 12.3 Engine block diagram.

the positive feedback loops, hence stabilizing the system at high speed. An extensive modelling (Thirurooran 1979) and identification exercise (Wellstead, Thirurooran and Winterbone 1978) has shown that, in keeping with the above arguments, the diesel engine can be modelled by a simple first order model relating the fuel rack position $u(t)$ to the engine speed $y(t)$, thus

$$\frac{y(s)}{u(s)} = \frac{k}{s + \alpha} \tag{1}$$

where α is a function of speed and is (i) negative at low speed (unstable), (ii) zero at medium speed (integrator), (iii) positive at high speed (stable).

A Remark on Model Order and Controller Order

The transfer function model given by equation (1) is a gross approximation to the dynamical behaviour of a diesel engine. However, it works very well in practice when used for *control system design*. A full model of a diesel engine involves characterising wave and flame propagation and is, by its very nature, unsuitable for control purposes. Even a linearized finite order model (Thirurooran, 1978) of the engine (involving thirty states and time-varying dynamics) is inappropriate for controller design for the speed loop. In an exhaustive study Thirurooran (1978) showed that for all practical purposes a low order model gave the most satisfactory controller design. Although we should note that satisfaction is measured here in terms of the design time required and controller complexity, as well as the final closed loop response. This point is relevant to the selection procedures for controller order in self-tuning. Except in very simple cases, the order of the underlying dynamical system is not the most appropriate for controller design, and the best way to select controller order is by cut and try in a self-tuning interaction with the system (Wellstead and Zanker 1978).

The approximate model (equation 1) therefore offers a suitable starting point from which to select the structure of a self-tuning system identifier and hence the controller. Notice that the transfer function of a reciprocating engine involves a small but significant time delay associated with the mean time between firing. In addition, the computation time for the digital control law must be accounted for, such that a revised system model looks like

$$\frac{y(s)}{u(s)} = \frac{k\, e^{-s\tau}}{s+\alpha} \tag{2}$$

where τ is the net system transport delay, which is generally much smaller than the control interval T. Applying the z transform tables appropriate to systems with partial time delays (Wellstead, Prager and Zanker 1979), and assuming a zero order hold is present, a simple discrete time model of the diesel engine is

$$\frac{y_t}{u_t} = \frac{b_1 z^{-1} + b_2 z^{-2}}{1 + a_1 z^{-1}} \tag{3}$$

where u_t, y_t are the input and output signals at the discrete times ($t = 0,1,2,...$) and z^{-1} is interpretted as the backward shift operator. The zero in the discrete time model is associated with the partial time delay, in this particular case the delay is not sufficient to place the zero outside the z-plane unit disc and thus warrant a pole-assignment regulator. Moreover, in the self-tuning tests the model structure of equation (3) was used as an initial guide to the desired controller structure.

12.4 *The Test Bed*

12.4.1 The engine (Judge 1967).

The engine used in the control experiments was a Ruston and Hornsby 6YEX Mk II 6 cylinder (in line) 4 stroke turbocharged diesel. The power delivered was 900 Nm at 1800 rev/min, which was also the maximum engine speed. The minimum speed was 900 rev/min, and over and under speed trips were provided which coupled to the fuel supply and the load. The loading was provided by a conventional water filled dynamometer. The load due to the dynamometer could be varied by adjusting the water level.

12.4.2 The test bed (Benson and Pick 1974).

The engine test bed comprised the turbocharged diesel engine coupled to a controllable-load hydraulic brake. Fuel flow to the engine was regulated via the position of the fuel rack. This was normally driven by a mechanical governor, although the rack position could also be controlled electrically and was equipped with a local rack position controller. Fuel rack position was measured potentiometrically and tachometers measured engine and turbocharger turbine speeds. A differential pressure cell measured the excess of inlet manifold pressure over ambient and an electro-optical device measured the opacity of the exhaust gas. Exhaust temperature was measured using a thermocouple, and the engine/load torque was measured using a load cell fixed between the dynamometer rotor and the casing. All signals were filtered and scaled before being fed to the computer analogue to digital converters.

12.4.3 The computer.

The control computer linked to the test bed was a DEC PDP15 machine equipped with floating point arithmetic hardware. Peripherals included a 32 channel multiplexed ADC, converting to 15 bits in 40 μS, and two 15 bit digital to analogue converters.

12.5 *Self-Tuning Trials*

Here we discuss the method of setting up and executing the self-tuning experiments on the diesel engine.

12.5.1 Setting-up the self tuner

(a) The system model and control objective

The general form of the system model in the self-tuning identifier was taken as (cf. equation (3)):

$$(1+\hat{A}(z^{-1}))y_t = z^{-k}\hat{B}(z) \quad \nabla u_t + e_t \tag{4}$$

where
$$\left.\begin{aligned} \hat{B}(z^{-1}) &= \beta_1 z^{-1} + \dots + \beta_{n_b} z^{-n_b} \\ \hat{A}(z^{-1}) &= \alpha_1 z^{-1} + \dots + \alpha_{n_a} z^{-n_a} \end{aligned}\right\} \tag{5}$$

$y(t)$ is the measured speed output and ∇u_t is the control increment, thus:

$$\nabla u_t = u_t - u_{t-1} \tag{6}$$

and u_t is the demanded fuel rack position.

Because of the incremental control action and the implied digital integrator which is cascaded with the system, the simple model becomes in terms of equation (4):

$$y_t = \frac{b_1 z^{-1} + b_2 z^{-2}}{(1-z^{-1})(1+a_1 z^{-1})} \nabla u_t = \frac{\beta_1 z^{-1} + \beta_2 z^{-2}}{1+\alpha_1 z^{-1} + \alpha_2 z^{-2}} \nabla u_t \tag{7}$$

and $\quad \alpha_1 = a_1 - 1, \; \alpha_2 = -a_1, \; \beta_1 = b_1, \; \beta_2 = b_2$

Since no reason could be found for the system being non-minimum phase at reasonable sample intervals, the detuned minimum variance (Wellstead, Edmunds, Prager and Zanker 1979) regulator algorithm was applied. In this instance, and with a time delay coefficient $k = 0$ in the model (equation (4)), the control algorithm under self-tuning is:

$$\nabla u_t = \frac{\hat{A}(z^{-1}) - T(z^{-1})}{\hat{B}(z^{-1})} (y_t - y_r) \tag{8}$$

where y_r is the reference command for the desired engine speed.

The use of incremental control effectively cascades an integrator in the loop and assures steady state correspondence between y_t and the demanded speed. The control law (equation 8) leads to a closed loop system whose poles are given by the roots of

$$1 + T(z^{-1}) = 1 + t_1 z^{-1} + \dots + t_{n_t} z^{-n_t} \tag{9}$$

Thus, $T(z^{-1})$ in equations (8) and (9) can be specified to obtain any desired closed loop speed of response by selecting the coefficients of $T(z^{-1})$ appropriately in conjunction with the sample rate. Typically, n_t is selected to be unity and t_1 is initially related to a continuous time pole closed loop pole thus,

$$t_1 = - e^{-aT}$$

where the desired pole at s = -a is set initially slightly faster than the open loop system. In the case of the engine, a closed loop pole at s = -4 was used initially.

(b) Forgetting factor λ

The self-tuning estimator is a recursive least squares algorithm of the standard form (Clarke, chapter two). Apart from system order, the most important practical part of setting up the self-tuning estimator is the "forgetting factor" λ which defines the rate at which old information is discounted from the estimator and hence from the controller design. Such considerations are clearly important in diesel regulation where, in addition to the initial tuning-in, the controller must adapt to the perceived changes in dynamics at the various operating speeds (see section 12.3). For most of the trials a fixed forgetting factor of λ = 0.99 was used, although in retrospect a variable forgetting factor which is geared to the *rate of change of the reference command signal*, is more appropriate to systems whose dynamics change with operating level (Fortescue et al 1978).

(c) Sampling Rate

The sampling rate selection is usually limited by the speed of the computer and physical considerations of the system. As the sample rate increases faster system dynamics become more apparent, forcing higher order system models and controllers. Additionally, wide band measurement noise becomes progressively more significant. The control is made excessively active correcting for the fastest dynamics and noise, at the expense of the slower dynamics which were the original control objectives. High order models, excessive controls and frequently limiting make tuning-in of the estimator slow and/or inexact.

In the case of the diesel engine the minimum time between firings was 11.1 mS (at 1800 r.p.m.) and the maximum 25 mS (at 800 r.p.m.). The computer could not easily be made to sample faster than 20 mS. Knowledge of the engine's stalling time and guesses at its dynamics suggested 200 mS as a reasonable value for sampling interval.

Notice an alternative sampling strategy would be to link the sampling to crank angle as is necessary in many other engine control loops. However, this was not tried.

(d) Initial Conditions

The start-up of the self-tuner involves initialization of the estimator, the model and hence the control. Inappropriate decisions about any of these could lead to large tuning-in transients. The problem was tackled by running the self-tuner initially on a simple first order digital simulation (stored in the control computer) until the control had settled, and then switching to the actual engine.

12.5.2 The Test Cycle

As stated in section 12.3 the dynamics of the diesel engine vary significantly with operating conditions, with speed and load being the most important factors. Accordingly, three speed and load settings were chosen to span the operating regime: 900, 1400 and 1800 rev/min were selected as low, medium and high speeds; 200, 500 and 800 Nm were the corresponding loads.

A test cycle was chosen to operate between two loads and two speeds, and include all 12 possible step changes in the two variables. A 5 bit PRBS generator was set up to form two sequences of 16 bits (15 switches) which provided the test cycle, additionally the operator was required to specify the number of sampling steps between each test cycle switch (typically 50 steps at 0.2 sec sampling, 80 or 100 steps at 0.1 sec sampling). Step changes in speed and load required different control actions over the operating range (cf. section 12.3). In particular, the engine was open loop unstable at low speeds but stable at high speed high load points.

12.5.3 Control Limits

As in fixed term control, self-tuning controllers must take regard of the physical limitations of the systems they control, else system non-linearities can cause them to detune. Control saturation is the most usual such factor. Software control limits must fall within the systems limits and must be reflected back to the estimator. The control output limiting sequence was as follows:

(i) calculate the differential control ∇u_t

(ii) integrate $u_t = \nabla u_t + u_{t-1}$

(iii) limit u_t and apply

(iv) recompute $\nabla u_t = u_t - u_{t-1}$

(v) store ∇u_t and u_t for the next estimation and control step.

The engine speed reference y_r was slew-rate-limited such that demanded accelerations were within the engine's capabilities. Rate limits were typically in the range 250 to 500 rev/min/second.

12.6 *Experimental Results*

In all, some thirty self-tuning runs were made on controller order, control objective (i.e. the polynomial $T(z^{-1})$ in equation (9)), forgetting factor, sample rate, slew-rate limits and the like. Throughout the trials the same basic test pattern was used and the following criteria applied in evaluating the control law:

(i) good disturbance rejection

(ii) satisfactory transient response

(iii) absence of violent control excursions

(C.1.)

(iv) small tuning-in and re-tuning transient
(v) rapid settling of the controller parameters } C.2.

The criteria C.1 are clearly those associated with conventional control laws. C.2 are the additional points which are needed for adaptive controller evaluation.

A detailed discussion of all the results is given in (Zanker 1980, Zanker and Wellstead 1978). Thus, for reasons of space economy, we merely pick out particular aspects of the more typical runs as they relate to self-tuning problems. In particular, we refer to a trial conducted with the model structure $n_a = 2$, $n_b = 2$ (i.e. as per equation (7)) and using detuned minimum variance with:

$$1 + T(z^{-1}) = 1 - 0.7z^{-1}$$

and a sample interval of 200 mS.

12.6.1 Tuning-in and re-tuning

Fig 12.4 shows the input/output behaviour of the engine during the initial tuning-in phase and for a re-tuning transient from medium to low speed. The initial transient is fairly oscillatory, but is not excessive for this kind of system. In addition, the supervisory control prevented the control saturating or changing at excessive rates.

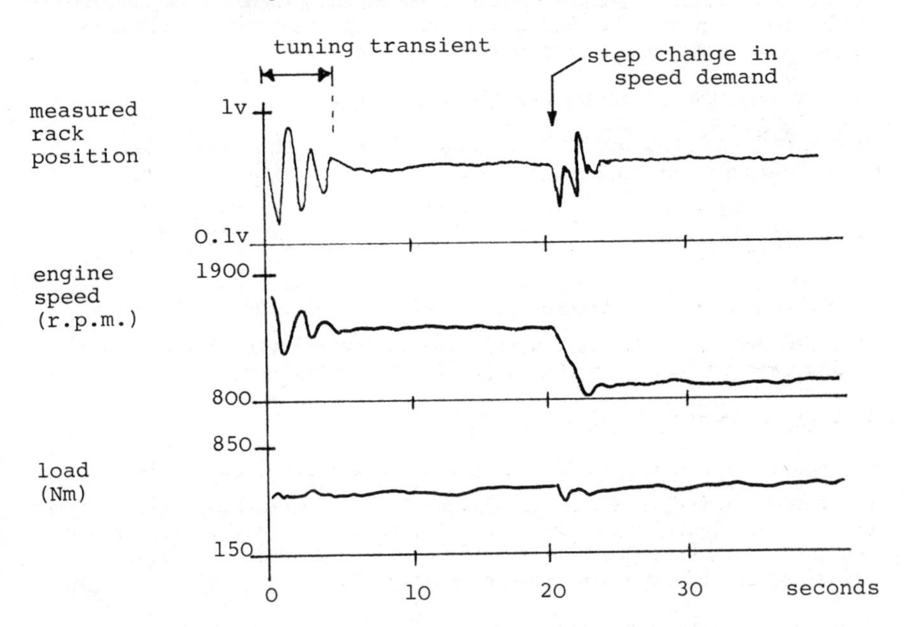

Fig 12.4 Showing the closed-loop response of a self-tuned diesel engine for a step change in demanded speed.

The re-tuning transient is more interesting, since in addition to producing an acceptable step response the controller has to retune from one suitable for a stable first order system (which is what the engine linearized dynamics are at medium speed) to cope with the open loop unstable nature of the engine at low speed.

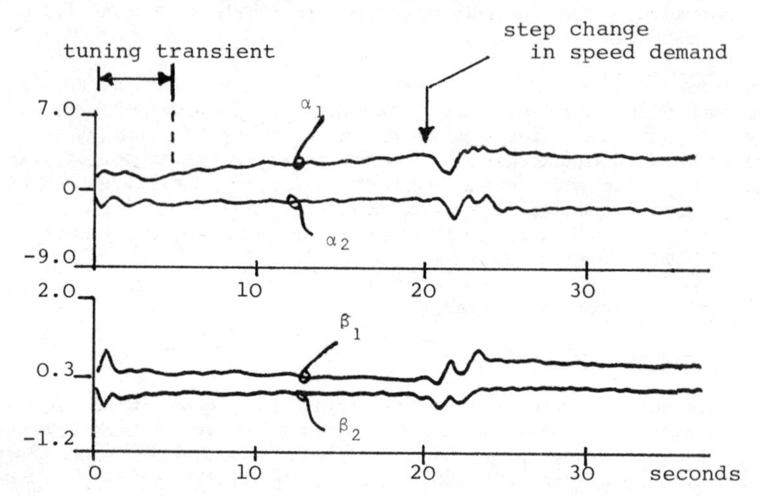

Fig 12.5 Showing the time evolution of the parameter estimates under closed loop control during tuning-in and a step change in speed demand (cf Fig 1a). Model used

$$y_t = \alpha_1 y_{t-1} - \alpha_2 y_{t-2} + \beta_1 u_{t-1} + \beta_2 u_{t-2}$$

Fig 12.5 illustrates the evolution of the system model estimates corresponding to the data shown in Fig 12.6. Notice the relatively

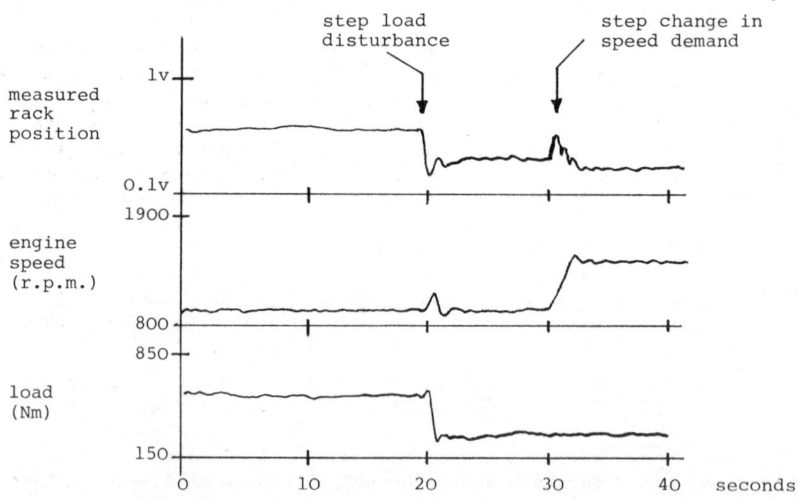

Fig 12.6 Showing the closed-loop response of a self-tuned diesel engine for a step change in load followed by a step change in demanded speed.

small change in parameters during tuning-in. In fact, this indicates that the tuning transient in Fig 12.4 is predominantly due to a signal level mis-match at the control handover point. The parameter variations at retuning are marked but nonetheless small. This is because of the relative insensitivity of the discrete time representation to s-domain transitions near $s = 0$. The changes are nevertheless crucial.

12.6.2 Disturbance Rejection

The load demand applied by the dynamometer represents a significant external disturbance. The ability of the self-tuned loop to reject such disturbances is indicated in Fig 12.6, which shows the response to a step change in load followed by a step increase in the reference engine speed. Again, the changed load also alters the perceived system dynamics so that controller re-tuning is required, and again the self-tuner complied in a satisfactory manner.

12.6.3 Other Experimental Observations

(i) Sampling Interval

Self-tuning experiments on the diesel engine covered three sampling intervals, 200 mS, 100 mS and 40 mS. Self-tuning control was most successful at the lowest sampling speed, both in terms of controls and parameter estimate behaviour. Faster sampling should, by simple physical reasoning, have allowed better control of the engine; however, in the self-tuning experiments faster sampling led to exaggerated re-tuning transients and oscillatory control actions. In fact, as the control sample interval is decreased, more system dynamics come into the bandwidth set by the sampling action. Accordingly, the system model order, and hence controller order, must be *increased* to account for these. This does not, however, mean the control is better at high sample rates.

(ii) Forgetting Factor

The forgetting factor λ sets the rate at which the self-tuner can weight out past information in favour of their data, and hence sets the speed of controller parameter adaption to new operating points. For the diesel engine various values of λ were tried, from low ($\lambda = 0.96$) to high ($\lambda = 0.999$). In all cases a fixed value of about $\lambda = 0.99$ gave the best compromise between oscillatory controls and inadequate controller re-tuning.

(iii) The Tailoring Polynomial $1+T(z^{-1})$

First order tailoring polynomials were introduced to detune the minimum variance self-tuning controllers, in order to damp out violent or oscillatory control actions. A single pole time constant τ would correspond to a z-domain tailoring polynomial $1-e^{-T/\tau}z^{-1}$ (sampling interval T), e.g. a 0.5 sec time constant is represented at 200 mS sampling by a tailoring polynomial $1-0.67z^{-1}$. A compromise had to be found between heavy detuning which left noticeable lags in the speed responses and light detuning which allowed excessive controller activity. Heavy detuning ($1+T(z) = 1-0.9z^{-1}$, 2 second time constant) showed smooth controls but clearly visible 2 second lags in the speed responses after step changes in load or speed demand. In contrast, a run with very light detuning ($1+T(z^{-1}) = 1-0.2z^{-1}$, i.e. a 50 mS time constant, almost minimum variance control) gave tight control but spiky excessive control action.

12.7 *Conclusion - Engine Management*

The self-tuning trials outlined here have been addressed simply to engine speed regulation. In practice, however, the goal is an integrated engine management system which accounts for all the various factors which influence engine performance. For stationary engine regulation (such as employed in stand-by power generation) the system outlined here is appropriate. However, for other applications (such as automotive and marine situations) a self-adaptive engine governor would fit into the context of a microprocessor based *supervisory control system* which itself would not be a simple closed-loop system. By the same token, the self-tuner should properly take account of such factors as exhaust, opacity, turbo-charger speed and so on. Moreover, as indicated earlier, this would involve a mix of feedback controllers and open loop scheduled controllers with the emphasis on the latter. In the specific context of engine management, the immediate future lies in the development of instrumentation such that engine variables can be sensed cheaply and accurately and hence feedback to the engine manager. In general, however, the future of self-tuning seems to lie in an integrated study of closed loop adaption and supervisory control. The problems which this raises are many, but such a combination of supervision and closed loop adaption could present a simple solution to convergence problems of adaptive algorithms. Indeed, in the trials discussed here the rudimentary supervision of the self-tuner (rate, limits, etc) were sufficient to ensure that the algorithm converged to a satisfactory engine speed regulator.

References

Laurance, N (1978). 'Development of an Automobile Engine Control System', Proc 7th IFAC World Congress, Helsinki, Finland.

Wellstead, P E and Zanker P M (1978). 'Techniques of Self-Tuning', Control Systems Centre Report 432, UMIST.

Thirurooran, C (1979). 'Modelling and Control of a Turbo-Charged Diesel Engine', PhD Thesis, UMIST.

Wellstead, P E, Thirurooran, C and Winterbone, D E (1978). 'Identification of a Turbo-Charged Diesel Engine', Proc 7th IFAC World Congress, Helsinki, Finland.

Wellstead, P E, Prager, D L and Zanker, P M (1979). 'Pole Assignment Self-Tuning Regulator', Proc IEE, Vol 126.8, pp 781-787.

Zanker P M (1980). 'Applications of Self-Tuning', PhD Thesis, UMIST.

Zanker, P M and Wellstead, P E (1978). 'On Self-Tuning Diesel Engine Governors', Control Systems Centre Report 422, UMIST.

Judge, A W (1967). 'High Speed Diesel Engines', Chapman & Hall, London.

Benson, R S and Pick, R (1974). 'Recent Advances in Internal Combustion Engine Instrumentation, with Particular Reference to High Speed Data Acquisition and Automated Test Beds', S.A.E. Paper 740695.

Wellstead, P E, Edmunds, J M, Prager, D L and Zanker, P M (1979). 'Pole-Zero Assignment Self-Tuning Regulators', Int J Control, 30.1, pp 1-26.

Wellstead, P E (1980). 'Self-Tuning Digital Control Systems - The Pole-Zero Assignment Approach', Control Systems Centre Report 490. (Also in SRC Vacation School Lecture Notes on Digital Control Systems, University of Sheffield).

Fortescue, T R, Kershenbaum, L S, Ydstie, B C (1978). 'Implementation of Self-Tuning Regulators with Variable Forgetting Factors', Internal Report, Dept of Chem Eng, Imperial College.

Oates, C (1980). PhD Thesis, UMIST.

Chapter Thirteen

Self-tuning controllers for surface ship course-keeping and manoeuvring

N MORT and D A LINKENS

13.1 *Introduction*

During the last decade, considerable research effort has been devoted to the design of ship autopilots using recently developed techniques of modern control theory. These developments are encouraging in view of the lack of progress in autopilot design since the first automatic steering control for ships was introduced by Sperry (1922). The operational performance of conventional autopilots at sea has been worse than that theoretically attainable by optimum steering. This discrepancy is mainly due to the fact that these conventional autopilots fail to take account of variations in both the ship's dynamic characteristics (course, speed, draught, loading condition etc) and the environment in which it operates (i.e. waves, wind, current etc). There has thus been a growing interest in autopilots that are adaptive in nature in order to overcome these known variations.

Adaptive control schemes applied to ship steering by Oldenburg (1975) and Schilling (1976) fall into the category of three-term PID controllers which are capable of automatically adjusting the control loop coefficients as a function of the external disturbances. Alternatively, model reference techniques, which have been extensively used in aircraft flight control systems, were used by Van Amerongen and Udink Ten Cate (1975) to design an adaptive ship steering system. It was found that this approach worked well for manoeuvring but was unsuitable for straight course-keeping because the model reference method does not account for disturbances. Modern control theory has been applied to ship steering by Millers (1973) who proposed on-line estimation of ship parameters via a Kalman filter cascaded with an optimal control law which minimised a quadratic performance index.

More recently, Källström et al (1977) and Brink et al (1978) have investigated the use of the self-tuning regulator in an autopilot context and the results of their work are encouraging. The purpose of this chapter is to describe, using digital simulation methods, the application of the more general self-tuning controller algorithm, first proposed by Clarke and Gawthrop (1975), to the control of surface ship course- and track-keeping. Linear and non-linear models of the rudder

Lt Cdr Mort is with the Royal Naval Engineering College, Manadon and Dr Linkens is with the Control Engineering Department, Sheffield University.

angle/yaw angle dynamics are considered together with stochastic disturbances due to sea wave effects. Finally the potential of the algorithm as an adaptive controller is investigated using shallow water effects to generate a time-varying linear model of the steering dynamics.

13.2 *Mathematical Models of Ship Dynamics*

13.2.1 Linear Models

The equations describing the motion of a ship are obtained from Newton's laws expressing conservation of linear and angular momentum (see Abkowitz (1964)). These equations are conveniently expressed using a coordinate system fixed in the ship. If the ship is considered as a rigid body then it has six degrees of freedom. The translational motions along the three axes are termed surge, sway and heave, and the rotational motions about the three axes are called roll, pitch and yaw. For large ships, it is commonly accepted that there is little coupling between the modes and the steering dynamics can be described by considering the horizontal plane motions only i.e. surge, sway and yaw. Thus, introducing the variables shown in Fig 13.1 to describe the horizontal motion of a ship;

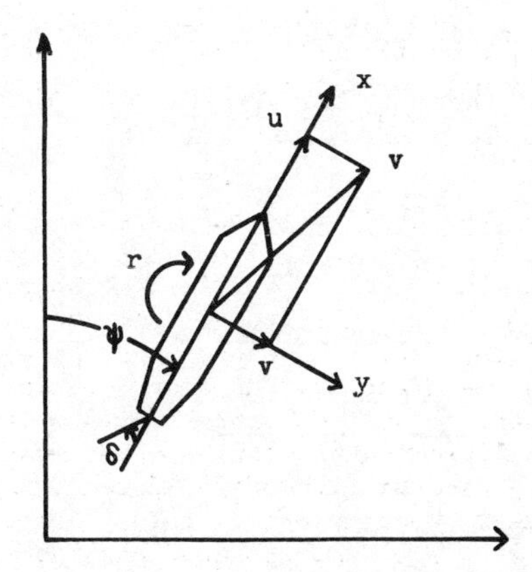

Fig 13.1 Coordinate system for ship manoeuvring equations.

the surge, sway and yaw motions can be written:

$$m(\dot{u} - vr - x_G r^2) = X$$

$$m(\dot{v} + ur + x_G \dot{r}) = Y$$

$$I_z \dot{r} + m x_G(\dot{v} + ru) = N \qquad (1)$$

where m is the mass of the ship, r is the rate of turn, x_G is the x-coordinate of the centre of gravity and I_z is the moment of inertia about the z-axis. X, Y, N on the right-hand side of (1) are the hydrodynamic forces and moments which are complicated functions of the ship's motion. The functions describing the forces and moments can be developed into a useful form for analysis purposes by means of the Taylor expansion of a function of several variables. The linearized equations of motion are derived by considering only the first order partial derivatives, called the "hydrodynamic derivatives", of the Taylor series. Thus using the notation $Y_v = \frac{\partial Y}{\partial v}$ etc for these derivatives, the equations of motion (1) can be linearized around the stationary values $r = 0$, $v = 0$ and, for constant forward speed u, reduce to

$$(Y_{\dot{v}} - m)\dot{v} + Y_v v + (Y_{\dot{r}} - mx_G)\dot{r} + (Y_r - mu)r + Y_\delta \delta = 0$$

$$(N_{\dot{v}} - mx_G)\dot{v} + N_v v + (N_{\dot{r}} - I_z)\dot{r} + (N_r - mx_G u)r + N_\delta \delta = 0 \qquad (2)$$

These equations can be converted into state space form by solving for $\dot{v}$ and $\dot{r}$ to give

$$\begin{bmatrix} \dot{v} \\ \dot{r} \\ \dot{\psi} \end{bmatrix} = \begin{bmatrix} a_{11} & a_{12} & 0 \\ a_{21} & a_{22} & 0 \\ 0 & 1 & 0 \end{bmatrix} \begin{bmatrix} v \\ r \\ \psi \end{bmatrix} + \begin{bmatrix} b_{11} \\ b_{21} \\ 0 \end{bmatrix} \delta \qquad (3)$$

where ψ, the heading angle, has been introduced as a state variable.

It follows from (3) that the transfer function relating rudder angle δ and heading angle ψ can be written as:

$$\frac{\psi(s)}{\delta(s)} = G(s) = \frac{b_1 s + b_2}{s(s^2 + a_1 s + a_2)} \qquad (4)$$

Equation (4) is commonly represented as:

$$\frac{\psi(s)}{\delta(s)} = G(s) = \frac{K(1 + sT_3)}{s(1 + sT_1)(1 + sT_2)} \qquad (5)$$

while Nomoto (1957) has suggested a second-order approximation of the form:

$$\frac{\psi(s)}{\delta(s)} = \frac{K}{s(1 + sT)} \qquad (6)$$

for stable ships and small δ (where $T = T_1 + T_2 - T_3$)

13.2.2 Non-Linear Models

The linearized model relating rudder angle to yaw angle is inadequate for some manoeuvring conditions. The non-linear higher order terms of the Taylor expansion of the forces and moments become increasingly important when large rudder deflections are demanded. These models are generally quite complex and to simplify the analysis of such systems a class of simpler, non-linear models has been developed by adding non-linear terms to the linear models given in equations (5), (6). Bech and Smitt (1969) suggested the following third order differential equation model:

$$\dddot{\psi} + \left(\frac{1}{T_1} + \frac{1}{T_2}\right)\ddot{\psi} + \frac{K}{T_1T_2}H(\dot{\psi}) = \frac{K}{T_1T_2}(\delta + T_3\dot{\delta}) \tag{7}$$

where $H(\dot{\psi})$ is a non-linear function of $\dot{\psi}$ and can be found experimentally from the relationship between δ and $\dot{\psi}$ in the steady state.

13.2.3 Parameter Variations

It is well known that the coefficients of the manoeuvring equations of surface ships are influenced by water depth. Fujino (1968) has conducted planar motion mechanism and oblique tow tests for a "MARINER" type ship. The parameters and eigenvalues of a third order transfer function (equation (5)) for varying depth-to-draft ratio (H/T) are given in Table 13.1 and correspond to a speed of 12 kts for the full scale ship.

Table 13.1 Shallow water parameters for "MARINER" type ship at 12 kts

	H/T			
	∞	2.50	1.93	1.50
T_1	102.8	83.45	70.47	27.23
T_2	8.92	9.57	9.39	9.44
T_3	19.51	17.71	13.77	8.68
T	92.22	75.32	66.08	26.94
K	-0.102	-0.095	-0.101	-0.047
λ_1	-0.101	-0.012	-0.014	-0.036
λ_2	-0.112	-0.105	-0.107	-0.106

13.2.4 Sea Wave Disturbance Effects

Any course control system should be designed with the knowledge that the ship's heading angle is likely to be subject to disturbances caused by the environment, e.g. wave motion. An accurate description of the forces and moments generated by waves is difficult to obtain. An approximation is used in this study whereby the heading angle disturbance is taken as the output of a filter driven by a wave record which is itself derived from the ITTC Wave Spectrum (see Price and Bishop (1974)). The variance of the filter output is adjusted to a value that is considered realistic with due regard to the type of ship and magnitude of the wave record. White and Linkens (1978) have published results for an Ocean Survey Ship of the Royal Navy predicting a yaw angle standard deviation of $\sigma = 5^o$ for the ship at 10 kts with a 'significant wave height' of 10 m and a wave encounter angle of 45°. However, it must be remembered that an Ocean Survey Ship is a small ship ($\sim$ 2900 tons) compared to most merchant vessels and it can be assumed that for larger ships such variations are attentuated.

13.3 *Self-Tuning Control of Ship Steering*

The self-tuning regulator, first proposed by Åström and Wittenmark (1973) was designed for the regulation of systems with constant but unknown parameters and subject to random disturbances. A refined version of this algorithm has been applied to ship steering control. The self-tuning controller is now examined as an autopilot design strategy making use of the ship models discussed in section 13.2 with a view to assessing its suitability given the non-linear, time-varying, stochastic nature of ship dynamics.

13.3.1 The Self-Tuning Controller (STC) Algorithm

The theoretical foundation for the STC algorithm is treated in chapter two by Clarke and as such is omitted here. One feature of the algorithm is that certain parameters have to be selected by the user at start-up e.g. estimated system order and time delay, initial parameter estimates, cost function weighting polynomials etc. The implications of such selections have been discussed by Clarke in chapter six and are not repeated here - the values chosen in each example are merely stated without justification.

13.3.2 Optimal vs Self-tuning Control

Inherent in the design of an optimised control system for ship steering is the formulation of a suitable criterion for "good" steering. A simple quadratic criterion has been suggested by Motora and Koyama (1968) and is of the form

$$J = \frac{1}{T}\int_0^T \{\psi_e^{\,2} + \lambda\delta^2\}dt \tag{8}$$

where ψ_e - heading error, δ - rudder angle and λ is a weighting factor. Let the course control system have the structure shown in Fig 13.2.

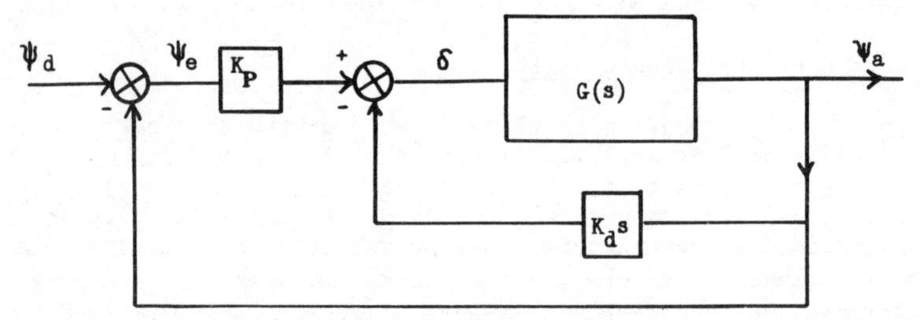

Fig 13.2 Simple course control system.

In this case ψ_d, ψ_a are the desired and actual headings and K_p, K_d are the gains of a PD controller.

Assuming the ship dynamics can be modelled in the simplest form:

$$G(s) = \frac{K}{s(1+sT)} = \frac{K'}{s(s+a)} \tag{9}$$

then the control law is

$$\delta = -K_p\psi_e - K_d\dot{\psi}_a \tag{10}$$

The optimal values of K_p, K_d such that (8) is minimised can be found using well established techniques of linear optimal control theory and are, in this case:

$$K_p = \frac{1}{\sqrt{\lambda}}$$

$$K_d = \frac{a}{K'}\left(\sqrt{1 + \frac{2K'}{a^2\sqrt{\lambda}}} - 1\right) \tag{11}$$

As a numerical example, consider the data given in Åström et al (1975) for the car and passenger ferry BORE 1. The model parameters of equation (9) are

$$K' = -0.00449 \text{ sec}^{-2}, \qquad a = 0.064 \text{ sec}^{-1}$$

The optimal values of the controller gains for the system when $\lambda = 1$ are $K_p = 1.0$, $K_d = 11.21$. The heading angle and rudder responses to a 20° demanded course change are shown in Fig 13.3(a). If the self-tuning controller is applied under the same conditions then the responses depicted in Fig 13.3(b) are obtained. Fig 13.3(c) shows the estimated values of the coefficients of the $\overline{F}$, $\overline{G}$ and $\overline{H}$ polynomials in the control law.

In this example, the following parameter values were selected:

Cost Function Weighting Polynomials	$P = 1.0$, $Q = 1.0$, $R = 1.0$.
Initial Covariance Matrix	$P_o = 10I$
Initial Parameter Estimates	$\hat{\theta}_o = [\underbrace{0.0\ 0.0}_{\overline{F}}\ \ \underbrace{1.0\ 0.0}_{\overline{G}}\ \ \underbrace{-1.0\ 0.0}_{\overline{H}}]$
Estimated Orders of A,B,C = 2,1,1	System description of the form:
Estimated time delay = 1	$Ay_t = z^{-k}Bu_t + C\xi_t$
Exponential weighting factor	$= \beta = 0.995$
Sampling time	= 1 sec

These results show that, for simple linear ship models, the self-tuning controller can be used in a ship steering control loop as a very good approximation to the optimal state feedback controller.

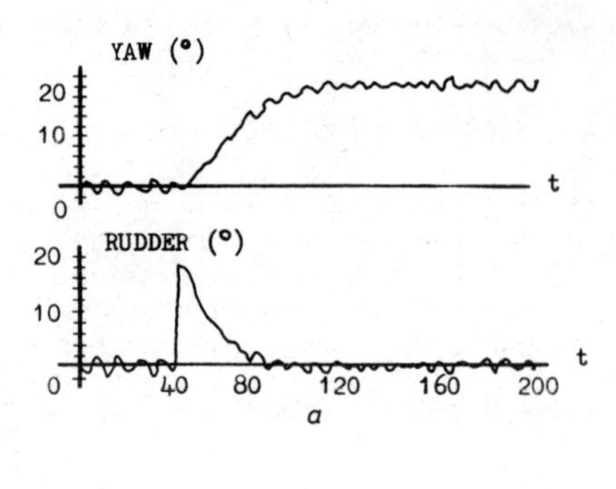

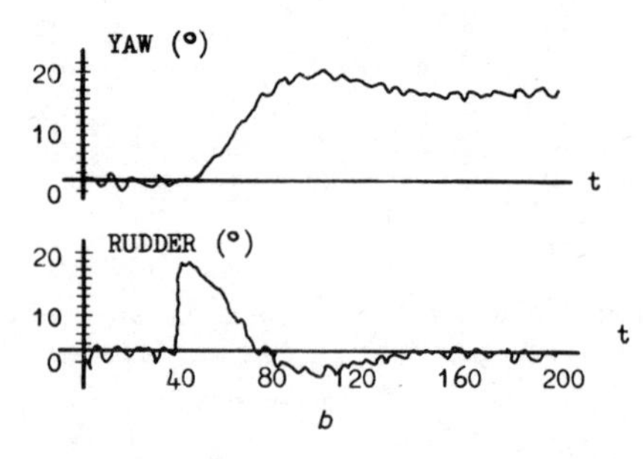

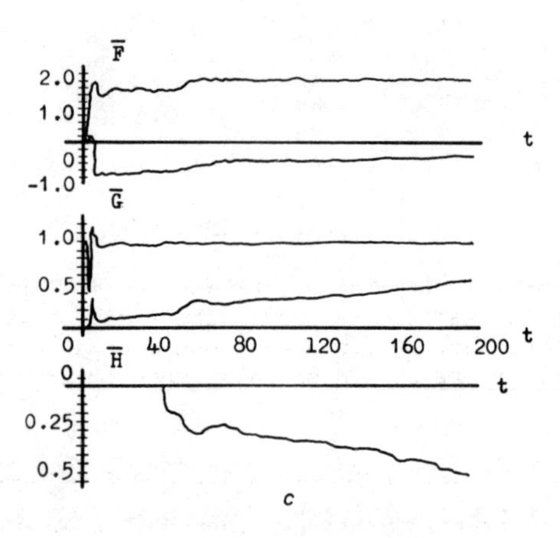

Fig 13.3 Optimal vs Self-Tuning Control for linear ship model.

13.3.3 Self-Tuning Control applied to Non-Linear Ship Models

We now widen the class of models used to describe ship dynamics to include the third order non-linear differential equation model of equation (7). Strom-Tejsen (1965) has published parameter values for a 'MARINER' class fast cargo ship corresponding to a speed to 12 kts:

$$K = -0.187 \text{ sec}^{-1},\ T_1 = 118.8 \text{ sec},\ T_2 = 7.71 \text{ sec},\ T_3 = 18.53 \text{ sec}$$

together with spiral test data leading to a third order polynomial for $H(\dot{\psi})$ of the form:

$$H(\dot{\psi}) = -1.942\dot{\psi} - 1.805\dot{\psi}^2 + 93.708\dot{\psi}^3 - 0.943$$

Also included in this example is the finite response time of the rudder since the instantaneous response indicated in the previous example is clearly not feasible. The maximum rudder angular velocity is taken as 3°/sec. The heading angle and rudder responses to a 30° course change are shown in Fig 13.4 together with the estimated $\overline{G}$ and $\overline{H}$ parameters. The STC parameters included at the start of the simulation are the same as the previous example except $\beta = 0.99$ in this case.

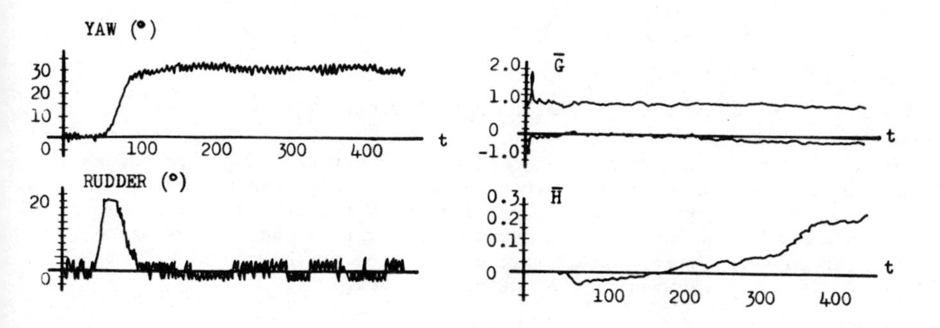

Fig 13.4 STC applied to non-linear "MARINER" model.

The effect of the non-linearity is to introduce more damping into the loop, and the yaw angle response is good. The small oscillations in the rudder are not satisfactory since they would lead to excessive wear on the steering gear. Steps should be taken to eliminate these in any final design.

13.3.4 Self-Tuning Control applied to Time-varying Ship Models

The final example presented considers the problem of the control of a ship whose dynamic characteristics are changing with time. In particular, the effect of shallow water is used as the mechanism for generating parameter variations and the results due to Fujino (1968) are used in the simulation. An interesting extension of Fujino's work is to investigate the design of an autopilot having the structure of Fig 13.2 given the variations in the transfer function parameters listed in Table 13.1. Consider the effect on the closed-loop response of the changing dynamics when the optimal gains are fixed at those values corresponding to $H/T = \infty$ i.e. deep water. Fig 13.5 depicts the variation in the eigenvalues as the ship enters shallow water (H/T decreasing) at a speed of 12 kts. Also shown on this graph are the eigenvalue positions when the optimal gains are varied in order that they remain optimal as the ship's dynamics change with varying depth. Two values of cost function weighting factor λ are considered.

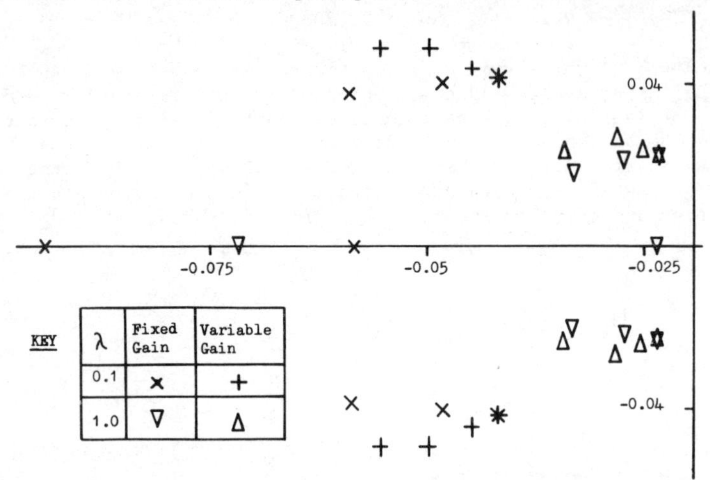

Fig 13.5 Eigenvalue variation with H/T for "MARINER" model.

From this graph, it can be seen that the deviation of the eigenvalues from the optimal deep water values is much reduced when the gains of the controller are allowed to vary along with the ship dynamics. This is an important point in the argument for adaptive controllers since, in this case at least, the overall performance of the course control system is improved when varying gains are used.

To examine the behaviour of the self-tuning controller in the context of an adaptive controller consider a ship approaching a land mass such that the depth-to-draft ratio (H/T) varies with time according to Fig 13.6(a). Then, using the values contained in Table 13.1, Fig 13.6(b) indicates the time variation in the parameters of the ship's transfer function expressed as a percentage of the deep water values.

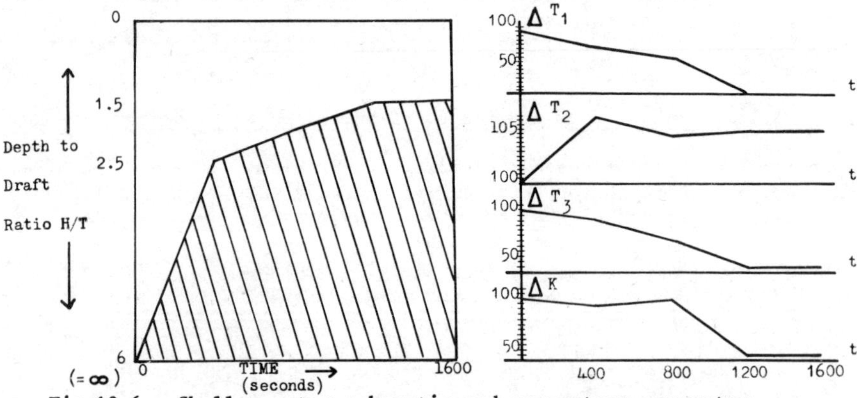

Fig 13.6 Shallow water schematic and percentage parameter variations.

The parameters for initialising the algorithm have been chosen as before. If a sequence of course changes such as those shown in Fig 13.7(a) is demanded, the response of the ship under self-tuning control is shown in Fig 13.7(b).

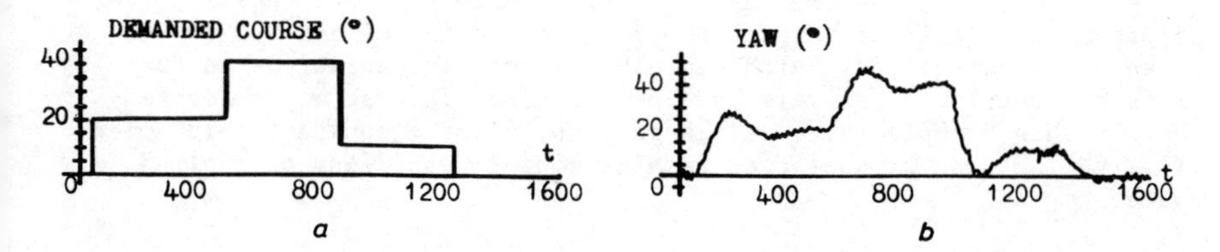

Fig 13.7 STC applied to "MARINER" in varying depth of water.

To illustrate the effectiveness of the self-tuning controller in monitoring the changing ship parameters consider Fig 13.8. This figure shows one of the estimated parameters ($\hat{g}_1$) when the course changes given in Fig 13.7(a) are demanded for the following three conditions of water depth:

(i) Constant deep water H/T = ∞.

(ii) Constant shallow water H/T = 1.5.

(iii) Varying depth of water as given by Fig 13.6(a).

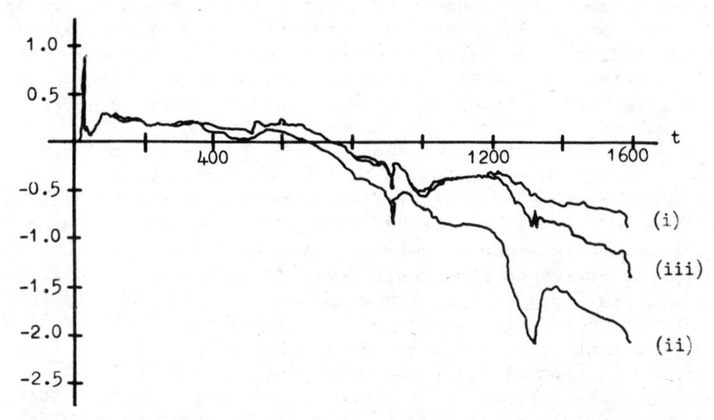

Fig 13.8 Variation of g_1 estimate in different conditions of water depth.

The rate at which the self-tuning controller "tunes-in" to the changing parameters is governed by several factors. Ideally, the value of β, the exponential weighting factor in the estimation algorithm, should be linked to the rate of parameter variations in the system. However, this information is rarely available so the selection of β is invariably dependent on the past experience of the user. Another factor that influences the rate of convergence of the estimates is the nature of the set point function. The example of a manoeuvring ship in shallow water has been used (Mort and Linkens (1980)) to demonstrate that the succession of course changes shown in Fig 13.7(a) encourages greater convergence of the estimated parameters than does a single course change.

To sum up, this example must be viewed in the correct context, i.e. it provides a qualitative picture of the potential of the STC to act as an adaptive controller. Although clearly defined variations in the transfer function parameters have been quoted, it must be remembered that in this example these values only apply for a constant ship speed of 12 kts and for a model that remains linear regardless of applied rudder angle.

13.4 *Conclusions*

The results presented in section 13.3 have indicated that the STC is a suitable candidate in the design of controllers for surface ship manoeuvring.

The algorithm compared very favourably with an optimal state feedback controller which was designed assuming complete knowledge of the linear second order transfer function parameters. Although the estimation procedure intrinsic in the algorithm is linear in form, satisfactory results were obtained when the system to be controlled was non-linear. Further work is necessary, however, to gain a fuller understanding of the performance of the algorithm when applied to different classes of non-linear system.

It is well known that a recursive estimation procedure should be capable of monitoring slowly-varying parameters in a dynamic system. The evidence produced in the example considered here suggests that the self-tuning controller can be applied to this class of system although no quantitative definition of "slowly-varying" has been attempted.

The dynamics of a surface ship can be modelled as a time-varying, non-linear, stochastic system which implies that controlling such a system in some "optimum" fashion is far from straight-forward. The self-tuning controller is one technique that has become available in recent years and which shows promise in attempting to solve the problem. However, further research is necessary to identify possible difficulties that may exist in applying the algorithm to the design of ship autopilots. For example, unlike large tankers, warships of the Royal Navy possess significant cross-coupling effects between rudder and fin stabilizers which implies that the control of yaw and roll motion cannot be considered separately. Hence, a multivariable structure is suggested for the ship model and it would seem prudent to consider this in any autopilot design using self-tuning methods.

References

Abkowitz, M A (1964). 'Lectures on ship hydrodynamics, steering and manoeuvrability', Report Hy-5, Hydro- and Aerodynamics Laboratory, Lyngby, Denmark.

Åström K J, Källström C G, Norrbin N H and Byström L (1975). 'The Identification of Linear Ship Steering Dynamics using Maximum Likelihood Parameter Estimation', SSPA Report No 75, Gothenburg, Sweden.

Åström K J and Wittenmark B (1973). 'On Self-Tuning Regulators', Automatica 9 pp 185-199.

Bech M and Smitt L W (1969). 'Analogue Simulation of Ship Manoeuvres'. Report Hy-14, Hydro- and Aerodynamics Laboratory, Lyngby, Denmark.

Brink A W, Bass G E, Tiano A and Volta E (1978). 'Adaptive Automatic Course-keeping Control of a Supertanker and a Container Ship - A Simulation Study', Proc 5th Ship Control System Symposium, Annapolis, USA.

Clarke D W and Gawthrop P J (1975). 'Self-tuning controller', Proc IEE Vol 122, No 9, pp 929 - 934.

Fujino M (1968). 'Experimental Studies on Ship Manoeuvrability in Restricted Waters - Part I'. International Shipbuilding Progress Vol 15, No 168, pp 279-301.

Källström C G, Åström K J, Thorell N E, Eriksson J and Sten L (1977). 'Adaptive Autopilots for Steering of Large Tankers', Report TFRT-3145, Department of Automatic Control, Lund Institute of Technology, Sweden.

Millers H T (1973). 'Modern Control Theory Applied to Ship Steering', Proc 1st IFAC/IFIP Symposium on Ship Operation Automation, Oslo.

Mort N and Linkens D A (1980). 'Self-tuning Controllers for Surface Ship Course and Track-keeping', International Symposium on Ship Steering Automatic Control, Genoa, Italy.

Motora S and Koyama T (1968). 'Some Aspects of Automatic Steering of Ships', Japan Shipbuilding and Marine Engineering, July issue.

Nomoto K (1957). 'On the Steering Qualities of Ships', International Shipbuilding Progress, Vol 4, No 35.

Oldenburg J (1975). 'Experiments with a new adaptive autopilot intended for controlled turns as well as for straight course steering', Proc 4th Ship Control Systems Symposium, The Hague.

Price W G and Bishop R E D (1974). 'Probabilistic Theory of Ship Dynamics', Chapman and Hall, London.

Schilling A C (1976). 'Economics of autopilot steering using an IBM/7 Computer', Proc 2nd IFAC/IFIP Symposium on Ship Operation Automation, Washington.

Sperry E (1922). 'Automatic Steering', Trans SNAME.

Strom-Tejsen J (1965). 'A digital computer technique for prediction of standard manoeuvres of surface ships', David Taylor Model Basin Report 2130.

Van Amerongen J and Udink Ten Cate A J (1975). 'Model Reference Adaptive Autopilots for Ships', Automatica, 9 pp 441 - 449.

White A D and Linkens D A (1978). 'An Adaptive Kalman Filter for Marine Navigation and Gravity Measurement', Proc IEE Vol 125, No 12, pp 1311-1317.

Chapter Fourteen

Self-tuning control of ship positioning systems

P T K FUNG and M J GRIMBLE

14.1 *Introduction*

Dynamic ship positioning control is used to maintain the position and heading of a ship, or floating platform, above a pre-selected fixed position on the seabed by using the vessels' thrusters. This form of control is also employed for tracking pipe lines or cables at a fixed speed. Dynamic positioning (DP) systems have the advantage of not involving anchors or mooring lines, however, they are sometimes combined with these position holding devices. Dynamic positioning systems are particularly suitable for drilling in deep water, typically 300-3000 metres.

This type of vessel can cost one hundred thousand pounds per day and thus it is an advantage that no delay in the laying or retrieving of anchors is involved with DP systems. Also no damage can be caused to oil pipelines on the seabed. Several position measurement systems are employed including radio systems, acoustic beacons and taut wire systems. Radio systems operating at 9.2 GHz can be used to measure the position of an accuracy of one metre at a range of six kilometres. The measurements are often combined in some optimal manner. An advantage of Kalman filtering based DP systems over the older notch filter systems is that the measurements can be lost for a number of sample periods without upsetting the filtering action. Thus, if the measurements are disturbed by, say fish movements, the input to the filter can be removed temporarily. This advantage is maintained in the self-tuning scheme to be described.

14.2 *Systems Description*

The basic components in a dynamic ship positioning system are shown in Fig 14.1. The motions of a vessel are assumed to consist of low frequency (LF) motions and high frequency (HF) motions. The LF motions (less than 0.25 rad/sec) are mainly due to thrusters, current, wind and second order wave forces. The last three forces can cause the vessel to drift from its station and these forces must be counteracted by using the thrusters. The first order oscillatory component of the wave force causes the HF motions (0.3-1.5 rad/sec) which cannot be counteracted effectively with the thrusters. Any attempt to do so would cause excessive wear and energy loss.

P T K Fung and M J Grimble are with the Department of Electrical and Electronic Engineering, Sheffield City Polytechnic.

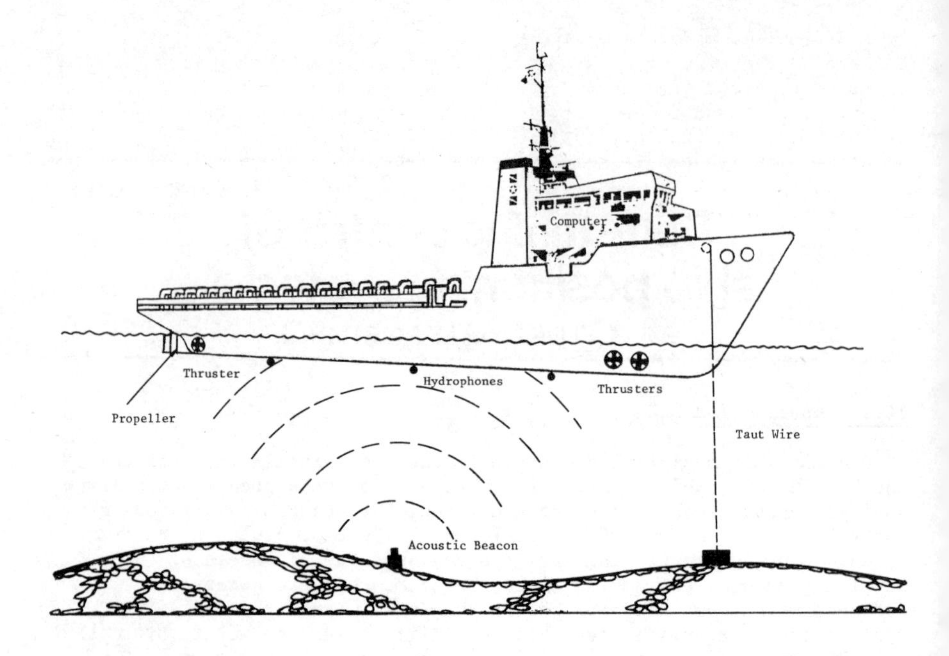

Fig 14.1 Measurement Systems of Dynamic Ship Positioning System.

14.2.1 Low Frequency motion

The vessel dynamics are governed by the following non-linear equations, English and Wise (1975):

$$\left.\begin{aligned} (M - X_{\dot{u}})\dot{u} - (M - Y_{\dot{v}})rv &= X_A + X_H(u, v, r) \\ (M - Y_{\dot{v}})\dot{v} + (M - X_{\dot{u}})ru &= Y_A + Y_H(u, v, r) \\ (I^2_{zz} - N_{\dot{r}})\dot{r} \qquad\qquad &= N_A + N_H(u, v, r) \end{aligned}\right\} \quad (1)$$

where
- u: surge velocity
- v: sway velocity
- r: yaw velocity
- X_A: applied surge direction force due to thrusters and environment
- Y_A: applied yaw direction force due to thrusters and environment
- N_A: applied turning moment on the vessel
- X_H, Y_H, N_H: the hydrodynamic forces and moment due to relative motion between the vessel and water
- $X_{\dot{u}}, Y_{\dot{v}}, N_{\dot{r}}$: added masses and added inertia which depend on the nature of the body motion and the resulting flow pattern
- M: mass of vessel
- I_{zz}: radius of gyration

A linearised model is normally used for design purposes. The linearised dynamics of a ship is dependent upon the ship motion relative to current, speed, sea conditions and the angles of inclination of the thrusters to the ship body axes. In this study the sway model of Wimpey Sealab (Wise and English, 1975), is considered with Beaufort No 8 sea condition (equivalent to a meanwind velocity of 19 m sec^{-1}). The current speed is difficult to measure, and for simplicity it is assumed to be zero. The normalised linearised model is:

$$\begin{aligned} \dot{\underline{x}}_\ell(t) &= \underline{A}_\ell \underline{x}_\ell(t) + \underline{B}_\ell u(t) + \underline{D}_\ell \omega(t) + \underline{E}_\ell \eta(t) \\ \underline{y}_\ell(t) &= \underline{C}_\ell \underline{x}_\ell(t) \\ \underline{z}_\ell(t) &= \underline{y}_\ell(t) + \underline{v}(t) \end{aligned} \tag{2}$$

where for one axes of motion (sway) the system matrices become:

$$A_\ell = \begin{bmatrix} -0.0546 & 0 & 0.5435 \\ 1 & 0 & 0 \\ 0 & 0 & -1.55 \end{bmatrix}$$

$$B_\ell^T = \begin{bmatrix} 0 & 0 & 1.55 \end{bmatrix}$$

$$C_\ell = \begin{bmatrix} 0 & 1 & 0 \end{bmatrix}$$

$$D_\ell^T = \begin{bmatrix} 0.5435 & 0 & 0 \end{bmatrix}$$

$$E_\ell^T = \begin{bmatrix} 0.384 & 0 & 0 \end{bmatrix}$$

$$\underline{x}\,(t) = \begin{bmatrix} x_{\ell_1}(t) \\ x_{\ell_2}(t) \\ x_{\ell_3}(t) \end{bmatrix} = \begin{bmatrix} \text{sway velocity} \\ \text{sway position} \\ \text{thruster state} \end{bmatrix}$$

Here $u(t) \in R^1$ represents the control input to the thruster, $\omega(t) \in R^1$ is a white noise sequence representing the random force on the vessel, $\eta(t) \in R^1$ represents the steady state and wind gust disturbance. Other disturbances, such as wave drift and current forces, cannot be measured and can be considered to produce an unknown mean value on the signal $\omega(t)$. Let $y(t) \in R^1$ denote the low frequency motion, $v(t) \in R^1$ the measurement noise and $z(t) \in R^1$ the total sway motion. For Wimpey Sealab, the estimated standard deviation for $\omega(t)$ $\sigma_p = 0.00228$, and that for $v(t)$ is $\sigma_m = 0.0033$.

14.2.2 High Frequency Motion

The assumption which is fundamental to the development of models for the HF motion of a vessel is that the sea state is known and can be described by a spectral density function. It is also assumed that in the worse case the HF motions of the vessel are equivalent to the first order wave motions. This implies that the effect of the vessel dynamics in attenuating the wave motions is neglected. A commonly used wave spectral density function (Pierson-Moskowitz 1964) is expressed as:

$$S_n(\omega) = \frac{A}{\omega^5} e^{-B/\omega^4} \tag{3}$$

where ω is the angular frequency in radians/second. $A = 4.894$, $B = 3.1094/(h_{1/3})^2$. The term $h_{1/3}$ is defined as the significant wave height. Typical power spectra for various Beaufort numbers are shown in Fig 14.2. The first order wave induced motions can be simulated

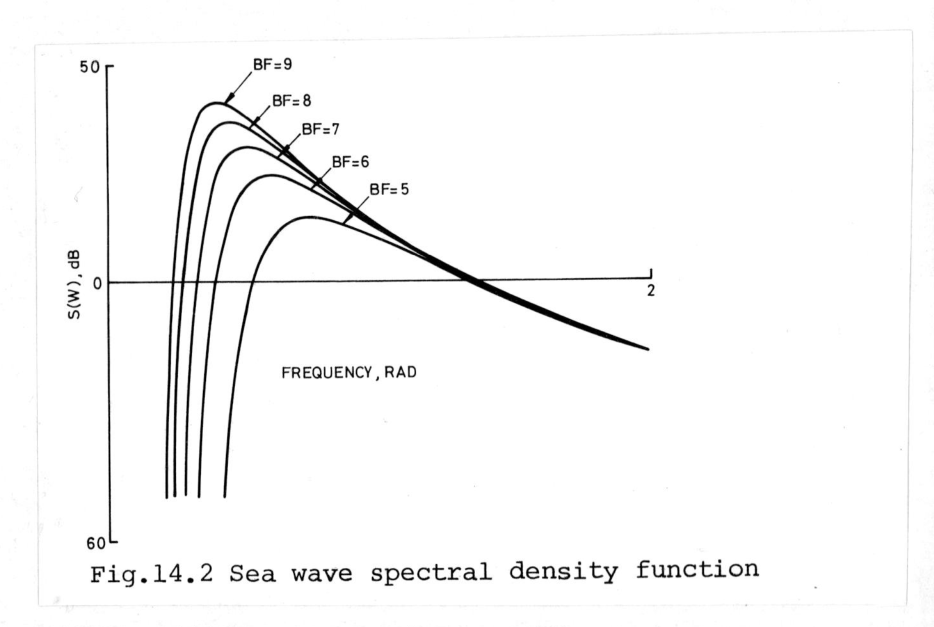

Fig.14.2 Sea wave spectral density function

using the following expression (Morgan, 1978):

$$y_h(t) = \sum_{i=1}^{20} y_o \sin(\omega_i t + \theta_i) \tag{4}$$

where θ_i are random numbers lying in $(0, 2\pi)$; y_o and ω_i are selected to approximate the Pierson-Moskowitz wave spectrum.

For a fully developed sea,

$$\omega_i = \frac{0.6990525}{\sqrt{h_{1/3}}} [\ln(\frac{2M}{2i-1})]^{-\frac{1}{4}} \tag{5}$$

$$y_o = h_{1/3} (M/2)^{-\frac{1}{2}} \tag{6}$$

where M is the number of equal parts into which the wave spectrum is divided. It is assumed the HF motion can be modelled by a second order transfer function.

$$y_h(t) = \frac{C(z^{-1})}{A(z^{-1})} \xi(t) \tag{7}$$

$$z_h(t) = y_h(t)$$

where

$$C(z^{-1}) = c_1 z^{-1}$$

$$A(z^{-1}) = 1 + a_1 z^{-1} + a_2 z^{-2} \tag{8}$$

The parameters of the polynomials $A(z^{-1})$ and $C(z^{-1})$ are unknown and $\xi(t)$ is an independent random sequence.

14.3 *The Filtering Problem*

In the DP control system, only LF motion needs to be controlled. Thus, the HF motion and LF motion must be estimated separately. The measured output consists of LF components, HF components and measurement noise. Balchen et al (1976) and Grimble et al (1979), (1980a), (1980b), have applied the Kalman filters to the problem of estimating the separate motions. The linear Kalman filter does not adapt to different sea conditions. The extended Kalman filter can obviate this problem but this increases the computing load. Moreover, some statistical properties such as the process noise and measurement noise covariances must normally be specified.

Some simplification of the extended filtering solution appears possible. For example, only LF motion states must be estimated for control purposes; not all of the system states. It is sensible to consider a replacement for the extended Kalman filter which may lead to a simpler adaptive scheme. A new philosophy for solving the filtering problem is proposed in the following. The basis of the scheme is illustrated in Fig 14.3.

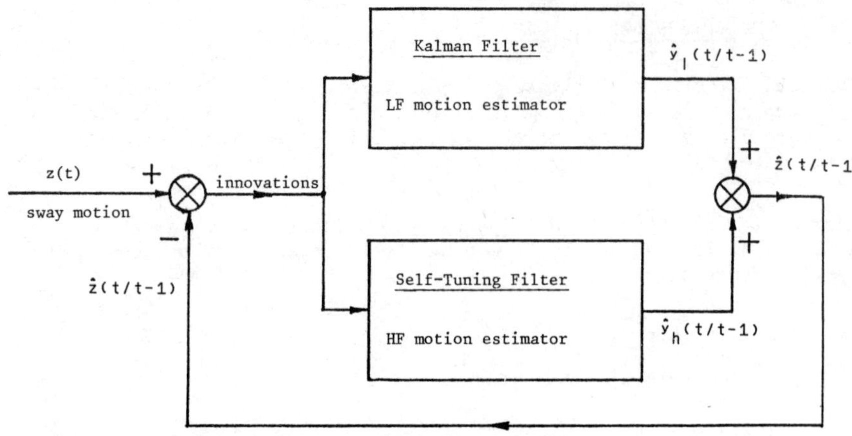

Fig 3 Adaptive and Kalman Filtering Scheme

14.3.1 Design of the Self Tuning Filter

Recall the model of HF motion in equation (7):

$$Y_h(t) = \frac{C(z^{-1})}{A(z^{-1})}\xi(t)$$

Let σ_ξ^2 be the covariance of $\xi(t)$. The low frequency model is assumed known and at time t, $\hat{y}_\ell(t|t-1)$ is known. The way in which this quantity is calculated will be discussed in the next section. Thus, the noise corrupted HF motion estimate becomes:

$$m_h(t) = z(t) - \hat{z}_\ell(t|t-1) \tag{9}$$

$$= z(t) - \hat{y}_\ell(t|t-1)$$

and

$$y_h(t) = z(t) - z_\ell(t) \tag{10}$$

where z(t) is the total measured motion. The one-step ahead prediction of $y_h(t)$ is the same as the one-step ahead prediction of $m_h(t)$, that is

$$\hat{m}_h(t|t-1) = \hat{y}_h(t|t-1) \tag{11}$$

It follows from equation (10) that

$$y_h(t) = z(t) - \hat{z}_\ell(t|t-1) - e(t) \tag{12}$$

where e(t) is the prediction error of $z_\ell(t)$ and is an independent sequence with zero mean. Substitute euqation (12) into equation (9) to yield

$$m_h(t) = y_h(t) + e(t) \tag{13}$$

Note that at time t, information for estimating $y_h(t)$ is available and e(t) can be treated as the measurement noise of the process. The corresponding innovation model (Kailath and Frost, 1968) can be represented by:

$$A(z^{-1})m_h(t) = D(z^{-1})\varepsilon(t) \tag{14}$$

where

$$D(z^{-1}) = 1 + d_1z^{-1} + \ldots + d_nz^{-n} \tag{15}$$

and the innovation follows from (11) as:

$$\varepsilon(t) = m_h(t) - \hat{y}_h(t|t - 1) \tag{16}$$

From equation (9), $\varepsilon(t)$ can also be expressed as:

$$\varepsilon(t) = z(t) - \hat{y}_\ell(t|t - 1) - y_h(t\ t - 1) \tag{17}$$

which is the same as the innovation for the total (LF + HF) filtering problem. The variance of $\varepsilon(t)$ is denoted by σ_ε^2.

The coefficients $\{d_i\}$ and the variance σ_ε^2 are given by the following spectral factorisation

$$\sigma_\varepsilon^2 D(z)D^*(z) = \sigma_\xi^2 D(z)C^*(z) + \sigma_e^2 A(z)A^*(z) \tag{18}$$

where $A^*(z)$, $C^*(z)$ and $D^*(z)$ are the reciprocal polynomials of $A(z^{-1})$, $C(z^{-1})$ and $D(z^{-1})$ respectively, and are defined using:

$$A^*(z) = z^n A(z^{-1}) \tag{19}$$

It is assumed that $D(z^{-1})$ has all its zeros outside the unit circle.

Hagander and Wittenmark (1977) have shown that the estimate of $\hat{y}_h(t)$ is given by

$$\hat{y}_h(t|t + k) = m_h(t) - \frac{\sigma_e^2}{\sigma_\varepsilon^2} F_k(z)[m_h(t) - \hat{y}_h(t|t - 1)] \tag{20}$$

where F_k is defined as:

$$F_k(z) = \sum_{i=o}^{k} f_i z^i \tag{21}$$

$$F(z) = \sum_{i=o}^{\infty} f_i z^i \tag{22}$$

and is computed from the following identity:

$$A(z) = D(z)F(z) \tag{23}$$

The smoothing delay is denoted by k but in the filtering problem k is zero, and $F_k(z)$ is found to be unity. Let σ_e^2 denote the covariance of e(t), then

$$\frac{\sigma_e^2}{\sigma_\varepsilon^2} = \frac{d_n}{a_n} \tag{24}$$

(n = 2 in the present system) and from equation (16) and equation (20):

$$\hat{y}_h(t|t) = m_h(t) - \frac{d_n}{a_n}\varepsilon(t) \tag{25}$$

14.3.2 Self-Tuning Filtering Algorithm

Step 1: Estimate the parameters a_i and d_i, i = 1, 2, ..., n, in the polynomials $A(z^{-1})$ and $D(z^{-1})$ using (14) and the extended least square technique (or recursive maximum likelihood). The estimated polynomials are denoted by $\hat{A}(z^{-1})$ and $\hat{D}(z^{-1})$.

Step 2: Compute the estimate from

$$\hat{y}_h(t|t) = m_h(t) - \frac{\hat{d}_n}{\hat{a}_n}\hat{\varepsilon}(t) \tag{26}$$

where

$$\hat{\varepsilon}(t) - \frac{\hat{A}(z^{-1})}{\hat{D}(z^{-1})} m_h(t) \tag{27}$$

and $m_h(t)$ is given from equation (9). The two steps of the algorithm are repeated at each sampling instant.

The above algorithm may be classified as an implicit self-tuning filter because the spectral factorisation stage is avoided. The innovation $\hat{\varepsilon}(t)$ and the quotient $\sigma_e^2/\sigma_\varepsilon^2$ can be computed directly from the estimated parameters.

14.3.3 The Kalman Filter

The Kalman filter for the LF motion estimation is obtained below. The LF model can be discretised and represented as:

$$\underline{x}_\ell(t + 1) = \Phi\underline{x}_\ell(t) + \Delta u(t) + \Gamma\omega(t) \tag{28}$$

$$y_\ell(t) = C_\ell \underline{x}_\ell(t) \tag{29}$$

$$z_\ell(t) = y_\ell(t) + v(t)$$

with

$$E\{\omega(t)\} = 0, \quad E\{\omega(k)\omega^T(m)\} = Q\delta_{km} = \sigma_p^2\delta_{km} \tag{30}$$

$$E\{v(t)\} = 0, \quad E\{v(k)v^T(m)\} = R\delta_{km} = \sigma_m^2\delta_{km} \tag{31}$$

where δ_{km} is the Dirac function. The wind disturbance can be measured and fedforward but is not considered here. The Kalman algorithm becomes:

predictor: $$\hat{\underline{x}}_\ell(t + 1|t) = \Phi\underline{x}_\ell(t|t) + \Delta u(t) \tag{32}$$

$$\hat{\underline{y}}_\ell(t + 1|t) = C_\ell\hat{\underline{x}}_\ell(t + 1|t) \tag{33}$$

estimator: $$\hat{\underline{x}}_\ell(t + 1|t + 1) = \hat{\underline{x}}_\ell(t + 1|t) + K(t + 1)(z(t + 1) - \hat{y}_\ell(t + 1|t) - \hat{y}_h(t + 1|t)) \tag{34}$$

From equation (17), equation (34) becomes:

$$\hat{\underline{x}}_\ell(t + 1|t + 1) = \hat{\underline{x}}_\ell(t + 1|t) + K(t + 1)\varepsilon(t + 1)$$

Now $\hat{\varepsilon}(t + 1)$ can be calculated from (27) since from equation (9) $m_h(t + 1) = z(t + 1) - \hat{y}_\ell(t + 1|t)$ and previous $m_h(t)$ values are known. Thence, the following approximation may be made for the estimator equation:

$$\hat{\underline{x}}_\ell(t + 1|t + 1) = \hat{\underline{x}}_\ell(t + 1|t) + K(t + 1)\hat{\varepsilon}(t + 1) \tag{35}$$

$$\hat{y}_\ell(t + 1|t + 1) = C_\ell\hat{\underline{x}}_\ell(t + 1|t + 1) \tag{36}$$

The Kalman gain matrix K(t + 1) can be calculated for a representative sea state using the usual procedure, (Grimble, Patton and Wise, 1979) for the total HF + LF system. However, the sea state varies and thus some approximation is involved in fixing this matrix. Fortunately, the effect of the high frequency subsystem on the low frequency gain calculation is small. Thus, it is more convenient here to neglect its effect and to calculate the gain using the low frequency equations only. The algorithm becomes:

$$P(t + 1|t) = \Phi P(t|t)\Phi^T + \Gamma Q\Gamma^T \tag{37}$$

$$K(t + 1) = P(t + 1|t)C^T[CP(t + 1|t)C^T + R]^{-1} \tag{38}$$

$$P(t+1|t+1) = [I - K(t+1)C]P(t+1|t)[I - K(t+1)C]^T + K(t+1)RK^T(t+1) \quad (39)$$

where P(.) is the error covariance matrix. If the disturbances are stationary and the system matrices are invariant, the gain K(.) will converge to a steady state value. The computing load may be reduced if K(.) is pre-calculated and maintained at this constant value. Thus, the only calculations involve equations (32), (33), (35) and (36) at each sampling instant. The structure of the combined filtering scheme is shown in Fig 14.4. For Wimpey Sealab, $\sigma_p^2 = 5.1984 \times 10^{-6}$, $\sigma_m^2 = 10^{-5}$. The computed steady-stage Kalman gains are given as:

$$K_{ss}(.) = \begin{bmatrix} 0.301351 \\ 0.776339 \\ 0.0 \end{bmatrix} \quad (40)$$

The thruster state has zero gain because the process noise is not fed into this state.

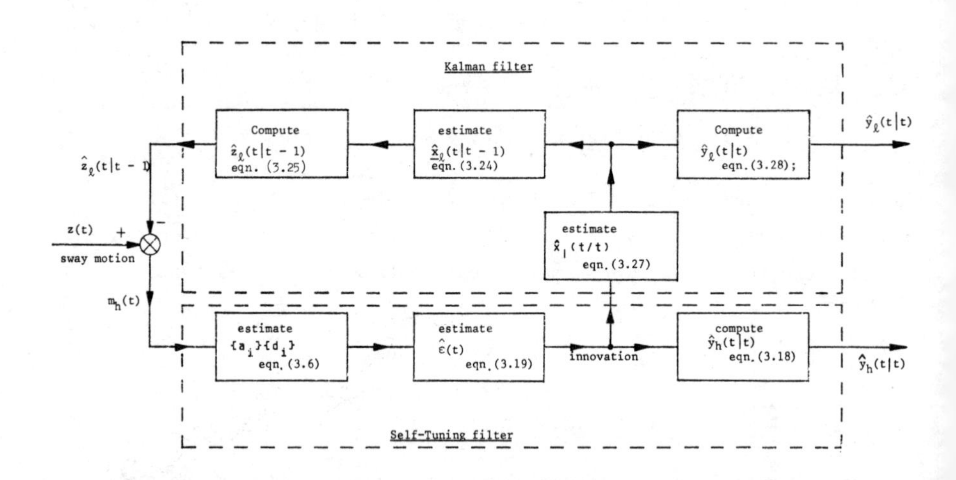

Fig 14.4 Structure of the Filtering Scheme

14.4 *Filtering Problem Results*

The estimated parameters are shown in Fig 14.5. The parameters converged to steady values after 350 seconds. It is, of course, well known that the convergence rate of any technique, where the innovations must be approximated, is very slow. Fig 14.6 shows the total sway motion and the estimated LF motion is shown in Fig 14.7. When the uncontrolled vessel was drifting away from station, the estimator

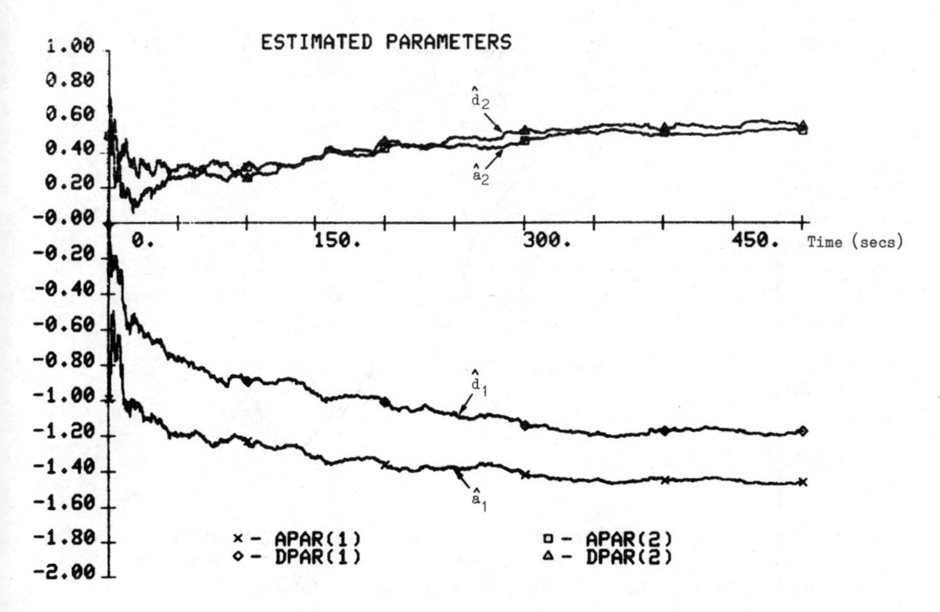

Fig 14.5 Estimated Parameters of Self-Tuning Filter

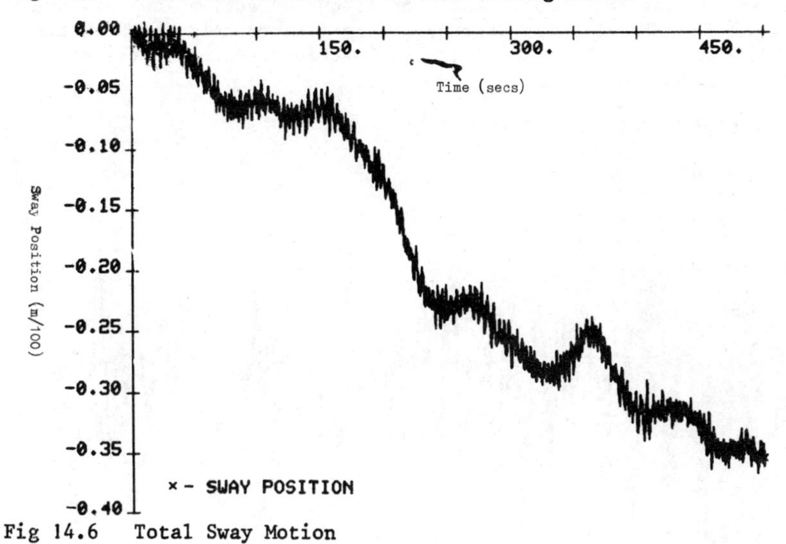

Fig 14.6 Total Sway Motion

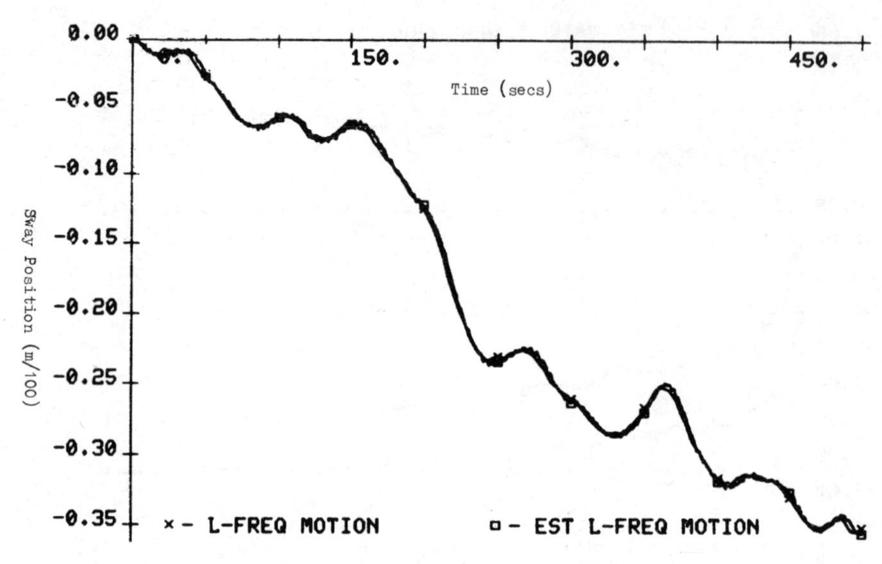

Fig 14.7 Actual and Estimated Low Frequency Motion

tracked the position well even though the parameters had not reached steady state (see Fig 14.5). The high frequency estimation is shown in Fig 14.8. The HF estimator started to track accurately the HF motion after 20 seconds. Before this time it tracked reasonably well except for a noisy envelope.

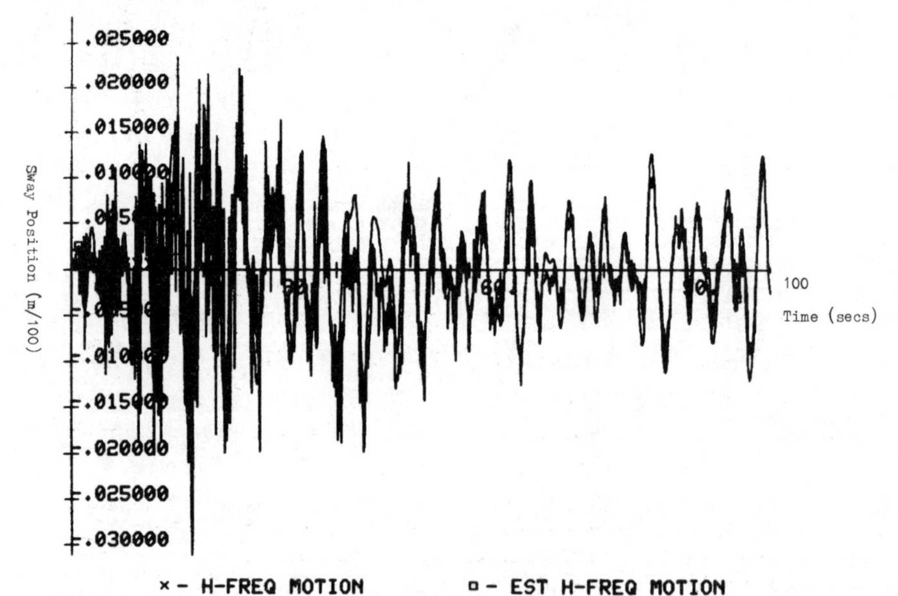

Fig 14.8 Actual and Estimated High Frequency Motion

For the innovation process (equation (14)) to be stable, the parameters a_1 and d_1 should lie in the interval [0, -2], and a_2 and d_2 should lie in the interval [0, 1]. Thus, the best guess of the initial values for a_1 and d_1 are both -1.0 and those for a_2 and d_2 are 0.5. When using the extended least square algorithm, the parameter gains and the estimation errors should be restricted so that $\{a_i\}$ and $\{d_i\}$ lie in the stable region otherwise it has been observed that the estimated parameters may blow up.

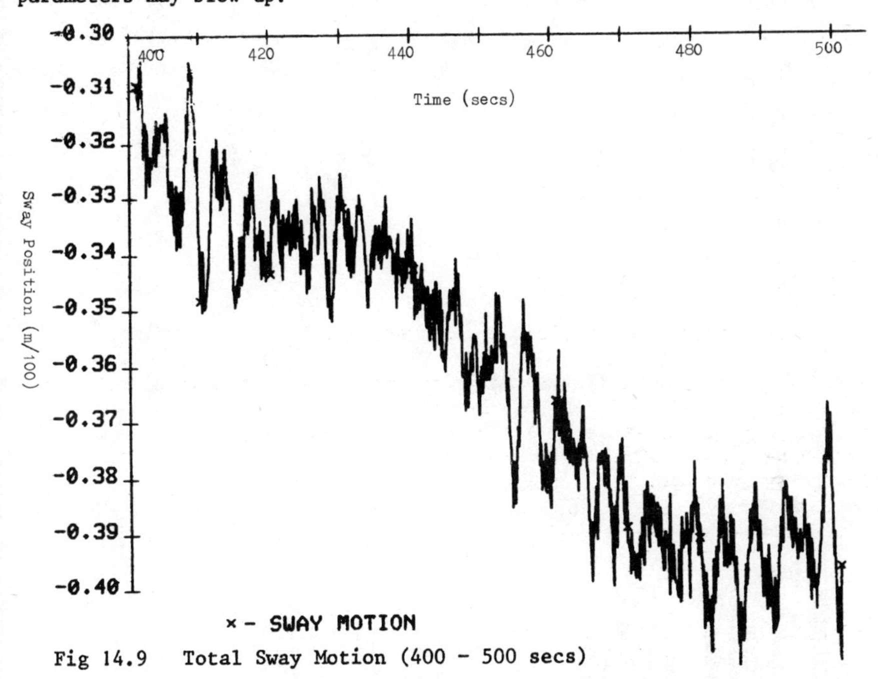

Fig 14.9 Total Sway Motion (400 - 500 secs)

It is important to consider how the filters behave in steady-state. Figs 14.9 to 14.11 show the motions between 400 to 500 seconds. Both LF estimator and HF estimator behaved well. The accumulative losses are shown in Fig 14.12. These tests were based on Beaufort No 8 ($h_{1/3}$ = 7.47 m) sea condition. Fig 14.13 shows the parameter estimation when the sea condition was changed to Beaufort No 5 ($h_{1/3}$ = 2.47 m). As the Beaufort No decreases the natural frequency of the wave spectrum increases, therefore the absolute values of the parameters $\{a_i\}$ should increase. The absolute values of the parameters $\{a_i\}$ do increase at the time when the sea condition is changed, and they converge to new steady state values. This demonstrated that, as required, the self-tuning scheme adapts to different weather conditions. The simulated frequency range of HF motion at Beaufort No 8 is {0.3, 1.2} radians per second, and that at Beaufort No 5 is {0.5, 1.9} radians per second.

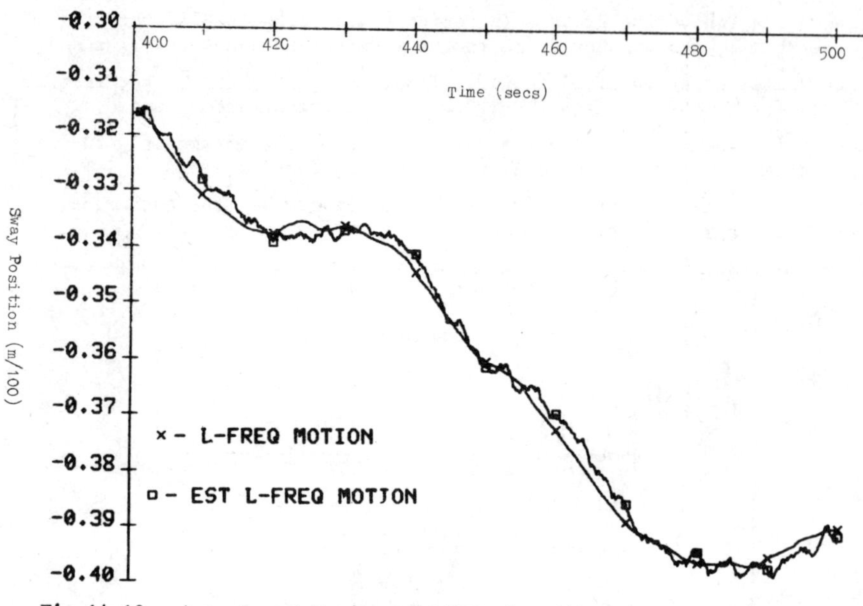

Fig 14.10 Actual and Estimated LF Motion (400 - 500 secs)

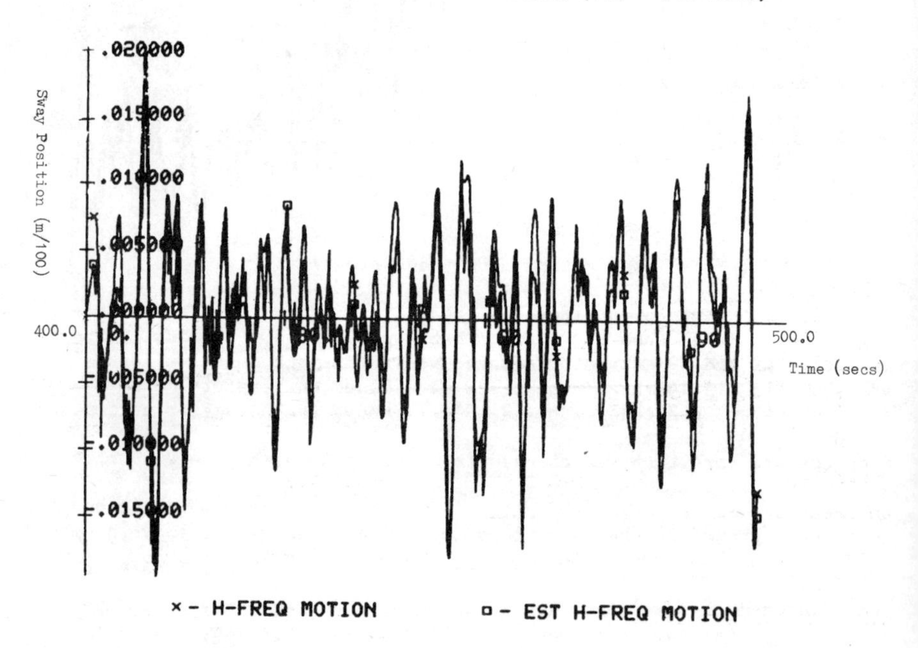

Fig 14.11 Actual and Estimated HF Motion (400 - 500 secs)

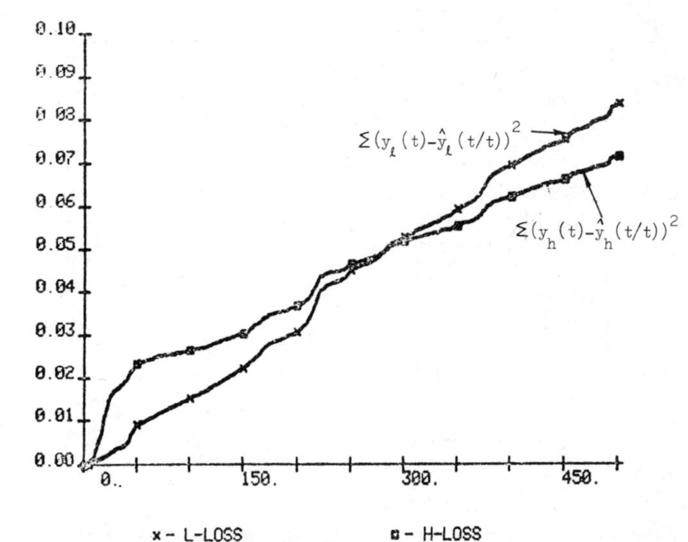

Fig 14.12 Accumulative Losses

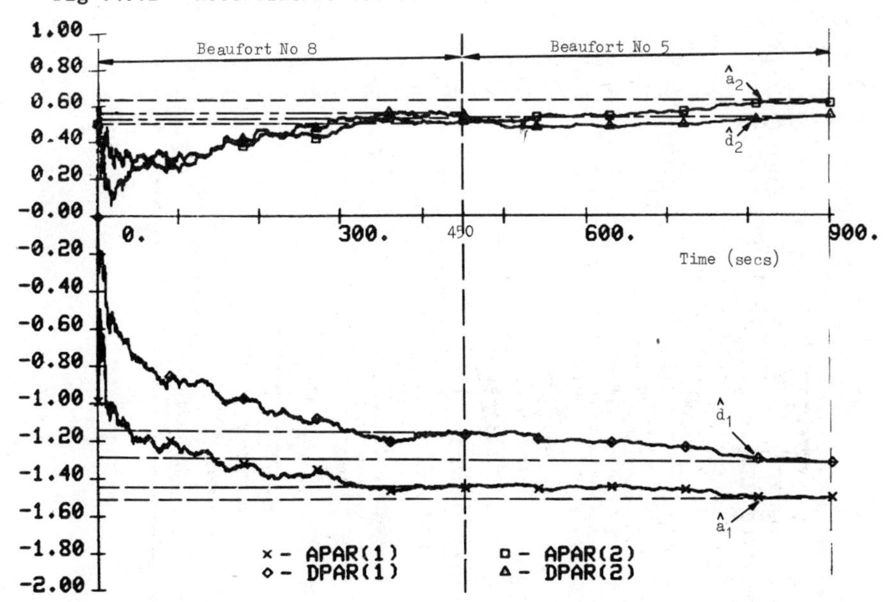

Fig 14.13 Estimated Parameters (with sea condition changed)

14.5 *Control Problem*

As was mentioned, only low frequency motion needs to be controlled. A state estimate feedback controller is employed based on the quadratic criterion:

$$J = E\{\lim_{T\to\infty} \frac{1}{2T} \int_{-T}^{T} < (\underline{x}_\ell(t) - \underline{x}_{\ell o}(t)), Q_c(\underline{x}_\ell(t) - \underline{x}_{\ell o}(t)>E_m$$
$$+ <\underline{u}(t), R_c\underline{u}(t)>E_r \; dt \tag{41}$$

where $\underline{x}_{\ell o}(t)$ is the reference state. The matrices Q_c and R_c are positive definite. For the Wimpey Sealab model let

$$Q_c = \begin{bmatrix} 300 & 0 & 0 \\ 0 & 10^4 & 0 \\ 0 & 0 & 10^3 \end{bmatrix} \tag{42}$$

and

$$R_c = [\, 50 \times 10^3 \,] \tag{43}$$

The low frequency subsystem is time invariant and the computed LQG steady-state gain matrix becomes:

$$K_c = [\, 1.4089,\ 0.4472,\ 0.4170 \,] \tag{44}$$

The optimal feedback controller is obtained from the separation principle as:

$$u(t) = K_c\hat{\underline{x}}_\ell(t|t) \tag{45}$$

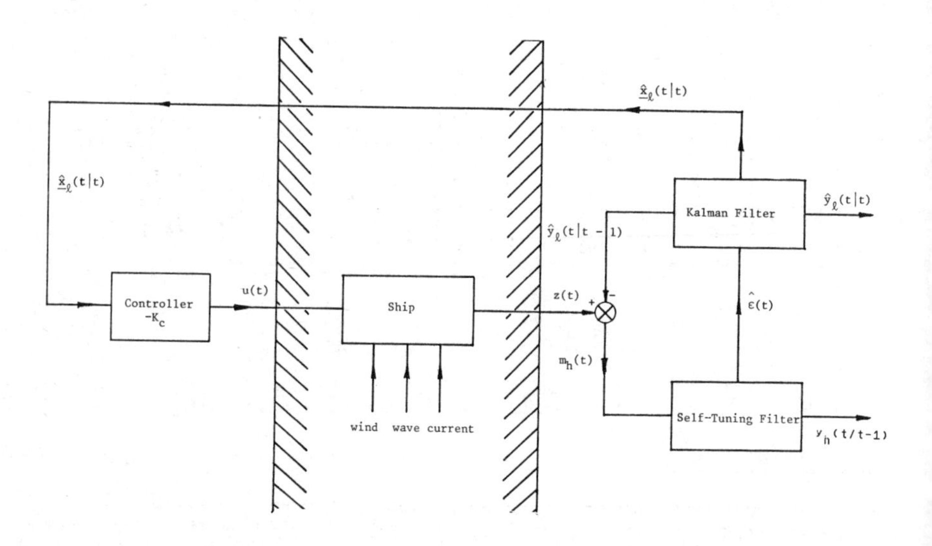

Fig 14.14 Control and Filtering Scheme

14.6 *Control Results*

The step response of the controlled system is shown in Fig 14.15.

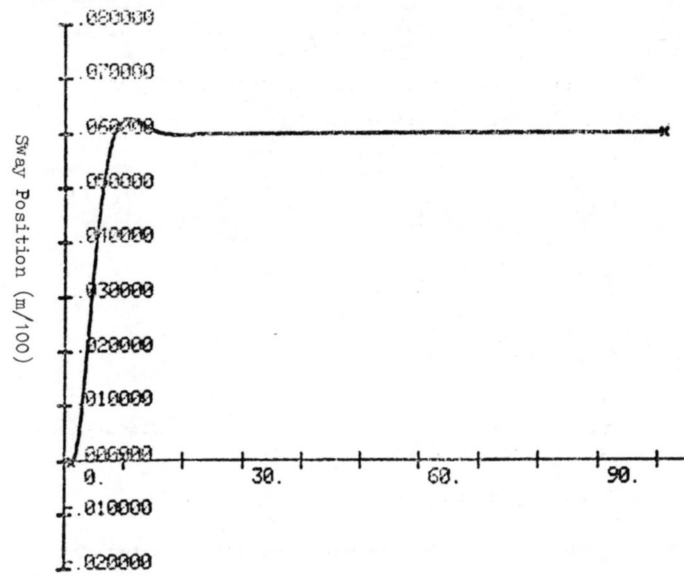

Fig 14.15 Step Response

The total sway motion with the reference at 0.06 per-unit is shown in Fig 14.16, and the LF and HF motions and their estimators are shown

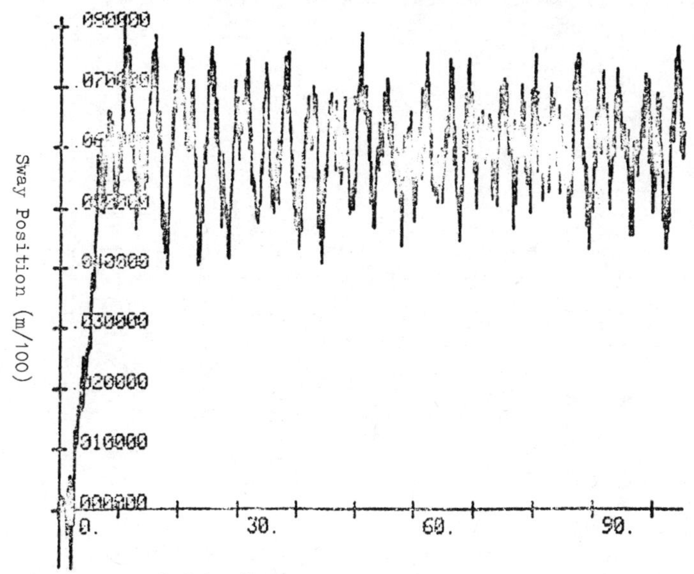

Fig 14.16 Total Sway Motion

in Fig 14.17 and 14.18. Both estimators track the motions well. The HF estimator gives similar results with and without control. The estimated parameters (Fig 14.19) and the accumulative losses (Fig 14.20) are similar to those without control. The control signal is shown in Fig 14.21. The hard limit of the controller for Wimpey Sealab is ±0.002 per unit. The generated control signal was within this limit after 10 seconds.

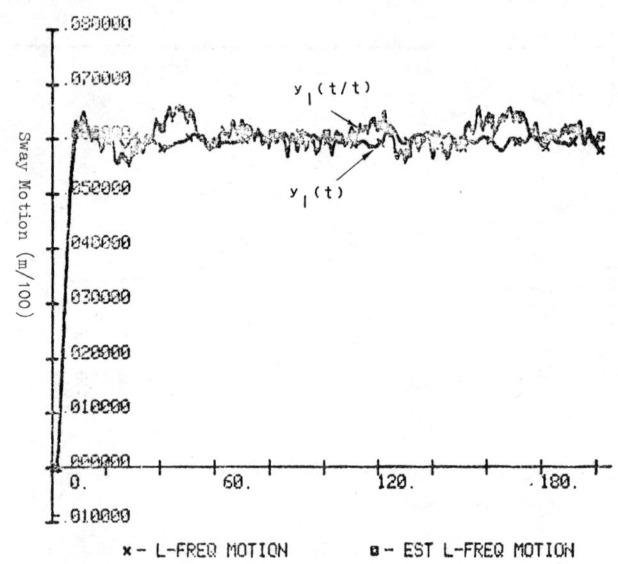

Fig 14.17 LF Motion and Estimate (with control)

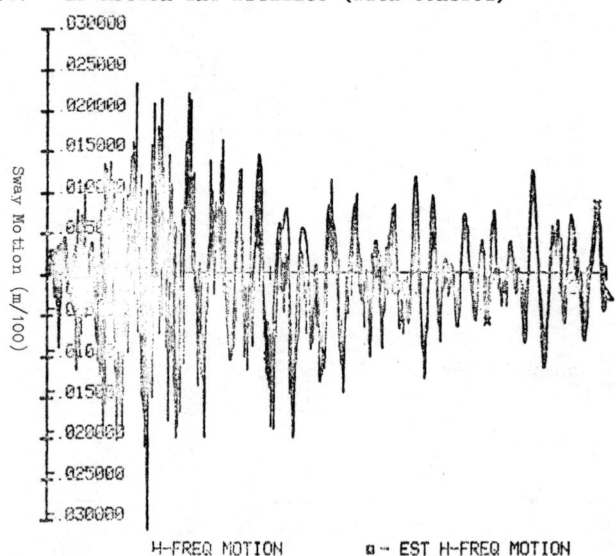

Fig 14.18 HF Motion and Estimate (with control)

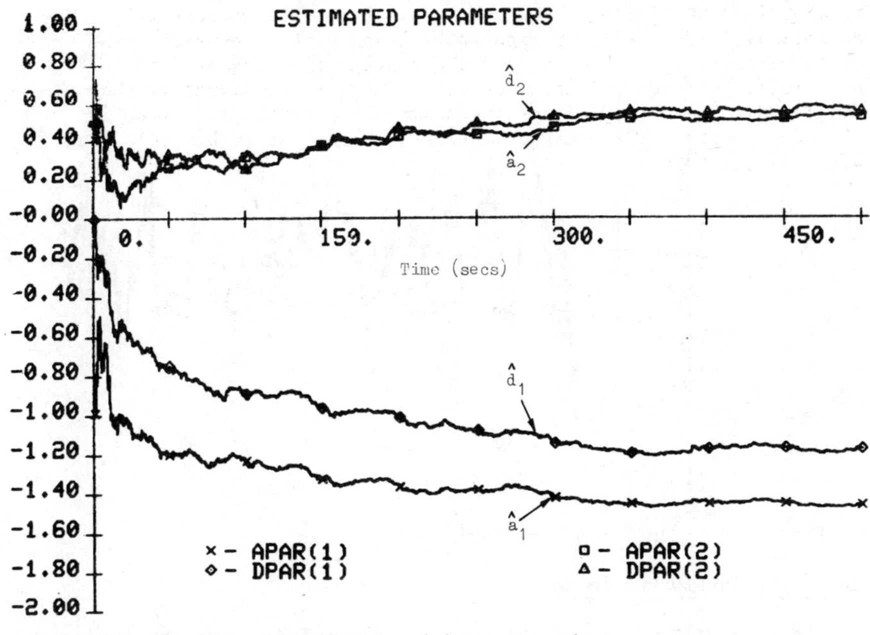

Fig 14.19 Estimated Parameters (with control)

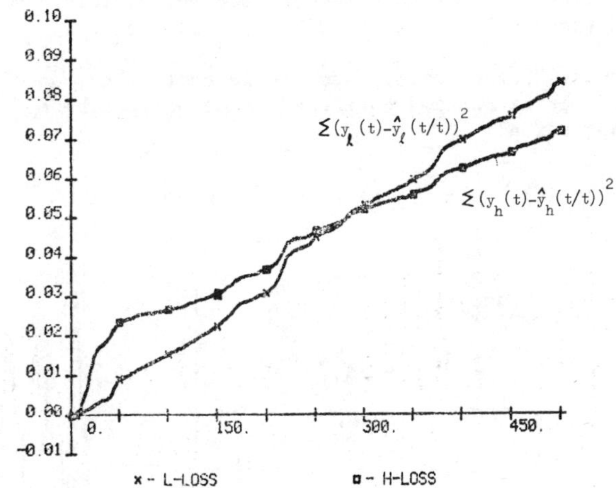

Fig 14.20 Accumulative Losses

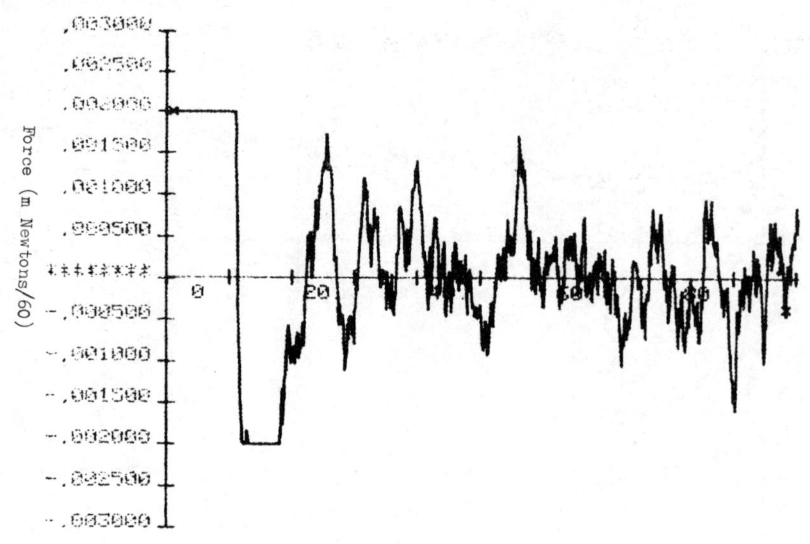

Fig 14.21 Control Signal Variations

14.7 *The Multivariable Case*

The HF models of the DP system are not coupled. Thus, individual self-tuning filters can be used to estimate the different motions. The structure of the algorithm is described briefly below.

In a DP system three motions need to be controlled, namely, surge, sway and yaw. It is assumed that the HF motions can be modelled by the following expression:

$$\bar{A}(z^{-1})\underline{y}_h(t) = \bar{C}(z^{-1})\underline{\xi}(t) \tag{46}$$

where

$$\bar{A}(z^{-1}) = \begin{bmatrix} A^{su}(z^{-1}) & 0 & 0 \\ 0 & A^{s}(z^{-1}) & 0 \\ 0 & 0 & A^{y}(z^{-1}) \end{bmatrix} \tag{47}$$

$$\bar{C}(z^{-1}) = \begin{bmatrix} C^{su}(z^{-1}) & 0 & 0 \\ 0 & C^{s}(z^{-1}) & 0 \\ 0 & 0 & C^{y}(z^{-1}) \end{bmatrix} \tag{48}$$

and $A^{su}(z^{-1})$, $A^{s}(z^{-1})$ and $A^{y}(z^{-1})$ have the same form:

$$A^{su}(z^{-1}) = 1 + a_1^{su}z^{-1} + a_2^{su}z^{-2} \tag{49}$$

and $C^{su}(z^{-1})$, $C^{s}(z^{-1})$ and $C^{y}(z^{-1})$ have the form:

$$C^{su}(z^{-1}) = c_1^{su} z^{-1} \tag{50}$$

$$\underline{y}_h(t) = \begin{bmatrix} y_h^{su}(t) \\ y_h^{s}(t) \\ y_h^{y}(t) \end{bmatrix} = \begin{bmatrix} \text{surge HF motion} \\ \text{sway HF motion} \\ \text{yaw HF motion} \end{bmatrix} \tag{51}$$

$$\underline{\xi}(t) \quad \begin{bmatrix} \xi_1(t) \\ \xi_2(t) \\ \xi_3(t) \end{bmatrix} \tag{52}$$

and $\underline{\xi}(t)$ is an independent random variable vector.

The low frequency model is known, thus the one-step ahead predicted value $\hat{\underline{y}}_\ell(t|t-1)$ can be computed at time t by using the Kalman filter discussed in section 14.3.3. All of the interaction occurs in the low frequency model and the Kalman filter may easily be applied in this multivariable situation. Thus, there is no difficulty in estimating the states even allowing for the coupling in the LF model. Using the same

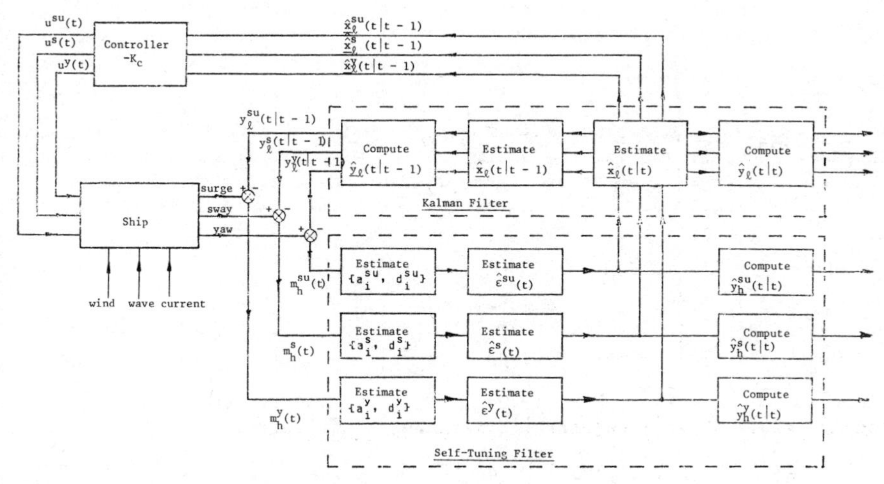

Fig 14.22 Multivariable Self-Tuning Control of Dynamic Ship Positioning

arguments as in Section 14.3.1, individual HF motions can be estimated with different self-tuning filters.

The control action can also be computed using the technique discussed in section 14.5. The structure of the multivariable system is shown in Fig 14.22.

14.8 *Conclusions*

The self-tuning filter has been shown to estimate the high frequency components of ship motion successfully. Its advantages over the usual extended Kalman filtering schemes are as follows:

(i) there is no need to specify the process noise and the measurement noise covariances (high frequency),

(ii) it does not estimate the HF states which are not needed for control and thus requires less computing time,

(iii) the structure of the multivariable system simplifies implementation and fault finding. The slef-tuning filter was found to adapt to various sea conditions effectively.

The new control scheme consists of a self-tuning filter, a constant gain Kalman filter and a state estimate feedback optimal control law. The latter sybsystems have been employed successfully on new DP vessels and the self-tuning filter involves small changes to the present system.

One of the major advantages of DP systems based upon Kalman filters is that when the measurements become heavily corrupted (or are lost) the inputs to the filter may be removed and the estimates will remain good for a reasonable period of time. The proposed system has this facility, but alternatively the self-tuning filter might be used to adapt to temporary measurement noise variations. The measurements are pooled from different measurement systems and as different devices are switched in or out it will also be useful to have a filter which adapts to the different noise characteristics.

Acknowledgments

We are grateful for the support of the United Kingdom Science Research Council and for the collaboration with GEC Electrical Projects Limited, at Rugby. We should like to thank Mr Dennis Wise of GEC and Dr Peter Gawthrop of Oxford University for their helpful comments.

References

Balchen, J G, Jenssen, N A and Saelid, S (1976). 'Dynamic positioning using Kalman filtering and optimal control theory', Automation in Offshore Oil Field Operation, pp 183-188.

English, J W and Wise, D A (1975). 'Hydrodynamic aspects of dynamic positioning', Trans North East Coast Inst of Engineers and Shipbuilders, V92, No 3, pp 53-72.

Grimble, M J, Patton, R J and Wise, D A (1979). 'The design of dynamic ship positioning control systems using extended Kalman filtering techniques', IEEE Oceans '79 Conf, San Diego, California, USA.

Grimble, M J, Patton, R J and Wise, D A (1980). 'The use of Kalman filtering techniques in dynamic ship positioning systems', Proc IEE, Vol 127, Pt D, No 3.

Grimble, M J, Patton, R J and Wise, D A (1980). 'The design of dynamic ship positioning control systems using stochastic optimal control theory', Optimal Control Applications and Methods, Vol 1, pp 167-202.

Hagender, P and Wittenmark, B (1977). 'A self-tuning filter for fixed lag smoothing, IEE Trans on Information Theory, Vol IT-23, No 3.

Kailath, T and Frost, P (1968). 'An innovation approach to least squares estimation: part II - linear smoothing in additive white noise', IEEE Trans on Aut Control, Vol AC-13, pp 655-668.

Morgan, M J (1978). 'Dynamic positioning of offshore vessels', Division of the Petroleum Publishing Company, Tulsa, Oklahoma.

Pierson, W J and Moskowitz, L (1964). 'A proposed spectral form for fully developed wind seas based on similarity theory of S A Kitagarodskii', J Geophysical Research, Vol 69.

Wise, D A and English, J W (1975). 'Tank and wind tunnel tests for a drill ship with dynamic position control', 7th Annual Offshore Tech Conf, Houston, Texas.

Index